Introduction to Optical Mineralogy

Second Edition

WILLIAM D. NESSE
University of Northern Colorado

New York   Oxford
OXFORD UNIVERSITY PRESS
1991

Oxford University Press

Oxford   New York   Toronto
Delhi   Bombay   Calcutta   Madras   Karachi
Petaling Jaya   Singapore   Hong Kong   Tokyo
Nairobi   Dar es Salaam   Cape Town
Melbourne   Auckland

and associated companies in
Berlin   Ibadan

Library of Congress Cataloging in Publication Data

Nesse, William D.
    Introduction to optical mineralogy / William D. Nesse.—2nd ed.
        p.   cm.
    Includes bibliographical references and indexes.
    ISBN 0-19-506024-5
    1. Optical mineralogy.   2. Mineralogy, Determinative.   I. Title.
QE397.N47   1991
549'.125—dc20   90-7328

Printing (last digit):   9 8 7 6 5 4 3 2 1

Printed in the United States of America
on acid-free paper

For Carl and Erik

Preface

Despite the major advances in the use of sophisticated analytical equipment that have taken place in mineralogy and petrology over the past several decades, the petrographic microscope still retains its essential role in rapid identification of minerals and interpretation of rock textures. The purpose of this book is to serve the needs of students in learning the procedures and theory required to use the petrographic microscope.

In the second edition, the text has been updated and a number of changes that have been suggested by students, colleagues, and reviewers have been incorporated. The net result of these changes should be to make the book more useful to students in learning how to identify minerals using the petrographic microscope. A number of additional photomicrographs of minerals are included as is a new determinative diagram that plots minerals as a function of index of refraction and birefringence.

As always, I will be delighted to hear from students, faculty, and other users of the book should errors or areas for improvement be identified.

Greeley, Colorado   W.D.N.
April 1990

Preface to the First Edition

This book is intended for use in an introductory optical mineralogy course. The objective in preparing the book was to present in a single volume of reasonable size both a thorough treatment of optical theory as it pertains to mineral identification with the petrographic microscope, and detailed mineral descriptions of the common rock-forming minerals.

The first seven chapters deal with optical theory and provide an introduction to the properties of light, a description of the petrographic microscope, and a discussion of the optical properties of isotropic and anisotropic materials. Detailed step-by-step procedures have been included to guide students through the measurement of optical properties in both thin section and grain mount. Selected spindle stage techniques also are included.

Chapter 8 provides an outline to guide students through the sometimes daunting process of identifying an unknown mineral with the microscope. A tabulation of the minerals likely to be found in common rocks has been included at the request of numerous students. There are obvious shortcomings to any such tabulation, which intructors will need to emphasize.

Chapters 9 through 15 contain detailed descriptions of the common rock-forming minerals. The minerals are grouped into oxides, sulfates, carbonates, orthosilicates, tectosilicates, and so forth, in the conventional manner. The optical data has been made as up-to-date as possible, but it may not reflect the tremendous advances made in other areas of mineralogy in recent years. Diagrams showing how optical properties vary as a function of composition are presented for many minerals. Since often the published data show a substantial amount of scatter, many of the diagrams were constructed with bands to show the common range of indices of refraction and so forth. The intent was to avoid the unwarranted impression of precision that single lines on the diagrams might give.

It is assumed that the reader of this book has the background normally acquired in a conventional introductory mineralogy course. Of particular importance is an understanding of the basics of crystallography including symmetry, crystal systems, crystal axes, and Miller indices. Students without this background, or whose grasp of crystallography is old and shaky, are encouraged to peruse a mineralogy textbook (e.g. Klein and Hurlbut's *Manual of Mineralogy*, 20th ed., John Wiley & Sons, New York) to acquire the needed background.

I would like to gratefully acknowledge Sturges W. Bailey, Stephen E. DeLong, Eugene E. Foord, Jeffrey B. Noblett, Howard W. Jaffe, Paul H. Ribbe, and John A. Speer for reviewing various portions of the manuscript. Their suggestions and criticisms are very much appreciated and materially improved the book. Numerous students suffered through the evolution of the manuscript. By their comments and reactions to it, they helped shape the end product probably more than they realized. Above all I owe a tremendous debt of gratitude to my wife, Marianne Workman-Nesse, who typed, edited, proofread, and generally helped and supported throughout. If, despite the aid and guidance of these people, errors, omissions, and inconsistencies still persist, they are solely my own responsibility, and I would appreciate being informed of them. I would also appreciate receiving any thoughts that readers might have concerning how the book might be made more usable in teaching optical mineralogy.

Greeley, Colorado   W.D.N.
January 1986

Contents

Introduction to
Optical Mineralogy

1

Light

The Nature of Light

In some ways light is an enigma. We know that it is a form of energy that is transmitted from one place to another at finite velocity and that it can be detected with the eye. In many ways it behaves as though it were composed of numerous tiny particles that travel bulletlike from one point to another. But it also behaves as though it were a wave phenomena in which the energy moves somewhat like the waves started by dropping a pebble in a pond. Because light behaves in these two seemingly contradictory ways, two different theories, the particle theory and the wave theory, have been developed to explain it.

In the particle theory, light is considered to be composed of subatomic particles called photons. When atoms are sufficiently heated, or otherwise excited, the outer electrons are forced into a higher-than-normal energy level. When the electrons revert to their normal energy level, a small amount of energy is released in the form of a photon, which is a small particle with essentially zero mass.

The wave theory considers light to be a form of radiant energy that travels wavelike from one point to another. These waves have both electrical and magnetic properties and are therefore called **electromagnetic radiation**. Light is just a small portion of a continuous spectrum of radiation ranging from cosmic rays at one end to radio waves and long electrical waves at the other (Figure 1.1).

Modern theories of matter and energy involving quantum mechanics have reconciled the seemingly contradictory particle and wave theories of light. Unfortunately, much of quantum mechanics does not lend itself to interpretation by simple analogs such as bullets or waves but can be understood only in abstract mathematical form. Both the particle and wave theories have been shown to be correct, it's just that in their simple forms, neither completely describes light. They are complementary theories and both can be used effectively in appropriate contexts. Because wave theory very effectively describes the phenomena of polarization, reflection, refraction, and interference—the meat of an optical mineralogy course—this book treats light as electromagnetic radiation.

Electromagnetic Radiation

All electromagnetic radiation, light included, is considered to consist of electric and magnetic vectors that vibrate at right angles to the direction in which the radiation is moving (Figure 1.2). For purposes of mineral optics it is necessary to consider only the vibration of the electric vector. It is the interaction of the electric vector with the electrical character of the atoms and chemical bonds in minerals that affects the behavior of light. Forces arising from the magnetic vector of light are generally very small and can be ignored for our purposes. It is important to note that **the vibration direction of the electric vector is transverse; it vibrates perpendicular to the direction in which the light wave is propagating**. The vibration direction of the electric vector is, in some ways, analogous to the movement of water in a water wave or the movement of the solid earth with the passage of an earthquake S wave. In both cases, the energy is propagated through the material, but the particles of water or earth are caused to move from side to side as the wave passes. The analogy is not complete, however, because with light it is not matter that vibrates from side to side but rather is an electric field that oscillates from side to side. Hereafter, when light's vibration direction is being discussed, it is the electric vector vibration that is being referred to.

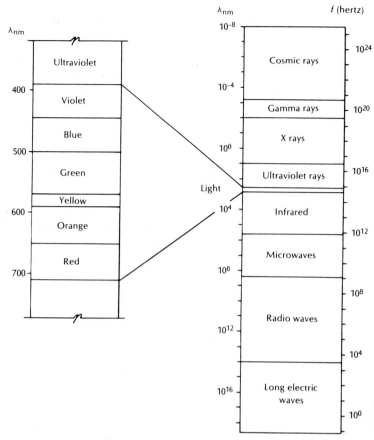

Figure 1.1 The electromagnetic spectrum. Visible light is a small portion of the electromagnetic spectrum and has wavelengths in vacuum between about 400 and 700 nm (1 nm = 10^{-7}cm).

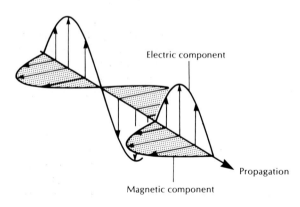

Figure 1.2 Electromagnetic radiation consists of electric and magnetic components that vibrate at right angles to each other and at right angles to the direction that the light is propagating.

A light wave can be described using the same nomenclature applied to any wave phenomenon. It has velocity, frequency, and wavelength (Figure 1.3), which are related by the equation

$$f = \frac{V}{\lambda}$$
1.1

where V is the velocity, λ is the wavelength or distance from one wave crest to another, and f is the frequency or number of wave crests per second that pass a particular point. Frequency is usually expressed as cycles per second or hertz (Hz). With some exceptions involving fluorescence that do not affect us here, **the frequency of light remains constant regardless of the material that the light travels through**. Hence, if the velocity changes, the wavelength also must change. Consider a wave train that is slowed when it passes through a piece of glass

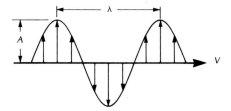

Figure 1.3 Wave nomenclature. The wave is traveling to the right with velocity V. The wavelength (λ) is the distance between successive wave crests. The frequency (f) is the number of wave crests that pass some point per second and is expressed as cycles per second or hertz (Hz). The amplitude (A) is the height of the wave. The intensity or brightness of the light is proportional to the square of the amplitude (A).

(Figure 1.4). The number of wave crests that enter the glass per second is the same as the number that exit the glass. Hence, the number of crests that pass a point inside the glass per second is the same as outside the glass, so the frequency remains constant. However, because the velocity in the glass is substantially slower than in the air, the waves bunch up and the wavelength decreases.

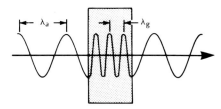

Figure 1.4 Passage of a light wave through a piece of glass (stippled). On entering the glass, the light is slowed down. Because the frequency remains the same, the wavelength in the glass (λ_g) must be shorter than the wavelength in the air surrounding the glass (λ_a).

The light passing through a mineral or through space does not consist of a single wave but rather can be considered to be composed of innumerable waves traveling together. It is, therefore, convenient to consider the waves en masse and introduce a few more terms. A **wave front** is a surface that connects similar points on adjacent waves. For example, wave fronts one wavelength apart can be drawn through each wave crest in Figure 1.5a. A line constructed at right angles to the wave front is called the **wave normal** and represents the direction that the wave is moving. A **light ray** is the direction of propagation

of light energy. In isotropic materials (light velocity the same in all directions), the light ray and wave normal coincide (Figure 1.5b). As we will see in Chapters 6 and 7, in anisotropic materials (light velocity different in different directions) the wave normal and light ray directions are usually not parallel (Figure 1.5c).

Phase

If two waves vibrate in the same plane and travel along the same path, they interfere with each other (Figure 1.6a). The distance that one wave lags behind the other is called the **retardation** (Δ). It can be described either in terms of the distance in nanometers that one wave lags the other, or in terms of the number of wavelengths that one wave lags the other. When the retardation equals an integral number of wavelengths (Figure 1.6b),

$$\Delta = i\lambda$$

where i is an integer, the two waves are **in phase** and they constructively interfere with each other. The resultant wave is the sum of the two. When the retardation equals $\frac{1}{2}$, $1\frac{1}{2}$, $2\frac{1}{2}$, etc., wavelengths,

$$\Delta = (i + \tfrac{1}{2})\lambda$$

the two waves are **out of phase** so they destructively interfere and cancel each other (Figure 1.6c). When the retardation is some intermediate value, the light is partially in phase (or partially out of phase, if you prefer) and the interference is partially constructive (or partially destructive) (Figure 1.6a).

If the two waves vibrate at an angle to each other, they can be resolved into a resultant vibration direction by means of vector addition. The resultant vibration direction R in Figure 1.7a is obtained by constructing a parallelogram whose sides are parallel to the vibration directions of waves A and B. Similarly, a component of a single wave may be resolved into any aribtrary vibration direction, as shown in Figure 1.7b. The component of wave X, which can be resolved into a new vibration direction V, is obtained by constructing a right triangle with X as the hypotenuse. The amplitude of V is given by the equation

$$V = X \cos \theta$$

where X is the amplitude of wave X and θ is the angle between the vibration direction of X and the

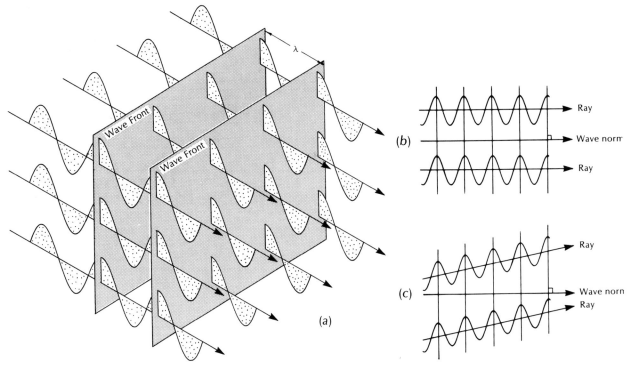

Figure 1.5 Wave fronts.   (*a*) Wave fronts are surfaces connecting equivalent points on adjacent waves. Successive wave fronts are one wavelength apart.   (*b*) In isotropic materials, the wave normal and light rays are both perpendicular to the wave front.   (*c*) In anisotropic materials, light rays are typically not parallel to the wave normal.

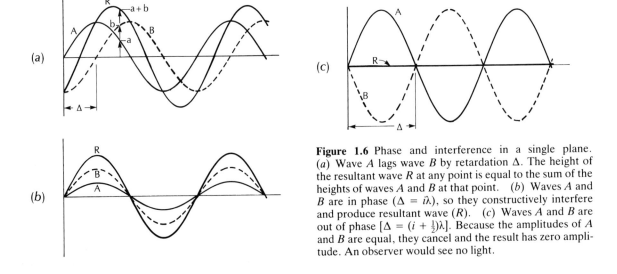

Figure 1.6 Phase and interference in a single plane. (*a*) Wave *A* lags wave *B* by retardation Δ. The height of the resultant wave *R* at any point is equal to the sum of the heights of waves *A* and *B* at that point.   (*b*) Waves *A* and *B* are in phase ($\Delta = i\lambda$), so they constructively interfere and produce resultant wave (*R*).   (*c*) Waves *A* and *B* are out of phase [$\Delta = (i + \frac{1}{2})\lambda$]. Because the amplitudes of *A* and *B* are equal, they cancel and the result has zero amplitude. An observer would see no light.

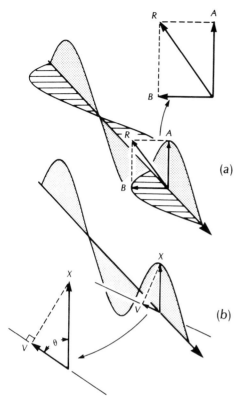

Figure 1.7 Vector resolution of light waves. (a) Waves A and B form a resultant R. (b) A component V of wave X can be resolved in a new direction at angle θ from X.

new vibration direction V. Note that if V is 90° to the original vibration direction, the resolved component must be zero. This is an important observation and accounts for a number of optical properties described in the following chapters.

The Perception of Color

The human eye is constructed so that it is able to discriminate the different wavelengths of light. Light whose wavelength in a vacuum is about 660 nm is perceived as red, light whose vacuum wavelength is about 600 nm is perceived as orange, and so forth. It would perhaps be better to talk about the different frequencies of light rather than wavelengths because frequency does not change on passing through different materials. However, the convention is to identify the different colors with their wavelengths

in a vacuum and that convention will be followed here.

If the light reaching the eye is essentially all one wavelength, it is **monochromatic light** and it is perceived as whatever wavelength is present. However, if **polychromatic light**, which consists of more than one wavelength, strikes the color receptors of the eye, the combination of wavelengths is still perceived as a single color, even though the wavelength associated with that color may not actually be present in the light. In fact, the sensation of all colors except those corresponding to wavelengths of about 420, 500, and 660 nm (violet, green, and red) can be produced by suitable combinations of two or more different wavelengths.

When all of the visible spectrum is present, the eye perceives it as white. The eye also will perceive as white various combinations of two colors called complementary light colors. There are an infinite number of complementary light color sets but none that include wavelengths in the green field. Other color sensations such as purple and brown have no counterpart in the visible spectrum and are formed by combining various wavelengths. The sensation of purple is produced by mixtures of red and violet light, while brown is formed by mixtures of red, blue, and yellow light.

About 4 percent of the population (mostly male) have forms of color blindness that affect their perception of color. For most day-to-day activities, these individuals have learned to adapt and the color blindness poses no significant difficulty. Unfortunately, the perception of color is important to certain areas of optical mineralogy and the inability to correctly perceive color may pose a hardship. The problems are not insurmountable. The individual can adapt by paying greater attention to the properties of minerals that do not require the accurate perception of color. The first step in dealing with the problem is recognizing that it exists. Not all people that have color blindness are aware of it because some forms are quite subtle. If there is any indication that an individual has color blindness, a vision specialist should be consulted.

Interaction of Light and Matter

Velocity

The velocity of light depends on the nature of the material that it travels through and the wavelength of the light. The maximum possible velocity is 3.0×10^{10} cm/sec (3×10^{17} nm/sec) in a vacuum. When light enters any other medium, it is slowed down. The detailed explanation of why the light is slowed is beyond the scope of this book, but it involves the interaction between the electric vector of the light and the electronic environment around each atom. Each atom consists of a positively charged nucleus surrounded by a number of negatively charged electrons. The nucleus is generally too heavy to respond to the forces imposed by the electric vector of light, but the electrons have low mass and can respond. When the light strikes an atom, the electron cloud around the atom is excited and is forced to vibrate or oscillate at the same frequency as the light. The excited electron cloud then re-emits the light. Because of the nature of the interaction between the electrons and the light, the re-emitted light is out of phase with the incident light. The re-emitted light then strikes the next atom along the path followed by the light, and the process is repeated. It can be shown through a series of derivations that the interference between the re-emitted light and the original light cancels the original light and produces a series of new light waves with the same frequency, but shorter wavelength and lower velocity (Eq. 1.1).

Reflection

When light reaches a boundary between two materials, some of it may enter the new material and the remainder is reflected at the boundary. The angle of incidence and the angle of reflection are identical (Figure 1.8).

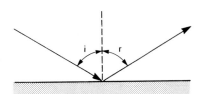

Figure 1.8 Reflection of light rays from a smooth surface. The angle of incidence (*i*) is equal to the angle of reflection (*r*).

Index of Refraction

It is well known that light is bent when passing from one transparent material to another at any angle other than perpendicular to the boundary (Figure 1.9). A measure of how effective a material is in bending light coming from a vacuum is called the index of refraction (or simply index) *n*:

$$n = \frac{V_v}{V} \qquad 1.2$$

where V_v is the velocity of light in a vacuum and V is the velocity of light in the material. The index of refraction of a vacuum is, therefore, 1.0 and for all other materials *n* is greater than 1.0. Most minerals have indices that fall in the range 1.4 to about 2.0.

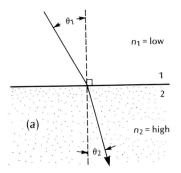

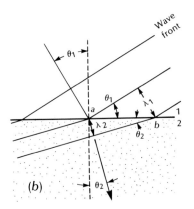

Figure 1.9 Refraction. (*a*) Light passing from material 1 (low index) to material 2 (high index) is bent as shown. The amount of refraction is given by Snell's law. (*b*) The wave fronts and wave normals must be bent at the interface because λ_2 is shorter than λ_1.

The velocity of light in air is almost the same as the velocity in a vacuum so n_{air} can be considered 1.0 for our purposes. Note that **a high index indicates low velocity and vice versa**.

The equation that allows the calculation of how much the light will be bent on traveling from one material to another is called **Snell's law**,

$$\frac{\sin\theta_1}{\sin\theta_2} = \frac{n_2}{n_1} \qquad 1.3$$

where n_1 and n_2 are the indices of materials 1 and 2, and θ_1 and θ_2 are the angles shown in Figure 1.9a. This equation holds whether the light passes from 1 to 2 or from 2 to 1. In general, light is refracted towards the normal to the boundary on entering a material with higher refractive index and is refracted away from the normal on entering a material with lower refractive index.

That the light must be bent on entering a material with a different index can be shown by referring to Figure 1.9b. Light in material 1 with wavelength λ_1 strikes the boundary at angle θ_1. On entering material 2, the light is slowed down. Because the frequency does not change, Equation 1.1 tells us that the wavelength λ_2 must be shorter than λ_1 because the velocity in 2 is lower than in 1. The only way for the wave fronts to be closer together in material 2 is to bend them at the boundary as shown. The wave normals, which are perpendicular to the wave fronts and indicate the direction the waves are moving, must also be bent, hence the light is bent on entering material 2.

Snell's law can be derived from Figure 1.9b. From Equation 1.1 we see that

$$f = \frac{V_1}{\lambda_1}$$

and

$$f = \frac{V_2}{\lambda_2}$$

or

$$\frac{\lambda_1}{\lambda_2} = \frac{V_1}{V_2} \qquad 1.4$$

where V_1 and V_2 are the velocities in materials 1 and 2 respectively. But

$$\lambda_1 = ab\sin\theta_1$$
$$\lambda_2 = ab\sin\theta_2$$

and, from Equation 1.2,

$$V_1 = \frac{V_v}{n_1}$$

$$V_2 = \frac{V_v}{n_2}$$

where n_1 and n_2 are the indices of refraction of materials 1 and 2 respectively. With substitution in Equation 1.4 we get

$$\frac{ab\sin\theta_1}{ab\sin\theta_2} = \frac{\dfrac{V_v}{n_1}}{\dfrac{V_v}{n_2}}$$

or

$$\frac{\sin\theta_1}{\sin\theta_2} = \frac{n_2}{n_1}$$

which is Snell's law.

Snell's law applies for both isotropic and anisotropic materials. However, in anisotropic materials, the angles θ_1 and θ_2 must be measured from the wave normals, not the rays. As we will see, rays and wave normals may not be coincident in anisotropic minerals.

Critical Angle and Total Internal Reflection

When light travels from a low-index material to a high-index material, at least some light can enter the higher-index material regardless of the angle of incidence. However, this is not true for light going from a high-index material to a low-index material where the angle of refraction is larger than the angle of incidence. Light with an angle of incidence greater than the **critical angle** (CA) cannot be refracted into the low-index material. **The critical angle is the angle of incidence that yields an angle of refraction of 90°.** Consider Figure 1.10. Rays *a* and *b* are refracted into the low-index material with angles of refraction that are larger than the angles of incidence according to Equation 1.3. For ray *b*, the angle of incidence is slightly less than the critical angle and the angle of refraction is almost 90°. For any angle of incidence larger than CA, the angle of refraction would have to be larger than 90°. This is not possible, however, because an angle greater than 90° would prevent the light from entering the low-index material. Instead of being refracted, light rays like *c*, reaching the boundary at angles of incidence greater than CA, exhibit **total internal reflection** because all of the light is reflected at the boundary. If n_1 and n_2 are known, the critical angle (CA) in material 1 can be calculated from Equation 1.3 as follows:

$$\frac{n_2 \; (\text{low})}{n_1 \; (\text{high})} = \frac{\sin \; CA}{\sin \; 90°}$$

which gives

$$\frac{n_2 \; (\text{low})}{n_1 \; (\text{high})} = \sin \; CA \qquad 1.5$$

If the low-index material (2) is air or vacuum ($n_2 = 1$), this becomes

$$\frac{1}{n_1} = \sin CA \qquad 1.6$$

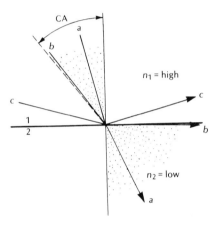

Figure 1.10 Critical angle and total internal reflection. Light such as rays *a* and *b* can only pass from a high-index to a low-index material if the angle of incidence is less than the critical angle (CA). Rays such as *c*, that have an angle of incidence greater than the critical angle, are entirely reflected.

Isotropic and Anisotropic Media

An optically isotropic material is one that shows the same velocity of light in all directions. The rock-forming materials that are isotropic include glasses and minerals belonging to the isometric system. In these materials, the chemical bonds are the same in all directions, at least on the average, so the electron clouds can oscillate the same in every direction. The electronic environment that the light "sees" is therefore independent of direction in the material.

An optically anisotropic material is one in which the velocity of light is different in different directions. The rock-forming materials that are anisotropic

include minerals in the tetragonal, hexagonal, orthorhombic, monoclinic, and triclinic systems. Minerals in these crystal systems have lower symmetry than those in the isometric system and show different strengths of chemical bonding in different directions. The electron clouds of the atoms or ions are not able to vibrate the same in all directions so the velocity of light is different for different directions. If normally isotropic materials are unevenly strained, they also may be anisotropic.

Dispersion

The index of refraction of a material is not the same for all wavelengths of light. This is easily demonstrated by passing white light through a prism (Figure 1.11). The light at the violet end of the spectrum is more strongly refracted than the light at the red end of the spectrum. This relationship, in which indices of refraction decrease for increasing wavelengths of light, is called **normal dispersion of the refractive indices**. Certain wavelength bands may have **abnormal dispersion of the refractive indices** and indices of refraction increase for increasing wavelengths. These terms are somewhat misleading because all materials show abnormal dispersion at certain wavelengths, but these wavelengths may be outside of the visible spectrum.

Dispersion is a consequence of the interaction of light with the natural resonant frequencies of the electron clouds around each atom. As was described earlier, the electric vector of the light causes the electron cloud around an atom to resonate at the frequency of the light. The atom then re-emits the light but it is not in phase with the incident light. The degree to which the re-emitted light is out of phase with the incident light depends on the degree to which the frequency of the incident light differs from the natural resonant frequency of the electron clouds. Through a complex set of equations, it can be shown that the index increases with increasing frequency (decreasing wavelength), producing normal dispersion if the frequency of the light is significantly different from a resonant frequency of the electron clouds (Figure 1.11*b*). If the frequency of the light is nearly the same as one of the natural resonant frequencies of the electron clouds, the light is strongly absorbed and the index of refraction sharply decreases with increasing frequency

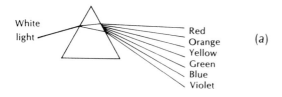

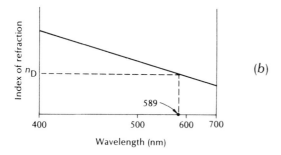

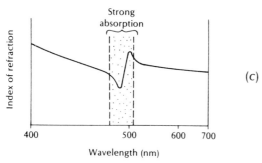

Figure 1.11 Dispersion of the refractive indices. (a) White light passed through a prism is split into its constituent wavelengths because the index of refraction increases from red to violet. (b) Normal dispersion of the refractive indices. Short wavelengths have higher indices of refraction. The index n_D for 589-nm light is the value usually reported as the index of refraction of a material. (c) Abnormal dispersion of the refractive indices. The mineral shows an increase in index for higher wavelengths In a wavelength band that is absorbed by the mineral. Minerals with abnormal dispersion are always colored.

(decreasing wavelength), producing abnormal dispersion (Figure 1.11c).

To describe the dispersion of the material, it is necessary to report the index of refraction at several wavelengths. By convention, indices usually are reported for light of 486 (n_F), 589 (n_D), and 656 nm (n_C). These wavelengths are used because they correspond to certain wavelengths called Fraunhofer lines, which are absorption lines in the sun's spec-

trum. It also happens that 589 nm is the wavelength produced by a sodium vapor lamp and is near the middle of the visible spectrum. In this book and in most other sources, it is n_D, the index for light of 589-nm wavelength, that is meant when the index of refraction of a material is reported. It may be described either as n_D or just n.

The coefficient of dispersion is defined as the value

$$n_F - n_C$$

where n_F and n_C are the indices for the F and C Fraunhofer lines at 486 and 656 nm, respectively. A large coefficient of dispersion means that the material shows a large change of index as a function of wavelength.

A related term, called the dispersive power, is defined as the value

$$\frac{n_F - n_C}{n_D - 1}$$

A large value for dispersive power means that the material shows a large change of index as a function of wavelength.

Light Absorption and Color

The color of a mineral or any other object is the color of light that is not absorbed on transmission or reflection. Usually the color of an object is the same in reflected and transmitted light, although there are exceptions. A white object looks white because it reflects essentially all of the visible spectrum. A clear mineral similarly transmits essentially all of the visible spectrum. A black object absorbs all wavelengths of light. If a mineral is colored, it is because it selectively absorbs certain wavelengths of light and transmits or reflects the remaining light to our eye. The color that is perceived depends on which wavelengths are transmitted to the eye and on how the eye interprets these wavelengths, as was described earlier. A number of different objects may appear to be the same color even though they each reflect a different complement of wavelengths to the observer.

Note that the perceived color of an object is not an inherent property but depends on the color of the incident light. An object that is white in sunlight is blue in blue light, yellow in yellow light, and so on, because these are the only wavelengths of light available to be transmitted to the observer. Colored objects may appear black in monochromatic light unless they are capable of reflecting or transmitting the wavelength of the monochromatic light.

On the atomic scale, the colors of light that a mineral absorbs depends on the interaction between the electric vector of the light and the natural resonances of the electron clouds around each atom. If the frequency of the light is significantly different from the natural resonance, then the light is trans-

mitted as previously described. However, if the frequency of the light is nearly the same as the natural frequency of the electron clouds, then the matter absorbs the light. The frequencies that show strong absorption also show abnormal dispersion. The absorbed light energy usually is converted to heat energy. Because dark-colored materials absorb more of the light than pale-colored materials, they heat up much faster when exposed to sunlight or other sources of light.

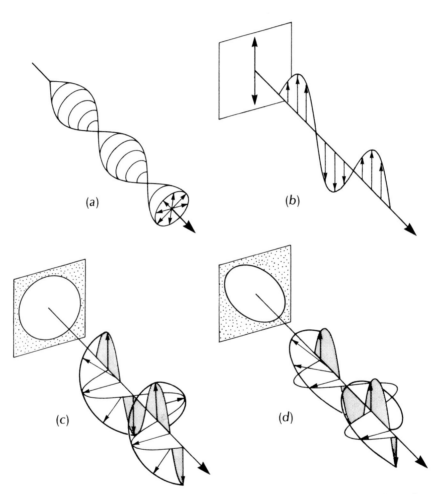

Figure 1.12 Polarization of light. (*a*) Unpolarized light. The light vibrates in all directions at right angles to the direction of propagation. (*b*) Plane or linear polarized light. The electric vector vibrates in a single plane. (*c*) Circular polarization. Two waves with equal amplitude vibrate at right angles to each other with one wave retarded $\frac{1}{4}$ wavelength relative to the other. The vector sum of these two waves is a helix whose cross section is circular. (*d*) Elliptical polarization. Two waves vibrate at right angles to each other with one wave retarded relative to the other by other than 0, $\frac{1}{4}$, or $\frac{1}{2}$ wavelength. The resultant is a helix whose cross section is an ellipse.

Polarized Light

Ordinary light, like that coming directly from the sun or an incandescent light bulb, vibrates in all directions at right angles to the direction of propagation (Figure 1.12*a*) and is unpolarized. The vibration direction of polarized light is constrained so that it is not uniformly distributed around the direction of propagation. There are three different but related types of polarization: **plane polarization, circular polarization**, and **elliptical polarization**.

In plane polarization (also called linear polarization), the electric vector vibrates in a single plane (Figure 1.12*b*). The light wave is a simple sine wave with the vibration direction lying in the plane of polarization. Plane polarized light, or simply plane light, is of primary interest in this book.

Circular polarized light is produced by two waves of plane polarized light with the same amplitude and whose vibration directions lie at right angles to each other (Figure 1.12*c*). One wave is retarded $\frac{1}{4}\lambda$ relative to the other. The two electric vectors can be added vectorially so that at any point along the wave path the two vectors produce a resultant vector. The resultant vectors sweep out a helical surface that resembles the threads on a screw. When viewed along the direction of propagation, the outline of the helix is a circle.

Elliptical polarized light is produced in the same manner as circular polarized light except the two waves that produce it are retarded relative to each other by a value different than $\frac{1}{4}\lambda$ (Figure 1.12*d*). The result is still a helix, but instead of being circular in cross section, the helix is elliptical.

Because circular and elliptical polarized light can be considered to be composed of two waves of plane polarized light that vibrate at right angles to each other. it is convenient in this book to treat circular and elliptical polarization in terms of the two component waves.

Methods of Polarization

Light may be polarized by selective absorption, double refraction, reflection, and scattering.

Selective absorption provides the basis for polarization with polarizing films. A variety of anisotropic materials, among them tourmaline, have the property of strongly absorbing light vibrating in one direction

and transmitting light vibrating at right angles more easily (Figure 1.13). This property is called pleochroism. When unpolarized light enters one of these anisotropic materials in the proper direction, it is split into two plane polarized rays that vibrate perpendicular to each other. About half the light energy goes into each of the two rays. If the material is thick enough and is strongly pleochroic, like tourmaline, one of the rays is absorbed and the other passes through. The ray that passes through retains its polarization on leaving the material.

The original polarizing film developed in 1928 by Edwin Land took advantage of the strong pleochroism of a material called herapathite, which forms long slender crystals. Millions of tiny grains of herapathite were embedded in plastic and aligned by extrusion of the plastic through a narrow slit. The resulting plastic sheets behaved as though they were large flat sheets of herapathite and did a credible job of polarizing the light that passed through them. More recent developments have dispensed with the herapathite crystals and have relied on long hydrocarbon molecules in the plastic. The result is a sheet of polarizing film with substantially better optical properties. Almost all modern petrographic microscopes use sheets of polarizing film to provide polarized light.

A second means of producing polarized light depends on the double refraction of anisotropic materials. The most commonly known device using this principle is the Nicol prism, which is constructed

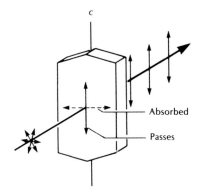

Figure 1.13 Polarization by selective absorption. Unpolarized light is split into two rays when entering tourmaline. One ray is strongly absorbed and does not pass through. The other ray is not absorbed and retains its polarization after exiting the mineral.

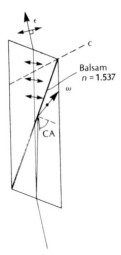

Figure 1.14 Polarization by double refraction, the Nicol prism. The prism is constructed of calcite, which has been cut on the diagonal and glued back together with balsam ($n = 1.537$). Light entering the bottom of the calcite prism is split into two rays (ω and ϵ) that have different indices of refraction ($n_\epsilon < 1.537 < n_\omega$). The ω ray encounters total internal reflection when it reaches the balsam, because the angle of incidence is greater than the critical angle (CA). The ϵ ray passes through the balsam, since its index is less than 1.537, and it retains its polarization after exiting the top of the prism. The dots on the ω ray indicate that it vibrates in and out of the page. The c crystallographic axis is indicated by the dashed line.

of clear calcite (Figure 1.14). A crystal of calcite is cut on the diagonal, as shown, and glued back together with balsam, which has an index of refraction of 1.537. As is discussed in detail in Chapters 5 and 6, when light enters the calcite, it is split into two plane polarized rays (ω and ϵ) with different velocities and that vibrate at right angles to each other. Because the two rays have different velocities, each can be assigned an index of refraction. It turns out that $n_\epsilon < 1.537 < n_\omega$. The cut through the crystal is oriented so that the ω ray strikes the boundary between the balsam cement and the calcite at greater than the critical angle, hence it is internally reflected and absorbed by black paint on the side of the prism. The ϵ ray, with its lower index of refraction, is able to enter the balsam, so it is transmitted through the prism to emerge at the top as plane polarized light. Nicol prisms, or other more optically efficient variants, were the preferred means of polar-

izing light in petrographic microscopes until replaced by polarizing film.

When unpolarized light strikes a smooth surface, such as a piece of glass, a smooth table top, or the surface of a lake, the reflected light is polarized so that its vibration direction is parallel to the reflecting surface (Figure 1.15a). The reflected light is not completely plane polarized unless the angle between the reflected and refracted rays is 90° (Figure 1.15b). The angle of incidence needed to produce the 90° angle between the reflected and refracted rays is called **Brewster's angle**. If the indices of refraction are known, Brewster's angle (θ_B) can be derived from Snell's law (Equation 1.3), which with substitution of trigonometric identities, gives

$$\frac{n_2}{n_1} = \tan \theta_B \qquad\qquad 1.6$$

The refracted light is polarized so that it vibrates at right angles to the reflected light, as shown in Figure 1.15b, but it is not completely plane polarized because it includes a substantial component of light that vibrates parallel to the surface.

The polarization direction of polarizing film can easily be identified by looking through the film

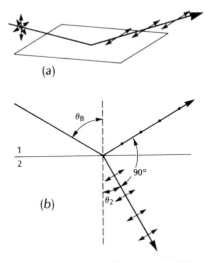

Figure 1.15 Polarization by reflection. (a) The reflected light is polarized parallel to the reflecting surface. (b) Complete polarization of the reflected light is achieved if the angle of incidence is Brewster's angle (θ_B), which places the reflected and refracted rays at 90° to each other (Equation 1.6).

at the glare reflected from a smooth horizontal surface such as a table top. Because the polarization of the reflected glare is horizontal, the polarizing film will transmit a maximum amount of glare when its polarization direction also is horizontal. If the polarizing film is rotated so that its polarization direction is vertical, it will absorb the glare. Sunglasses that use polarizing film have it oriented with the polarization direction vertical, so that it absorbs the reflected glare coming from lakes, automobile hoods, and other horizontal surfaces.

While polarization by scattering is of no particular interest to optical mineralogy, it is responsible for the blue color of the sky and the brilliant colors of sunsets. When light passes through the air, a certain amount of the light is scattered by the dust particles in the air and by the air molecules themselves. The scattered light is polarized so that it vibrates in a plane at right angles to the original path of the light before scattering. This can be demonstrated by looking at the bluest part of the sky with a piece of polarizing film whose vibration direction is known. The wavelengths that are most strongly scattered depend on the size of the scattering particles. For the atmosphere, it turns out that blue light is most strongly scattered, hence the sky looks blue from the blue light that is scattered down to us from light passing through the upper reaches of the atmosphere. Sunrises and sunsets take on colors in the red, orange, and yellow end of the spectrum because the light reaching us directly from the sun has had much of the blue end of the spectrum scattered away as it passes at a low angle through the atmosphere. Volcanic dust blown into the upper parts of the atmosphere is very effective in scattering light and may produce vivid sunrises and sunsets.

SUGGESTIONS FOR ADDITIONAL READING

Brown, E. B., 1965, Modern optics: Reinhold Publishing, New York, 645 p.

Cornsweet, T. N., 1970, Visual perception: Academic Press, New York, 475 p.

Ditchburn, R. W., 1976, Light (two volumes): Academic Press, London, 776 p.

Hecht, E., and Zajec, A., 1976, Optics: Addison-Wesley Publishing Co., Reading, Massachusetts, 565 p.

Jenkins, F. A., and White, H. E., 1976, Fundamentals of optics: McGraw-Hill, New York, 746 p.

LeGrand, Y., 1968, Light, colour, and vision: Chapman and Hall, Ltd., London, 564 p.

Welford, W. T., 1981, Optics, 2nd Edition: Oxford University Press, Oxford, 150 p.

2

The Petrographic Microscope

There are numerous petrographic microscope designs available. While each is different in detail, all have fundamentally the same design and construction (Figure 2.1). From the bottom up, they consist of an illuminator, substage assembly, stage, objective lenses, upper polar, Bertrand lens, and ocular lens. Most modern microscopes are equipped with a prism between the upper polar and ocular that allows the microscope tube to be tilted from the vertical, and all have a focusing mechanism.

Illuminator

Most modern microscopes are equipped with an incandescent light mounted in the base. The light from the bulb is directed upward with a combination of lenses and mirrors. The light is usually diffused somewhat by passing through a piece of ground glass and filtered with a piece of blue glass so that the color balance of the light more closely approximates natural sunlight. Light intensity is adjusted with a

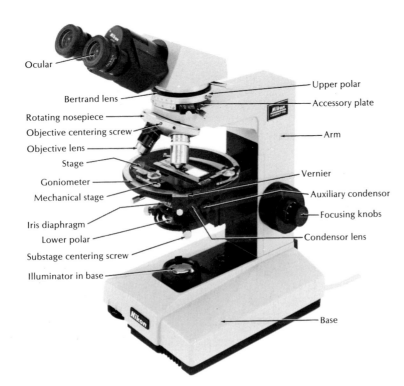

Ocular

Bertrand lens

Rotating nosepiece

Objective centering screw

Objective lens

Stage

Goniometer

Mechanical stage

Iris diaphragm

Lower polar

Substage centering screw

Illuminator in base

Upper polar

Accessory plate

Arm

Vernier

Auxiliary condensor

Focusing knobs

Condensor lens

Base

Figure 2.1 The petrographic microscope. Photo courtesy of Nikon Inc., Instrument Division, Garden City, New York.

rheostat control or with filters. Some microscopes are equipped with an iris diaphragm called a field diaphragm mounted in the base. The field diaphragm controls the size of the area on the sample that is illuminated.

In some microscopes, the illuminator may be replaced with a mirror. An external source of light is directed at the mirror, which is adjusted to direct the light beam upward. The mirror is used when an external source of monochromatic light, such as a sodium vapor lamp, is needed or when a built-in illuminator is not provided.

Substage Assembly

The principal parts of the substage assembly are the lower polar, aperture diaphragm, condensing lens, auxiliary condensing lens, and one or more filter holders. The assembly is commonly mounted in a mechanism which allows it to be raised or lowered.

The **lower polar** on most modern microscopes consists of a piece of optical-quality polarizing film mounted in a ring that can be rotated at least 90° and usually 180° or 360°. This allows the vibration direction of the polarized light passing through the microscope to be set in any desired orientation. The polars are sometimes referred to as Nicols, because Nicol prisms were used to provide polarized light on many early petrographic microscopes. In some cases, the lower polar is mounted on a pivot so that it can be swung out of the optical path.

The **aperture diaphragm** is an iris diaphragm mounted either above or below the fixed condensing lens. Its function is to adjust the size of the cone of light that passes up through the microscope. Closing the aperture diaphragm decreases the size of the cone of light and increases the contrast in the image seen through the microscope. The aperture diaphragm is not intended to be used to adjust the intensity of the illumination. The rheostat control or neutral gray filters should be used for this purpose.

The **condensor lenses** serve to concentrate the light onto the area of the sample immediately beneath the objective lens. The fixed condensor lens usually has a numerical aperture (discussed shortly) about the same as the numerical aperture of the medium-power objective lens. Because the light reaching the sample from the fixed condensor is only

moderately converging, the illumination provided is called **orthoscopic illumination**.

The **auxiliary condensing lens** is mounted on a pivot so that it can swing in or out of the optical path. Its function is to provide **conoscopic illumination**, which consists of strongly converging light. This lens is swung into the optical path to allow production of optical phenomena called interference figures, which are examined with the high-power objective. Not all microscopes are equipped with auxiliary condensors. If an auxiliary condensor is not present, the fixed condensor is usually designed to provide somewhat more strongly convergent light and provides a compromise between the needs of orthoscopic and conoscopic illumination. Some variation in the degree of convergence can be accomplished by adjusting the aperture diaphragm or by raising or lowering the condensor, but this arrangement generally is not very successful.

Colored or gray filters can be placed in slots or swing-out holders at the bottom of most substage assemblies. Gray filters are used to adjust the intensity of the illumination, and colored filters can be used to adjust the color balance of the light or to produce roughly monochromatic light.

Microscope Stage

The circular stage of the petrographic microscope is mounted on bearings so that it can be rotated smoothly. The **stage goniometer** on the outside edge of the stage is marked in degrees so that angles of rotation can be measured accurately. Sometimes a vernier is provided by the index mark, although vernier accuracy is rarely needed. On some microscopes, a thumb screw or lever can be engaged to lock the stage and prevent it from rotating.

Objective Lenses

The objective lenses provide the primary magnification of the optical system and are, in effect, the heart of the microscope. Most student-model microscopes are equipped with three objectives with magnifications of around 2.5, 10, and 40X. They are either mounted on a rotating nosepiece, or may be detached and interchanged by releasing a small catch.

The **numerical aperture** (NA) of a lens is a measure of the size of the cone of light that it can accommodate. It is given by the equation

$$NA = n \sin \frac{AA}{2}$$

2.1

where n is the index of refraction of the medium between the objective lens and the item being examined, and AA is the **angular aperture** (Figure 2.2).

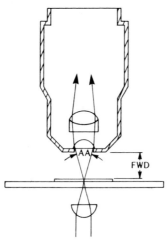

Figure 2.2 Objective lens construction. The free working distance (FWD) is the distance between the end of the lens and the top of the sample. The angular aperture (AA) is the size of the cone of light that the lens can accept.

The resolution of a lens is a measure of the ability to reveal fine detail. The limit of resolution (d) is the smallest distance between two points that can still be distinguished. A small limit of resolution means the lens has a high resolving power. The theoretical limit of resolution is given by the equation

$$d = \frac{\lambda}{2NA}$$

2.2

where λ is the wavelength of the light. We can see that as the numerical aperture increases, the limit of resolution does go down and the resolving power of the lens goes up. High numerical apertures are practical only with high-magnification lenses. If the material between the lens and the object being examined is air, $n = 1$, and the numerical aperture can, in principle, be no more than 1. In practice, the NA will not exceed 0.95, because the angular aperture must be somewhat less than 180°. It should be apparent that the numerical aperture and, therefore,

the resolving power, could be increased substantially by placing something with a higher index of refraction between the lens and the object being examined. Oil immersion lenses are designed so that a drop of mineral oil can be placed between them and the microscope slide to increase the numerical aperture. The oil usually used has an index of 1.515, so the maximum numerical aperture usually obtainable is about 1.4. From Equation 2.2, we can see that the limit of resolution with an oil immersion lens using light from the middle of the visible spectrum is about 200 nm. The high-power lens on most student-model microscopes has a numerical aperture around 0.7 and is used without oil, so the theoretical limit of resolution is about 400 nm. In practice, the limit of resolution will be substantially worse than the theoretical limit.

The free working distance is the distance between the top of the slide and the bottom of the objective lens. High-power objectives have a free working distance of less than a millimeter, while low-power objectives have a free working distance of several centimeters. When using the high-power objective it is necessary to exercise caution, because it is quite easy to damage either the lens or the microscope slide by forcing the slide against the lens while focusing. Most high-power objectives are constructed so that the lower part of the lens can be pushed up into the body of the lens against spring tension to reduce the hazard.

An objective lens can be focused precisely at only one point. However, objects above and below the point of precise focus also are in reasonably sharp focus. The distance between the lower and upper limits of reasonably sharp focus is called the depth of field. Lenses with low magnification and low numerical aperture have a relatively large depth of field, and those with high magnification and high numerical aperture have a small depth of field.

Most objective lenses (Figure 2.3) are marked with the magnification, numerical aperture, length of the microscope tube it is designed for (usually 160 or 170 mm), and the thickness of cover slip that provides the greatest optical efficiency (usually 0.17 mm). Cover glass thickness is not very important with lower-power objectives but may be quite important with high-power objectives. If the objective is designed as an oil immersion lens it will be marked "oil" or "oel." Oil should never be used with lenses that are not designed for it.

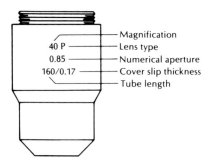

Figure 2.3 Typical objective lens markings. The P indicates that the lens is constructed of strain-free lens elements and is intended for use with polarized light.

Upper Polar

The upper polar or analyzer is located above the objective lens and is mounted on a slide or pivot so that it can be easily inserted or removed from the optical path. It usually is constructed of a piece of optical quality polarizing film, although Nicol prisms or equivalent are used on many older microscopes. The vibration direction of the upper polar may be adjusted on some microscopes, but on others there is no provision for adjustment. The vibration direction of the lower and upper polars are conventionally set so that they are at right angles to each other. When the upper polar is inserted, the polars are said to be **crossed**; with nothing on the microscope stage the field of view is dark because the plane polarized light that passes the lower polar is absorbed by the upper polar. If the upper polar is removed, the polars are said to be **uncrossed** or, alternately, the view through the microscope is with **plane light** because the light from the lower polar is plane polarized.

Bertrand Lens

The Bertrand lens (also called the Amici–Bertrand lens) is a small lens just below the ocular. It is mounted on a pivot or slide, so that it can be easily introduced or removed from the optical path. Its function is to allow the observer to view optical phenomena called interference figures that are seen near the top surface of the objective lens. Interference figures are described in detail later. In some cases, the Bertrand lens is equipped with an iris diaphragm or a pinhole to restrict the field of view.

Ocular

Oculars (eyepieces) are lenses that slide into the upper end of the microscope tube. They magnify the image provided from the objective lens and focus the light so that it can be accepted by the human eye. The usual magnification of oculars is between 5 and 12X, and many microscopes come equipped with both lower- and higher-power oculars. The total magnification of a microscope is the magnification of the objective lens times the magnification of the ocular. While a high-power ocular increases the magnification, it does not increase the resolution which is controlled by the numerical aperture of the objective. The image coming from the objective can be magnified as much as desired, but the amount of detail in the image does not change. The fuzziness gets magnified along with the image. For practical purposes, the maximum useful magnification is about 1000 times the numerical aperture of the objective.

Crosshairs or other markings are mounted in the ocular. The upper part of the ocular lens is mounted so that it can be screwed in or out to bring the crosshair into crisp focus. Detents are provided in the top of the microscope tube that match a small peg on the side of the ocular. They allow the crosshairs to be accurately oriented N–S and E–W or in the 45° positions. For routine work, oculars usually are oriented N–S and E–W. On binocular microscopes, only the right ocular is equipped with the crosshair or other marking.

Some oculars are equipped with a micrometer scale to allow measurement of the size of grains on the microscope slide (Figure 2.4). The scale must be calibrated for each different objective lens by comparison with a stage micrometer, which is a microscope slide on which a millimeter scale has been etched. The calibration is done by focusing on the stage micrometer and determining how many millimeters is represented by each unit on the ocular micrometer.

Conventional oculars are designed for use without eyeglasses. If eyeglasses are worn, the field of view through the microscope will be restricted, because the eyes are kept too far from the ocular(s). If the eyeglasses correct just for near- or farsightedness, the microscope may be used without the eyeglasses and the focus is adjusted to compensate. If the eyeglasses correct for significant amounts of astigmatism, they should be used to prevent eyestrain. Many newer microscopes are equipped with high eye-point oculars. They are intended for use with eyeglasses and usually have rubber eye cups that can be flipped up for use without eyeglasses.

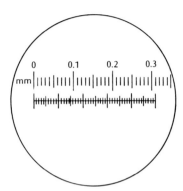

Figure 2.4 Calibration of an ocular micrometer scale. The 50 divisions of the ocular micrometer scale (bottom scale) subtend 0.31 mm. Each ocular micrometer division therefore equals 0.0062 mm. A different calibration must be made for each objective lens.

Focusing Mechanism

Focusing is accomplished by raising or lowering the stage, or by raising or lowering the microscope tube, by means of a screw or gear mechanism. Both coarse and fine focusing usually are provided for. The coarse focusing allows for rapid changes in the distance between objective lens and stage, and the fine focus allows for very precise adjustment. The mechanism may use two knobs, one for coarse and one for fine focus, or may use a single knob that incorporates both functions.

Accessories

A slot is provided immediately above the objective lenses for insertion of accessory or compensator plates. The usual accessory plates are the **gypsum plate, mica plate**, and **quartz wedge**. Each consists of a metal or plastic frame with the optical element mounted in a hole or slot in the center. Their function is described in later sections.

The gypsum plate also is known as a one-wavelength or first-order red plate. The optical element is usually a piece of quartz or gypsum, ground to a precise thickness and mounted between two pieces of glass. Depending on the manufacturer, it may be marked Gips, Gyps, rot I, Quartz sensitive tint, 1λ, $\Delta = 550$ nm, or $\Delta = 537$ nm.

The mica plate (quarter-wavelength plate) consists of a piece of muscovite or quartz that is mounted between two pieces of glass, just like the gypsum plate. Depending on the manufacturer, it may be marked Mica, Glimmer, $\frac{1}{4}\lambda$, or $\Delta = 147$ nm.

The quartz wedge in its simplest form is a piece of quartz ground into a wedge shape and cemented between two pieces of glass. It is mounted in a frame so that the thick end of the wedge is located nearest the handle.

Additional Equipment

A mechanical stage (Figure 2.5) is mounted on the stage of the microscope and grasps a slide so that the slide can be moved in a systematic way. It is most commonly employed when it is desired to determine the abundance of the different minerals in a rock thin section.

A spindle stage (Figure 2.6) consists of a fine wire spindle on which a single mineral gain is cemented (Appendix A). The spindle is mounted on a base plate so that it privots around a horizontal axis while holding a grain immersed in oil between a glass window and a cover slip. Selected procedures for using the spindle stage are described in the following chapters. For more information on the spindle stage, consult Bloss (1981).

A universal stage (Figure 2.7) is an apparatus that allows a thin section (Appendix A) to be rotated about several axes so that selected mineral grains can be precisely oriented in the optical path. Procedures for using the universal stage are beyond the scope of this book. Interested readers are referred to Hallimond (1972) or Wahlstrom (1979).

General Care of the Microscope

Microscopes must be kept as clean and dust free as possible. They should be picked up by the base and arm only, never by the microscope tube, stage, or substage, and the lenses should not be touched. Great care should be taken when cleaning the

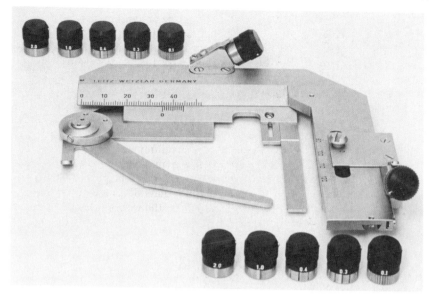

Figure 2.5 Mechanical stage. A mechanical stage is mounted on the microscope stage (cf. Figure 2.1) and securely holds a microscope slide. The small knobs are turned to move the slide. Detentes are provided on the knobs to allow the slide to be moved in uniform increments and the size of the increments can be selected by using different knobs. Photograph courtesy of Wild Leitz USA, Inc.

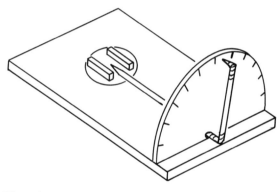

Figure 2.6 Spindle stage. A mineral grain is mounted on the spindle which can be rotated about a horizontal axis while in immersion oil.

lenses, because optical glass generally is softer than common glass and scratches easily. Dust should be removed only with compressed air and a camel hair brush. Oil, fingerprints, and related soil should be removed with lens tissue or other clean soft cloth intended for cleaning optical equipment. Be sure that all dust and grit is removed before cleaning with the lens tissue to avoid rubbing the dirt around on the lens. If a little liquid is needed, just breathe on the lens or dampen the lens tissue with distilled water. Solvents such as acetone or alcohol should not be used unless approved by the manufacturer because they may dissolve the cement holding the lenses together.

Lenses and other parts of the microscope should be disassembled only by a qualified technician.

Adjustment of the Microscope

Focusing

The microscope is focused by turning the focusing knobs mounted on the microscope base. Usually, if the microscope is focused with one of the lenses, it will remain approximately focused when another lens is inserted. If the sample is not visible with the high-power objective, the following procedure will allow the sample to be brought into focus without risking damage from inadvertently raising the stage and sample up against the objective lens.

1. While looking at the objective lens and sample from the side, raise the stage so that the sample almost touches the high-power objective lens. Avoid getting index of refraction oil on the lens if it is being used.
2. While looking through the microscope, lower the stage to bring the sample into focus. If focus cannot be achieved, the thin section is probably upside down.

On binocular microscopes, the spacing between the oculars may be adjusted to match the distance between the observer's eyes. Also, one of the ocu-

Figure 2.7 Universal stage. A thin section, mounted between glass hemispheres which are attached to the center of the universal stage, can be rotated to orient a sample so that selected optical, crystallographic, or textural features can be examined. Photograph courtesy of Nikon Inc., Instrument Division, Garden City, New York.

lars may be independently focused while the other is fixed. The ocular with the crosshair usually is the fixed ocular. To bring both oculars into focus, the microscope is first focused for the fixed ocular and then the focus of the adjustable ocular is manipulated to match.

Adjusting the Oculars

The crosshair or other marking in the ocular is brought into focus by turning the upper part of the lens in or out of the tube. This adjustment is best made with the microscope image well out of focus or with nothing on the stage. This allows the focusing

to be done with the eye in a relaxed condition and will avoid eyestrain.

Adjusting the Illuminator

Ideally, the intensity of illumination should be adjusted by inserting or removing neutral gray filters to preserve the same color balance at all intensities. However, it usually is more convenient to adjust the intensity of light with a rheostat control. The color balance of the light changes as a consequence, but this poses no problem for most work. The intensity of the light should be adjusted so that the field of view is well illuminated but not uncomfortably bright. On some microscopes, the position of the light bulb can be adjusted to provide uniform illumination. If a field diaphragm is provided, it is adjusted per the manufacturer's instructions to provide illumination to an area on the sample just slightly larger than the field of view.

Centering the Objectives

The objective lenses must be centered so that they coincide with the axis or rotation of the stage. When the objectives are centered, the crosshairs are centered on the point about which the image rotates as the stage is turned. The adjustment is accomplished in most cases by moving the lenses from side to side with two small machine screws located either in the lens collar or adjacent to the lens mounting hole in the rotating nosepiece. Some lenses are centered by rotating rings built into the lens body. The procedure is as follows (Figure 2.8).

1. Focus on a thin section and rotate the stage. Find the center of rotation, which is the stationary point about which the image rotates. Move the slide (if needed) to get a distinctive feature at the center of rotation.
2. Adjust the centering screws to move the crosshairs to the center of rotation (like aiming a rifle at a target).

It will generally be necessary to recenter the lenses periodically.

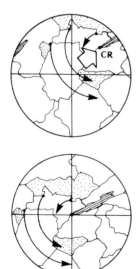

Figure 2.8 Centering the objectives. The centering mechanism is adjusted to move the cross hairs to the center of rotation (CR) seen when the stage is rotated.

On some microscopes, the location of the lenses cannot be adjusted, and the stage is moved to provide the necessary alignment. With this system, it is generally possible to have correct alignment for only one lens. The high-power objective lens should be accurately aligned and some misalignment tolerated in the other lenses. The procedure is the same as described in Figure 2.8 except that the stage, rather than the lens, is moved to produce alignment.

Adjusting the Substage

In many microscopes, the substage can be adjusted so that its axis coincides with the axis of rotation of the stage. This adjustment can be made only after the objectives have been centered. The procedure is to insert the auxiliary condensor and stop the aperture diaphragm down until it is almost closed. With the low-power objective, only a small spot of light will be seen somewhere near the center of the field of view. Move the substage with its adjusting screws until the spot of light is centered on the crosshairs.

For normal viewing through the microscope, the best quality image generally is produced when the aperture diaphragm is adjusted so that the field illuminated by the substage condensor is just slightly larger than the field of view seen through the ocular. This is most easily done by closing the diaphragm until the image just starts to dim. For optimum performance, the diaphragm should be readjusted for each change of objective lenses.

If the substage can be raised and lowered, it should be raised all the way to its upper stop so the auxiliary condensor just clears a slide placed on the stage. If the mechanism is loose or worn, the substage may creep downward because of vibration and gravity and may therefore require periodic attention.

Alignment of Polars

Depending on the manufacturer, the lower polar may pass light which vibrates either east–west or north–south. (The convention will be to refer to the field of view in terms of compass directions. North is at the top, south at the bottom, and east and west to the right and left, respectively.) The simplest way to check this alignment is with a thin section containing biotite (Figure 2.9). Biotite strongly absorbs light that vibrates parallel to its cleavage. In plane light

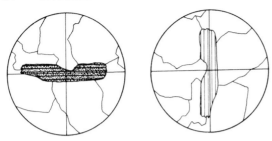

Figure 2.9 Checking alignment of the lower polar. As shown in the figure, if the lower polar passes light vibrating east–west, a grain of biotite will be darkest in plane light (upper polar removed) if the cleavage is aligned east–west and lightest when north–south. If the biotite is darkest on the north–south, then the lower polar passes light vibrating north–south.

(upper polar removed), biotite should be darkest when the cleavage is oriented parallel to the vibration direction of the lower polar and lightest when the cleavage is at right angles. Tourmaline also may be used, but it is lightest when the long dimension is parallel to the vibration direction of the lower polar. On most microscopes, the lower polar will be correctly oriented when the index mark on the polar mounting ring is placed at zero.

The upper polar may not be adjustable on student-model microscopes and is set so that its vibration direction is at right angles to the lower polar. When the upper polar is inserted into the optical path, the polars are crossed and the field of view should be black if nothing is on the microscope stage. If the field of view is dark, but not entirely black, then the polars are not oriented at exactly 90° from each other. Unless the upper polar has been broken loose or rotated in its mount, the problem generally is that the lower polar has been misadjusted. Check to ensure that the ocular is properly installed and the crosshairs are N–S, E–W. Also check to see that the substage assembly is properly installed. If readjusting the lower polar does not produce an entirely dark field of view with crossed polars, then a qualified technician should be consulted.

It is essential that the microscopist know the vibration directions of the lower and upper polar on his or her instrument because measurement of a number of optical properties depends on that knowledge. The major microscope manufacturers have adopted the convention of orienting the lower polar east–west and that convention is followed here. However, many older microscopes have the lower polar oriented north–south. The reader using a microscope whose lower polar is north–south may either adapt by bearing in mind the convention used on his or her microscope, or have a technician reorient the polars.

General Considerations

Extended periods of looking through the microscope may lead to eyestrain or headaches for some individuals. This problem can be minimized by carefully adjusting the microscope, focusing accurately, and taking a few simple precautions. The seating should provide a comfortable head position. Avoid looking down the microscope for long, unbroken periods of time. If a monocular microscope is used, keep both eyes open, and alternate looking down the microscope with the left and right eyes. The illumination reaching both eyes should be the same. It is sometimes useful to cut a hole in a piece of neutral-colored tagboard or paper and slide it over the monocular microscope tube. This will block the view and provide some illumination for the eye not looking down the microscope.

SUGGESTIONS FOR ADDITIONAL READING

Bloss, F. D., 1981, The spindle stage: principles and practice: Cambridge University Press, New York, 340 p.

Bradbury, S., 1984, An introduction to the optical microscope: Oxford University Press, Oxford, 85 p.

Hallimond, A. F., 1972, The polarizing microscope: Vickers Ltd., Vickers Instruments, York, England, 302 p.

Wahlstrom, E. E., 1979, Optical crystallography, 5th Edition: John Wiley and Sons, New York, 488 p.

3

Refractometry

While there are a variety of ways available to measure the refractive indices of minerals, by far the simplest and most convenient is the **immersion method**. With this method, the index of refraction of the mineral is compared to the known index of an immersion oil (Figure 3.1). If the indices of oil and mineral are not the same, the light passing from the oil into the mineral is refracted and the mineral grains appear to stand out. If the indices of oil and mineral are the same, the light passes through the oil–mineral boundary unrefracted and the mineral grains do not appear to stand out. If the oil and mineral are both colorless, the mineral grains may be almost indistinguishable from the oil. This can easily be demonstrated by placing a piece of common glass in water ($n = 1.33$) and in carbon tetrachloride ($n = 1.52$). Because the index of common glass is about 1.52, it stands out in water but is nearly invisible in carbon tetrachloride. By comparing the index of an unknown mineral with the known index of a variety of oils, it is possible to determine the index of the mineral. In practice, grain mounts of the unknown mineral are prepared as described in Appendix A, and examined with plane light (upper polar removed). The indices of the mineral and oil are compared. If they do not match, additional mounts are made until a match is obtained.

Relief

The degree to which mineral grains stand out from the mounting medium is called relief (Figure 3.2). If the mineral stands out strongly, the relief is high, and the mineral and oil indices differ by roughly 0.12 or more. If the mineral grains do not appear to stand out from the mounting medium, the relief is low, and the indices of refraction are within roughly 0.04 of each other. Moderate relief is intermediate and

indicates that the indices of oil and mineral differ by roughly 0.04 to 0.12. Because perception of relief is subjective to some degree, and varies depending on the adjustment of the microscope, these ranges should be used only as approximate guides.

If the index of the mineral is higher than the oil, the mineral has **positive relief**. If the index of the mineral is lower than the oil, the mineral has **negative relief**. Because positive and negative relief look alike, a method is required to determine whether the index of refraction of the oil is higher or lower than the

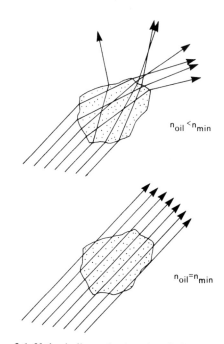

$n_{oil} < n_{min}$

$n_{oil} = n_{min}$

Figure 3.1 If the indices of mineral and oil are different, light is refracted at the mineral–oil boundary and the mineral grain stands out. If the indices of mineral and oil are the same, the light passes unrefracted and the mineral grain does not stand out.

index of refraction of the mineral. There are two convenient methods: the Becke line method and the oblique illumination method.

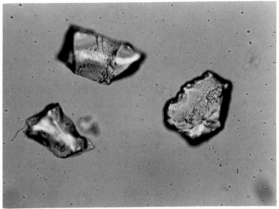

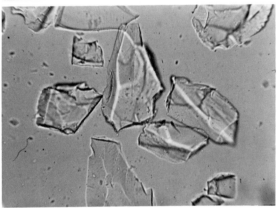

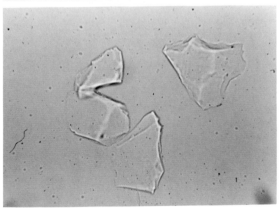

Figure 3.2 Relief. (*Top*) High relief. (*Middle*) Moderate relief. (*Bottom*) Low relief.

Becke Line Method

When the mineral grains in a grain mount are slightly out of focus, a band or rim of light called the Becke line should be visible along the grain boundaries in plane light (Figure 3.3). The rim of light may be either on the inside or the outside of the grain boundaries depending on exactly how the microscope is focused. The Becke line is usually most distinct if the aperture diaphragm is stopped down somewhat and the intermediate power objective (e.g., 10X) is used. If the focus is raised so that the distance between the sample and objective lens is increased, the Becke line appears to move into the material with the higher index of refraction. The production of the Becke line involves the lens effect and the internal reflection effect.

The lens effect depends on the observation that most mineral fragments are thinner on the edges than in the middle, so they act as crude lenses (Figure 3.4). If the mineral has a higher index of refraction than the oil, the grains act as converging lenses and the light is concentrated toward the center of the grain. If the grains have a lower index than the oil, the grains act as diverging lenses and the light is concentrated in the oil.

The internal reflection effect depends on the requirement that the edges of the grains must be vertical at some point. Moderately converging light from the condensor impinging on the vertical grain boundary is either refracted or internally reflected, depending on the angles of incidence and the indices of refraction. As can be seen in Figure 3.5, the result of the refraction and internal reflection is to concentrate light into a thin band in the material with the higher index of refraction.

Both the internal reflection effect and the lens effect tend to concentrate light into the material with the higher index of refraction in the area above the mineral grain. It is convenient to consider the Becke line to be a cone of light propagating up from the edges of the mineral (Figure 3.6). If $n_{mineral} > n_{oil}$, the cone converges above the mineral, and if $n_{mineral} < n_{oil}$, the cone diverges. If the microscope is crisply focused on the grain then the Becke line is coincident with the edge of the grain or it may disappear. If the stage is lowered so that light near the top or above the grain is brought into focus, the Becke line appears in the mineral if $n_{mineral} > n_{oil}$, or in the oil if $n_{mineral} < n_{oil}$. Hence, **as the stage is lowered, the**

Figure 3.3 The Becke line. (*Top left*) Microscope focused on grains of fluorite. (*Bottom left*) Microscope stage lowered. The Becke line (B) has moved into the mineral grains indicating that the fluorite has a higher index of refraction than the immersion oil. (*Top right*) Microscope focused on grains of fluorite. (*Bottom right*) Microscope stage lowered. The Becke line (B) has moved into the immersion oil indicating that the immersion oil has a higher index of refraction than the fluorite. Field of view of all photographs is 0.5 mm wide.

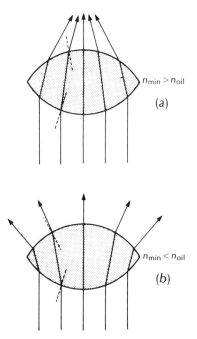

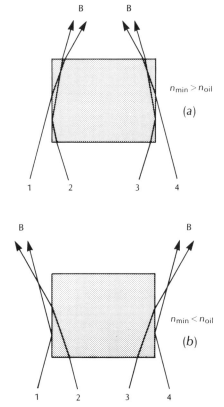

Figure 3.4 Formation of the Becke line by the lens effect. (*a*) The mineral has a higher index, so it acts as a crude converging lens concentrating the light above the mineral. The normals to the boundaries are shown for one ray to illustrate refraction towards the normal on entering the mineral and away from the normal when exiting, as predicted by Snell's law. (*b*) The mineral has a lower index, so it acts as a crude diverging lens and concentrates light around the margins of the grain. The light is refracted away from the normal on entering the mineral and towards the normal on exiting.

Becke line moves toward the material with the higher index of refraction. The reverse is true if the stage is raised.

The careful observer may notice that the Becke line often is paired with a dark band or shadow that moves in the opposite direction as the stage is raised or lowered. The dark band exists because the light that might otherwise be present there is reflected or refracted to form the Becke line.

Dispersion Effects

The dispersion of immersion oil is greater than the dispersion of most minerals, so it is possible to produce a match for the indices at only one wavelength (Figure 3.7). Ideally, the object should

Figure 3.5 Formation of the Becke line by internal reflection. (*a*) The mineral has a higher index than the oil. Rays 1 and 4 are refracted into the minerals. Rays 2 and 3 are internally reflected because they strike the boundary at greater than the critical angle. The Becke line (B) is formed by the concentration of light inside the mineral boundary. (*b*) The mineral has a lower index than the oil. Rays 2 and 3 are refracted out of the mineral grain. Rays 1 and 4 are internally reflected because they strike the boundary at greater than the critical angle. The Becke line (B) is formed by the concentration of light outside the mineral boundary.

be to produce a match at the wavelength of the yellow sodium vapor lamp (589 nm) (Figure 3.7a), because that is the wavelength for which indices of refraction are usually reported.

If the dispersion curves intersect in the visible spectrum, the oil will have higher indices of refraction for wavelengths shorter than the match, and the mineral will have higher indices of refraction for the longer wavelengths. This results in the formation of two Becke lines. One is composed of the shorter

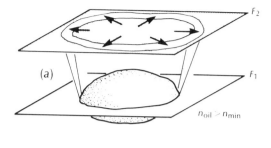

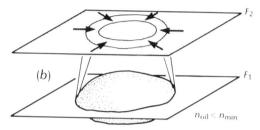

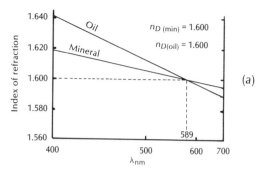

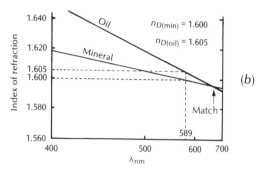

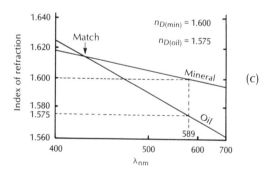

Figure 3.6 Movement of the Becke line as the stage is lowered. The Becke line can be considered to be formed of a cone of light that extends upwards from the edge of the mineral grain. (*a*) A mineral grain is immersed in oil where $n_{oil} > n_{min}$. The cone of light diverges above the mineral grain. If the stage is lowered so the plane of focus changes from F_1 to F_2, the Becke line will appear to move out of the grain and into the oil. (*b*) If $n_{oil} < n_{min}$, the cone of light forming the Becke line converges above the mineral. If the stage is lowered so the plane of focus changes from F_1 to F_2, the Becke line appears to move into the mineral grain.

wavelengths and it moves into the oil as the stage is lowered. The other is composed of the longer wavelengths and it moves into the mineral grain as the stage is lowered. The color of the lines depends on the wavelength at which the dispersion curves cross. The relationships described below are summarized in Table 3.1.

If the match is at 589 nm so that $n_{D(oil)} = n_{D(min)}$ (Figure 3.7a), then the mineral has the higher index of refraction for wavelengths longer than 589 nm. The light with these longer wavelengths, which will be perceived as yellowish orange, will form a Becke line that moves into the mineral as the stage is lowered. For wavelengths shorter than 589 nm, the oil has the higher index of refraction, and this light will form a bluish white Becke line that moves into the oil as the stage is lowered.

Figure 3.7 Formation of colored Becke lines. (*a*) Indices of mineral and oil matched for light whose wavelength is 589 nm. The longer wavelengths form a yellowish orange Becke line that moves into the mineral as the stage is lowered. The shorter wavelengths form a bluish line which moves into the oil. (*b*) Indices of mineral and oil matched at the red end of the spectrum. The mineral has higher indices only for red light, so a red line moves into the mineral, while a bluish-white line moves into the oil as the stage is lowered. (*c*) Indices of mineral and oil matched near the violet end of the spectrum. The oil has higher indices only for blue and violet light, so a blue-violet line moves into the oil and a yellowish white line moves into the mineral as the stage is lowered.

Table 3.1. Becke line relationships seen when the microscope stage is lowered

Condition	Observation	Intepretation[a]
n_{oil} higher for all wavelengths	White line into oil	$n_{D(oil)} \gg n_{D(min)}$
$n_{oil} = n_{min}$ for orange/red light	Red line into mineral, bluish white line into oil	$n_{D(oil)} > n_{D(min)}$
$n_{oil} = n_{min}$ for yellow light	Yellowish orange line into mineral, pale blue line into oil	$n_{D(oil)} = n_{D(min)}$
$n_{oil} = n_{min}$ for blue light	Blue-violet line into oil, yellowish white line into mineral	$n_{D(oil)} < n_{D(min)}$
n_{oil} lower for all wavelengths	White line into mineral	$n_{D(oil)} \ll n_{D(min)}$

[a] n_D is the index for yellow light (589 nm).

If the match is near the red (long wavelength) end of the spectrum, so that $n_{D(oil)} > n_{D(min)}$ (Figure 3.7b), then only red light whose wavelength is longer than the match can form a Becke line that moves into the mineral as the stage is lowered. The remaining portion of the spectrum, which will be perceived as bluish white, will form a Becke line that moves into the oil as the stage is lowered. The bluish white line will be significantly brighter than the red line.

If the match is near the violet (short wavelength) end of the spectrum, so that $n_{D(oil)} < n_{D(min)}$ (Figure 3.7c), then the oil has the higher index of refraction only for violet and some blue light, and the bluish violet Becke line formed of that light will move into the oil as the stage is lowered. The mineral has the higher index of refraction for longer wavelengths, which will be perceived as yellowish white, so a yellowish white Becke line will move into the mineral as the stage is lowered. The yellowish white line will be significantly brighter than the blue line.

If the dispersion curves for mineral and immersion oil do not intersect in the visible spectrum, all of the different visible wavelengths are refracted/reflected the same at the mineral–oil boundary and form a single white Becke line.

If a sodium vapor lamp or other monochromatic light source is used, then only one Becke line can be produced and it is composed of whatever wavelength of light is used. If the indices of the mineral and oil are matched for this particular wavelength, then no Becke line is produced. If the indices of the mineral and oil are not matched for this wavelength, then a Becke line is produced and it moves into the material with the higher index as the stage is lowered.

Oblique Illumination Method

The oblique illumination method of comparing the indices of a mineral and immersion liquid involves examining the "shadows" cast by the grains when part of the light coming up through the microscope is blocked. The light can be conveniently blocked by inserting a piece of cardboard (fingers also work) between the illuminator and substage so that it blocks about half of the light, or by partially inserting one of the accessory plates into the accessory slot. As the cardboard or accessory plate is inserted, half of the field of view is darkened and the grains become light on one side and darker on the other (Figure 3.8). Commonly, if the grains are dark on the side facing the darkened part of the field of view, the index of the grains is lower than the index of the oil. If the grains are light on the side facing the darkened part of the field, then the index of the grains is higher than the index of the oil. Note that this relationship may be reversed on some microscopes depending on design and where the stop is inserted. The petrographer should determine what the relationship is on his or her particular microscope using a grain mount where the relative indices of grains and oil are known.

The principles involved in producing shadows in the oblique illumination method are similar to those involved in producing the Becke line except that only light coming from one side of the field of view reaches the eye (Figure 3.9). Much of the light reflected or refracted on one side of the grains is deflected far enough that it misses the objective lens. Therefore, that side of the grain is dark. If the dispersion curves of grain and oil cross somewhere in the visible spectrum, two sets of shadows will be formed: one side of the grain will be illuminated by light of wavelengths greater than the match, and the other side will be illuminated by light of wavelengths less than the match.

Practical Considerations

Determining the index of refraction of an unknown mineral with grain mounts can be a tedious and frustrating task unless it is approached systematically. The method recommended here for routine work involves successively bracketing the index of the unknown mineral until a match can be obtained. With practice, a match can usually be obtained with the preparation of as few as four or five grain mounts as detailed in the following steps.

1. Prepare the first grain mount (see Appendix A) using an immersion oil whose index of refraction is in the middle of the available range. Examine the grains with the microscope to determine the relief and whether the mineral has a higher or lower index than the oil. Either the Becke line or oblique illumination method can be employed. If relief is very high, the oblique illumination method may give better results. If the index of the mineral is higher than the soil, then all indices less than the oil can be eliminated from consideration. If the index of the mineral is lower than the oil, then all the higher indices can be eliminated.
2. Prepare the second grain mount using an index oil somewhere in the middle of the remaining range of possible indices. Examine for relief and whether the mineral index is higher or lower than the oil. If the relief in the first mount was moderate or low, the second oil can be a little closer to the first, and if the relief in the first oil was very high, the second oil can be a little further away. The object is to bracket the index of the mineral.
3. Prepare a third grain mount using an index oil somewhere in the middle of the now much-reduced range of possible indices. Use the relief seen in the first and second mounts as a guide in selecting the oil. If the index of the mineral did not fall between the first and second oil and the mineral showed very high relief in the second oil, it may be advisable to use the oil at the extreme end of the available range (i.e., either the lowest or highest) to determine whether the mineral actually falls within the range of available oils.
4. Make a fourth grain mount, again selecting an oil to split the range of possible indices. With luck, the first three steps will have provided a bracket narrow enough to allow making a close match with the fourth

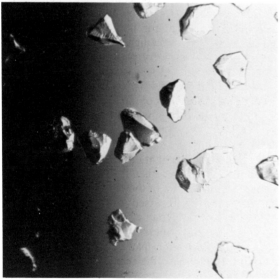

Figure 3.8 Oblique illumination. Part of the incident light is blocked, so half of the field of view is darkened. (*Top*) $n_{min} > n_{oil}$. Mineral grains are bright on the side facing the darkened portion of the field of view. (*Bottom*) $n_{min} < n_{oil}$. Mineral grains are dark on the side facing the darkened portion of the field of view. These relationships may be reversed, depending on where the light is stopped and on the construction of the microscope.

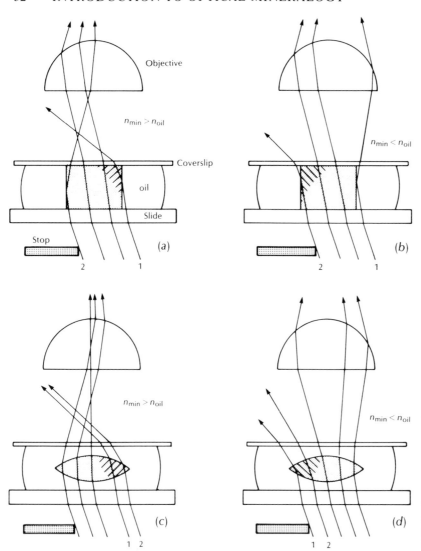

Figure 3.9 Production of shadows with oblique illumination. The location of the dark shadows are ruled. (*a*) Internal reflection effect for $n_{min} > n_{oil}$. A shadow is produced on the right side of the grain since ray 1 is refracted so it does not reach the objective lens. The left side is bright because ray 2 is internally reflected. (*b*) Internal reflection effect for $n_{min} < n_{oil}$. The shadow is on the left side of the grain because ray 2 is refracted away from the objective lens. (*c*) Lens effect for $n_{min} > n_{oil}$. The shadow is on the right side because rays 1 and 2 are refracted away from the objective lens. (*d*) Lens effect for $n_{min} < n_{oil}$. The shadow is on the left side because rays 1 and 2 are refracted away from the objective lens.

oil. If in doubt, select the fourth oil to further narrow the bracket, rather than to produce an exact match.
5. Repeat until a match is obtained.

If the index of refraction is being determined to confirm a tentative identification, then the first mount should be made using the oil having the index of the tentatively identified mineral.

If the unknown mineral is anisotropic, a number of additional considerations are involved. There are two or three indices of refraction to be determined and the index of refraction displayed by any particu-

lar grain on the slide depends on its orientation. These complexities are dealt with in later chapters. The practice of bracketing the indices of refraction still applies, however.

Accuracy of the Immersion Method

The accuracy of the immersion method depends on the accuracy to which the oils are calibrated, quality of the microscope optics, and whether white or monochromatic light is used. For routine work with white light, accuracy of about ±0.003 is possible. If a sodium vapor lamp is used to provide monochromatic yellow light, the accuracy can be improved to about ±0.001. Improved accuracy can be obtained by carefully controlling all the various factors involved but it is not necessary for most mineral identification purposes. The variation between different samples of the same mineral is often greater than the errors involved in routine measurements.

Commercially available immersion oils can be obtained in increments of 0.002, 0.004, or 0.005. Intermediate indices can be obtained by mixing two of the oils in various proportions. A simple and moderately accurate method is to place equal-sized drops of the different oils on a slide or watch glass and then mix. More precise methods of measuring the amounts of the two liquids to be mixed also can be employed. If great accuracy is required, the index of the mixed oil should be measured with a refractometer, which is an instrument used to measure refractive indices.

An important source of inaccuracy is contamination of the index oils. The usual culprits are interchanging droppers or bottle lids, or touching a dropper to a dirty slide. The index of refraction of an oil also may be altered if the container is left open so that some of the constituents of the oil are allowed to evaporate.

The indices of the oils also vary as a function of the temperature. Most are calibrated for 20°C, which is normal room temperature. If the temperature varies from that, a correction factor should be applied. The temperature correction (dn/dT) and temperature (T_r) at which the oil was calibrated will be printed on the bottle for commercially prepared oils. The equation that gives the index of the oil for temperatures other than T_r is

$$n_D(T) = n_D + (T - T_r)\left(\frac{dn}{dT}\right) \qquad 3.1$$

where n_D is the index at T_r and T is the temperature of the oil. The index of refraction decreases with increasing temperature. The correction factor is usually around $-0.0004/°C$, so a few degrees variation from 20°C is not significant for routine work.

If a heating stage is available, the temperature variation of the oil can be used to advantage. A sample is prepared so that $n_{D(oil)}$ at 20°C is higher than that of the mineral. The sample is then heated until the index of refraction of the oil is reduced to match that of the mineral. The index of refraction of the mineral, which varies little as a function of temperature, can be calculated from Equation 3.1 if the temperature at which the match is obtained and temperature correction factor are known. This method is particularly appropriate in situations where monochromatic light is used and a high degree of accuracy is desired.

Determining Indices in Thin Section

It is generally not possible to determine indices of refraction of minerals accurately in thin section, but it is possible to make estimates or to establish limits. The index of an unknown mineral can be compared to the index of the cement (balsam, epoxy, etc.) or to other known minerals in the thin section. The index of balsam is 1.537. The index of epoxy depends on manufacturer but is usually about 1.540. The relief gives an indication of how close the index of the mineral is to the cement. Becke lines form at mineral–cement and mineral–mineral boundaries and can be interpreted as described above.

SUGGESTIONS FOR ADDITIONAL READING

Allen, R. M., 1954, Practical refractometry by means of the microscope: R. P. Cargill Lab., Inc., New York.

Emmons, R. C., 1929, The double variation method of refractive index determination: American Mineralogist, v. 14, p. 482.

Emmons, R. C., and Gates, R. M., 1948, The use of Becke line colors in refractive index determinations: American Mineralogist, v. 33, p. 612–618.

Hurlbut, C. S., Jr., 1984, The jeweler's refractometer as a mineralogical tool: American Mineralogist, v. 69, p. 391–398.

Laskowski, T. E., Scotford, D. M., and Laskowski, D. E., 1979, Measurement of refractive index in thin sections using dispersion staining and oil immersion techniques: American Mineralogist, v. 64, p. 440–445.

Saylor, C. P., 1935, Accuracy of microscopial methods for determining refractive index by immersion: Journal of Research of the National Bureau of Standards, v. 15, p. 277–294.

4

Optics of Isotropic Materials

In isotropic materials, the velocity of light is the same in all directions. Materials that are optically isotropic include gases, liquids, glasses, and minerals in the isometric system. In each of these materials, the chemical bonding is the same in all directions (at least on the average), so the light passing through them sees the same electronic environment regardless of direction. However, if the solids are strained by bending them or by unevenly squeezing them, some chemical bonds will be stretched and others compressed and the normally isotropic materials will become anisotropic. The so-called liquid crystals used in the displays of calculators and other electronic equipment are anisotropic because they contain strongly aligned asymmetric molecules that produce an anisotropic structure.

Isotropic Indicatrix

A geometric figure that shows the index of refraction and vibration direction for light passing in any direction through a material is called an indicatrix (Figure 4.1). **An indicatrix is constructed so that indices of refraction are plotted on lines from the origin that are parallel to the vibration directions of the light.** Consider ray a traveling along the X axis in Figure 4.1 and vibrating parallel to the Z axis. The index of refraction for this ray is n_a and is plotted on both ends of the Z axis. For ray b which vibrates parallel to the X axis, the index of refraction n_b is plotted on the X axis. Because the material is isotropic, n_a must equal n_b, so the shape outlined in the X–Z plane is a circle. The light traveling any other direction through the material has the same index, so in three dimensions the indicatrix is a sphere.

To find the index of refraction for a light wave traveling in any particular direction, the wave normal is constructed through the center of the indicatrix

(Figure 4.2), and a slice is taken through the indicatrix perpendicular to the wave normal. The index of refraction of this light is the radius of the slice that is parallel to the vibration direction of the light.

It should be obvious that the indicatrix is not needed to tell that the index of refraction is the same in all directions through an isotropic material. The indicatrix is introduced here to prepare the reader for its application with anisotropic materials.

Distinguishing Between Isotropic and Anisotropic Minerals

Isotropic and anisotropic minerals can be quickly distinguished by crossing the polars. All isotropic minerals are dark between crossed polars. Anisotropic minerals are generally light unless they are in

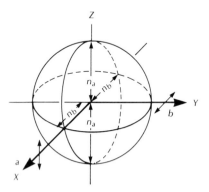

Figure 4.1 Isotropic indicatrix. The indicatrix is constructed by plotting indices of refraction parallel to the *vibration* direction of the light. Ray a vibrates parallel to the Z axis, so its index of refraction (n_a) is plotted along the Z axis. Ray b vibrates parallel to X, so n_b is plotted along the X axis. Since the indices of refraction for all vibration directions are the same, the isotropic indicatrix is a sphere whose radius is the index of refraction.

34

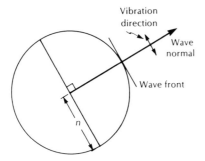

Figure 4.2 Use of the indicatrix (here shown in two dimensions). The wave normal direction is plotted and a section through the indicatrix is constructed perpendicular to the wave normal. The radius of the section (*n*) is in the index of refraction and the vibration direction is parallel to the radius. The section is parallel to the wave front.

certain orientations. Isotropic minerals are dark because they do not affect the polarization direction of the light coming up from the lower polar. The light that passes through the mineral on the stage of the microscope is absorbed by the upper polar. Occasionally, isotropic minerals may show bits of light along edges or cracks because the polarization of the light is modified somewhat by reflection (see Chapter 1).

Anisotropic minerals do affect the polarization of light that passes through them, so some light is generally able to pass through the upper polar. Because of a number of considerations that are discussed in following chapters, anisotropic minerals appear dark or extinct every 90° of rotation of the microscope stage. Anisotropic mineral grains that happen to be extinct become bright if the microscope stage is rotated a little. Also, if anisotropic mineral grains are placed in one or two specific orientations on the stage, they may behave as though they are isotropic. While a few grains of an anisotropic mineral may remain dark between crossed polars as the stage is rotated, most will not because it is highly unlikely that all will be so fortuitously oriented. If there are just a few grains, or doubt exists as to whether the mineral is isotropic, an interference figure, described in Chapters 6 and 7, can be obtained on a suspect grain. If the grain is actually anisotropic, but remains dark between crossed polars when the stage is rotated, an interference figure called an optic axis figure will be produced. In addition, some minerals may be just

slightly anisotropic and appear nearly black between crossed polars. Weak anisotropism can be detected by inserting the gypsum plate. If the suspect grain displays an interference color different than the normal magenta color of the gypsum plate, it is anisotropic. The development of interference colors and the function of the gypsum plate are discussed in Chapter 5.

Identification of Isotropic Minerals in Grain Mount

Determining the index of refraction is the primary means of identifying isotropic minerals in grain mount. Samples are prepared and the index of refraction determined as described in Chapter 3. Once the index is known, Appendix C and the mineral descriptions can be consulted to determine the identity. Unfortunately, the index of many minerals varies depending on chemical composition and different minerals may have the same index. Hence, other information is often required to confirm an identification.

If samples of the unknown mineral are large enough, it is important to determine the physical properties (color, luster, streak, hardness, cleavage or fracture, etc.) if possible. These properties are often overlooked in the rush to get a sample on the microscope. While none are diagnostic by themselves, they can help confirm a tentative identification.

Color is often useful in identifying a mineral but should be used with a certain amount of care. With some minerals, the color of different samples may be different because of chemical variation, the presence of minute inclusions, or incipient alteration. The color of isotropic minerals remains the same on rotation of the stage.

While it is generally not possible to measure the angles between cleavages accurately, it is easy to determine if the mineral has cleavage and, with careful observation, to ascertain which type of isometric cleavage is present (Figure 4.3). The cleavages commonly found in the isometric crystal system include cubic {001} (three at right angles), octahedral {111} (four cleavages that outline an octahedron), and dodecahedral {110} (six cleavages that outline a dodecahedron). If the mineral lacks cleavage, then the nature of the fracture should be noted. Isometric

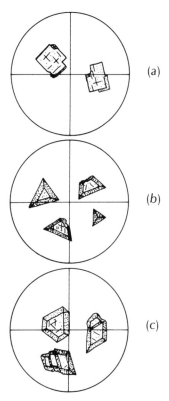

Figure 4.3 Cleavages in isometric minerals as seen in grain mount. (*a*) Cubic {001} cleavage. (*b*) Octahedral {111} cleavage. (*c*) Dodecahedral {110} cleavage.

minerals do not systematically fracture into elongate or splintery fragments.

Mineral characteristics described as structures include the presence of inclusions, color banding or variation, and the nature of the intergrowth with other minerals. Because grain mounts are made of crushed samples, textures or structures are usually not recognizable.

Mineral association is often a valuable aid in mineral identification. Even an introductory geology student soon finds that certain minerals are commonly associated, whereas others are rarely or never found together. Hence, educated guesses can often be made concerning the identity of an unknown mineral based on the identity of associated minerals. A granite, for example, will contain quartz, K-feldspar, and plagioclase but will rarely contain olivine. However, a certain amount of caution is called for. Undue reliance on association as a guide in mineral identification will blind the observer to unusual or new associations or may prevent recognition of a new mineral.

An estimate of hardness can sometimes be made by rubbing the grains between two glass slides and seeing whether the slides are scratched. Other materials also can be used.

Identification of Isotropic Minerals in Thin Section

It is not possible to determine refractive indices in thin sections accurately, but estimates can sometimes be made, as described in Chapter 3. Lacking a value for refractive index, other properties of the mineral must be used. In addition to the characteristics of the mineral described in the preceding section, it also is possible to get some idea of the crystal shape from the thin section. The shape seen in thin section is the shape of a random slice through the crystal. Allowance must be made for the fact that different slices may produce quite different shapes. Sections through cubic crystals commonly show as three- or four-sided shapes. Sections through octahedrons usually show as four- or six-sided shapes, and sections through dodecahedrons usually show as six- or eight-sided shapes. Cleavage in thin section usually shows as straight lines or cracks in the grains.

5

Optics of Anisotropic Minerals: Introduction

Anisotropic minerals are distinguished from isotropic materials because (1) **the velocity of light varies depending on direction through the mineral** and (2) **they show double refraction**. The light that enters anisotropic minerals in most directions is split into two rays with different velocities. The two rays vibrate at right angles to each other. Each anisotropic mineral has either one or two directions, called optic axes, along which the light behaves as though the mineral were isotropic. **Minerals in the hexagonal and tetragonal systems have one optic axis and are optically uniaxial. Minerals in the orthorhombic, monoclinic, and triclinic systems have two optic axes and are optically biaxial.**

We can demonstrate that the light is doubly refracted or split into two rays by placing a cleavage rhomb of clear calcite on a mark on a piece of paper (Figure 5.1). Two images corresponding to the two rays are produced. If the calcite is viewed through a piece of polarizing film whose vibration direction is parallel to the short axis of the rhomb, only one image is seen (Figure 5.1b). When the polarizing film is rotated 90° so that its vibration direction is parallel to the long axis of the rhomb, only the other image is seen (Figure 5.1c). **The two rays must therefore be plane polarized and vibrate at right angles to each other.** If the indices of refraction of the two rays are measured, it will be found that one is higher than the other. **The ray with the lower index is called the fast ray, and the ray with the higher index is called the slow ray.**

The reader needs to be cautioned for this and all following discussions to clearly keep in mind the difference between propagation direction and vibration direction. The propagation direction is the direction that the light is traveling. The vibration direction represents the side-to-side oscillation of the electric vector of the plane polarized light.

Electromagnetic theory provides an explanation

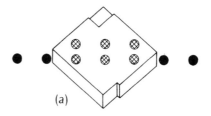

(a)

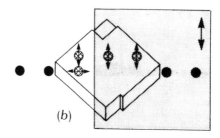

(b)

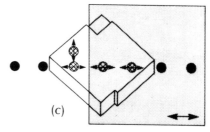

(c)

Figure 5.1 Double refraction and the calcite experiment. (a) A cleavage rhomb of clear calcite on a row of dots. Two images are produced because the light is split into two rays that vibrate at right angles to each other. (b) A polarizing film with its vibration direction parallel to the short diagonal of the rhomb passes one set of dots and absorbs the other. (c) If the polarizing film is rotated 90°, the first set of dots is absorbed and the other passes. In intermediate orientations, both sets will be visible with subdued brightness.

37

of why the velocity of light varies depending on direction through the anisotropic mineral. As described in Chapter 1, the strength of chemical bonding and the density of atoms are different in different directions. The light finds a different electronic environment depending on direction, and the electron clouds about each atom vibrate with different resonant frequencies in different directions. Because the light velocity depends on the interaction between the vibration of the electric vector and the resonant frequencies of the electron clouds, we can anticipate that velocity will vary with direction.

Electromagnetic theory also provides an explanation of why the light entering an anisotropic mineral is split into two rays vibrating at right angles to each other. Consider a wave front propagating through an anisotropic mineral (Figure 5.2). As always, the vibration vector of the light must lie in (or tangent to) the wave front. The wave front represents a planar section through the mineral. Within that plane the average density of the electron clouds and, therefore, the strength of the electric field varies with direction. With appropriate mathematical manipulations, it can be shown that a plot of the electronic field strength within the plane of the wave front is an ellipse. The axes of the ellipse, which represent the maximum and minimum field strengths, are at 90° to each other and correspond to the vibration directions of the two rays. Because the two rays "see" different electron cloud densities and field strengths with different associated resonant frequencies, their velocities and hence indices must be different. However, there will always be one (in uniaxial minerals) or two (in biaxial minerals) planes

through any anisotropic mineral that show uniform electron cloud densities (i.e., a plot of the electric field strengths is a circle). Lines at right angles to these planes are the optic axes that represent directions in the mineral along which light propagates without being split into two rays.

The mathematical rationalization of the foregoing is rather long. Because the important point is that the light *is* split into two rays with different velocities, the reader will be spared several pages of derivations. Readers wishing to pursue the topic are encouraged to read Ditchburn (1976), Jenkins and White (1976), or Phemister (1954).

Interference Phenomena

When an anisotropic mineral is placed between crossed polars, it is generally light and may show vivid colors. **The colors seen between crossed polars are called interference colors and are produced as a consequence of light being split into two rays on passing through the mineral.** For convenience of presentation, we begin our explanation of how the interference colors are formed with a description of interference phenomena produced with monochromatic light and then progress to a description of interference with polychromatic light.

Monochromatic Illumination

Consider a ray of plane polarized light that enters a plate of an anisotropic mineral (Figure 5.3). When light enters the anisotropic mineral, it is split into

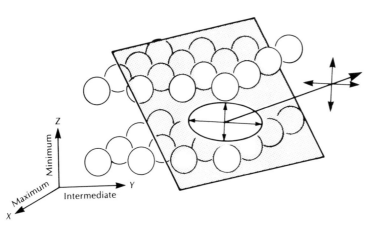

Figure 5.2 Separation of light into two rays. Atoms are closely packed along the *X* axis, moderately far apart along the *Y* axis, and widely spaced along the *Z* axis. The strength of the electric field produced by the electrons around each atom must therefore be maximum, intermediate, and minimum along the *X*, *Y*, and *Z* axes respectively. Within a random wave front (stippled), the strength of the electric field must have a maximum in one direction, and a minimum at right angles. Incident unpolarized light must interact with the electric field and is split so it vibrates parallel to the maximum and minimum electric field strengths within the plane of the wave front.

two rays that vibrate at right angles to each other and that have different velocities. The amplitude of each ray can be determined by vector addition as shown.

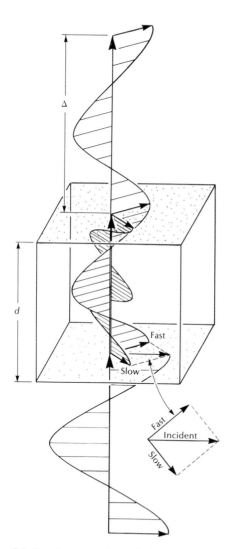

Figure 5.3 Development of retardation. The light entering the mineral with thickness d is split into slow and fast rays. In the time it takes the slow ray to pass through the mineral, the fast ray will have traveled through the mineral plus an additional distance Δ, which is the retardation.

Retardation

Because of the difference in velocity, the slow ray lags behind the fast ray. The distance that the slow ray is behind the fast ray after both have exited the crystal is called the retardation (Δ). The magnitude of the retardation depends on the thickness of the crystal plate (d) and the differences in the velocity of the slow ray (V_s) and the fast ray (V_f). The time it takes light to travel a particular distance is equal to the distance divided by the velocity. The time (t_s) it takes the slow ray to traverse the crystal is, therefore,

$$t_s = \frac{d}{V_s} \qquad 5.1$$

However, during the interval of time t_s, the fast ray not only went through the crystal but also went an additional distance equal to the retardation. Hence,

$$t_s = \frac{d}{V_f} + \frac{\Delta}{V} \qquad 5.2$$

where V is the velocity of light in air. By combining 5.1 and 5.2, we get

$$\frac{d}{V_s} = \frac{d}{V_f} + \frac{\Delta}{V}$$

which, by rearranging, gives

$$\Delta = d\left(\frac{V}{V_s} - \frac{V}{V_f}\right)$$

Because the value of V is essentially the speed of light in a vacuum, the two terms inside the parentheses are equal to n_s and n_f, which are the indices of refraction for the slow and fast rays, respectively. Hence,

$$\Delta = d(n_s - n_f) \qquad 5.3$$

This relationship is particularly important and forms the basis of a great deal of what follows.

Birefringence

The term $n_s - n_f$ is called **birefringence** (δ) and is the difference between the indices of the slow and fast rays. Its numerical value depends on the path followed by the light through the mineral. Some paths (i.e., along optic axes) show zero birefringence, others show a maximum, and most will show an intermediate value. The maximum birefringence is a characteristic of each mineral. Birefringence also may vary depending on the wavelength of the light.

The dispersion characteristics of the fast and slow rays may be different, so birefringence may be different for different wavelengths of light. Unless stated otherwise, numerical values of birefringence are for light whose wavelength is 589 nm.

Interference of the Two Rays
The microscope is arranged so that plane polarized light from the lower polar enters a mineral grain on the stage where it is split into a slow ray and a fast ray that vibrate at right angles to each other. When they exit the top of the mineral, the fast ray will be ahead of the slow ray by an amount equal to the retardation. Interference phenomena are produced when the two rays are resolved into the vibration direction of the upper polar. What happens when the two rays pass through the upper polar depends on whether they are in or out of phase.

In Figure 5.4*a*, the slow ray is retarded exactly one wavelength relative to the fast ray (remember that we are still dealing with monochromatic light) or

$$\Delta = i\lambda$$

where *i* is an integer. When the light reaches the upper polar (crossed polars) a component of each ray is resolved into the vibration direction of the upper polar. When the two rays are in phase with each other but vibrating at right angles, the resolved components are in opposite directions so they destructively interfere and cancel each other. No light passes the polar and the mineral grain appears dark.

In Figure 5.4*b*, the retardation is equal to one-half wavelength or

$$\Delta = (i + \tfrac{1}{2})\lambda$$

where *i* is an integer. As before, the rays are resolved into the vibration direction of the upper polar. However, both components are in the same direction, so the light constructively interferes and light passes the upper polar.

An alternate way of treating the same phenomena is to examine the vector sum of the two waves after they exit the top of the crystal plate. If the retardation is an integer number of wavelengths (Figure 5.4*a*), the vector sum of the two waves is a plane polarized wave vibrating parallel to the vibration direction of the incident light. Because there is no vector component of this wave that can be resolved into the vibration direction of the upper polar, no light is allowed to pass. If the retardation is $(i + \tfrac{1}{2})\lambda$

(Figure 5.4*b*), the vector sum of the two waves is a plane polarized wave vibrating parallel to the vibration direction of the upper polar, and light is allowed to pass with maximum intensity. For all other amounts of retardation, the light reaching the upper polar has either elliptical or circular polarization and some component of this light is allowed to pass.

If the sample is wedge shaped instead of flat (Figure 5.5), the thickness and retardation vary continuously along the wedge. When placed between crossed polars, the areas along the wedge where the retardation is equal to $i\lambda$ are dark and the areas where the two rays are out of phase are light. The brightest illumination is where the two rays are one-half wavelength out of phase. For a colorless mineral and ideal optical conditions with no losses from reflection or absorption, the amount of light that reaches the upper polar and is allowed to pass is given by (Johannsen, 1918)

$$T = [-\sin^2 \frac{180°}{\lambda} \frac{\Delta}{} \cdot \sin 2\tau \cdot \sin 2(\tau - 90°)] \cdot 100 \qquad 5.4$$

where *T* is percent transmission, Δ and λ are the retardation and wavelength of the light, respectively, and τ is the angle between lower polar vibration direction and the closest vibration direction in the mineral. If the mineral is placed so that $\tau = 45°$, which yields the brightest illumination, this becomes

$$T = \left(\sin^2 \frac{180°}{\lambda} \frac{\Delta}{}\right)100 \qquad 5.5$$

Figure 5.5*b* graphically illustrates this relationship for the monochromatic light passing through the quartz wedge shown in Figure 5.5*a*.

Confusion is sometimes encountered here because the light cancels when the rays are in phase, and the light passes the upper polar when the two rays are out of phase. This appears to contradict the discussion of interference in Chapter 1. Note, however, that in Figure 1.6 the two rays of light are vibrating in the same plane. When the two rays are in phase, they constructively interfere. In the present case, the two rays are in phase while vibrating at right angles to each other. The components of these two in-phase rays that are resolved into the vibration direction of the upper polar end up being out of phase (they are vibrating in opposite directions), hence the light destructively interferes.

If the upper polar is rotated so that it is parallel

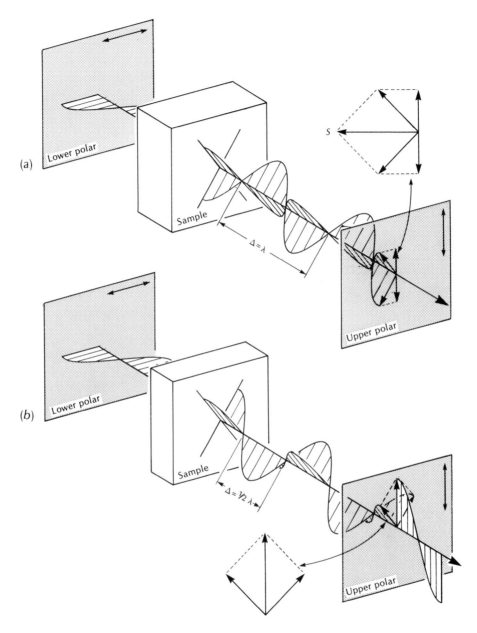

Figure 5.4 Interference at the upper polar.   (a) The retardation (Δ) is one wavelength. When the vector components of the two rays are resolved into the vibration direction of the upper polar, they are in opposite directions and cancel, so no light passes. The vector S is the vector sum of the two waves and is at right angles to the vibration direction of the upper polar.   (b) Retardation is one-half wavelength. Vector components of both rays resolved into the vibration direction of the upper polar are in the same direction, so they constructively interfere. Light passes the upper polar and the mineral appears bright.

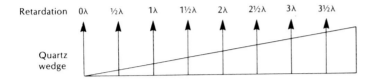

(a)

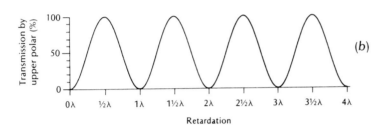

(b)

Figure 5.5 Interference pattern formed by a quartz wedge with monochromatic light. (*a*) Where the retardation is an integer number of wavelengths, the slow and fast rays destructively interfere at the upper polar and a dark band is seen. Where the retardation is $i + \frac{1}{2}$ wavelengths the two rays constructively interfere at the upper polar, and light passes with maximum intensity. (*b*) Percent transmission by the upper polar, assuming ideal optical conditions. Computed from Equation 5.5

with the lower polar, the relations described in this section are reversed. When the waves are in phase, they contructively interfere when resolved into the vibration direction of the upper polar, and they destructively interfere when they are one-half wavelength out of phase. The reader is encouraged to go through a construction similar to Figure 5.4 to demonstrate this.

Polychromatic Illumination

If white light is used instead of monochromatic light, all of the different wavelengths are present, and each is split into slow and fast rays. For a given thickness of mineral, approximately the same amount of retardation will be produced for all wavelengths. Because the wavelengths are different, some wavelengths reach the upper polar in phase and are canceled, and others are out of phase and are transmitted. The combination of different wavelengths that pass the upper polar produces the interference color. They are not an inherent property of a mineral but depend on the retardation between slow and fast rays and therefore on both the thickness and birefringence of the mineral (Equation 5.3).

If a quartz wedge is placed between crossed polars, a range of colors is produced (Plate 1). At the thin edge of the wedge, the thickness and retardation are essentially zero, so all wavelengths cancel at the upper polar ($\Delta \cong 0\lambda$) and the color is black or dark gray. As the thickness increases, the color changes from black to gray, white, yellow, red, and then a repeating sequence of blue, green, yellow, and red, with each repetition becoming progressively paler. The color produced at any particular point along the wedge depends on which wavelengths of light pass the upper polar and which are canceled (Figure 5.6). If the retardation for all wavelengths is 250 nm, then the slow and fast rays for all of the visible spectrum are substantially out of phase. Over 80 percent of every wavelength of light passes through the upper polar (Equation 5.5). The light appears white, with a slight yellow cast, because small amounts of both the red and violet ends of the spectrum are canceled at the upper polar (Figure 5.6*a*). If the retardation is 500 nm, then a substantial portion of the blue and green section of the spectrum is canceled at the upper polar and the light that passes is perceived as red (Figure 5.6*b*). If the retardation is 2500 nm, then a substantial part of each section of the spectrum is allowed to pass and the light is perceived as a creamy white (Figure 5.6c).

Because the thickness of the quartz wedge at various points can be measured and the birefringence is known, it is a simple matter to determine the amount of retardation or path difference between slow and fast rays that corresponds with each interference color. These values are plotted along the bottom edge of Plate 1.

Orders of Interference Colors

As we can see on Plate 1, the interference colors go through a repeating sequence, with the change from red to blue occurring at retardations of approximately 550, 1100, and 1650 nm. For convenience, these boundaries are used to separate the color sequence into orders. First-order colors are produced by retardations of less than 550 nm, second-order colors are for retardations between 550 and 1100 nm, and so forth. First- and second-order colors are the most vivid. The higher-order colors become progressively more and more washed out so that, above fourth order, the colors degenerate into a creamy white.

Anomalous Interference Colors

Certain minerals display anomalous interference colors not shown in the interference color chart. These colors are produced when birefringence and retardation are significantly different for different wavelengths of light. The combination of wavelengths that are in and out of phase is different than if retardation is uniform for all wavelengths (Figure 5.7). A different complement of wavelengths passes the upper polar and is perceived as a different interference color.

The color of a mineral can also influence the interference color, because some wavelengths of light are selectively absorbed by the mineral. For example, green minerals transmit green light and absorb other wavelengths to various degrees. The interference colors tend to look greenish as a consequence.

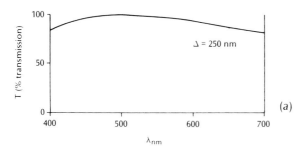

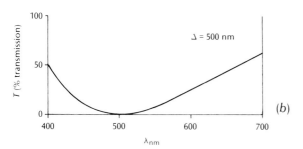

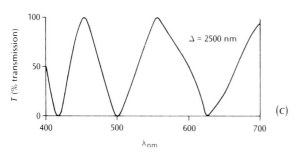

Figure 5.6 Formation of interference colors with polychromatic light. (*a*) The retardation for all wavelengths of light is 250 nm. The slow and fast rays are largely out of phase for all wavelengths, so all rays come through and form a first-order white interference color. (*b*) Retardation for all wavelengths is 500 nm. Only a portion of the red and violet ends of the spectrum are transmitted, and 500 nm light is completely blocked. The color is perceived as first-order red. (*c*) Retardation for all wavelengths is 2500 nm. Wavelengths around 417, 500, and 625 nm are canceled and wavelengths near 455, 555, and 714 nm are passed with maximum intensity. The combination of transmitted light is perceived as upper-order white.

Determining Thickness of a Sample

Equation 5.3 provides·the basis for determining the thickness of a sample. The equation contains three variables: thickness, birefringence, and retardation. If two can be determined, the third can be calculated. Retardation can be determined by examining the interference color displayed by a mineral in grain mount or thin section. The retardation correspond-

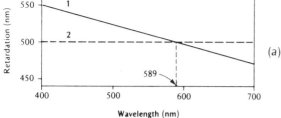

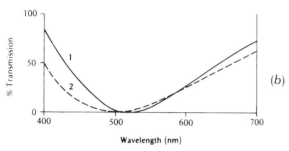

Figure 5.7 Formation of anomalous interference colors. (*a*) Mineral 1, which displays anomalous colors, his significant variation in birefringence and retardation for different wavelengths of light. Mineral 2, which shows normal interference colors, has the same birefringence and retardation for all wavelengths.   (*b*) The complement of wavelengths that are transmitted by the upper polar for mineral 1 is different than for mineral 2, despite its having the same retardation for 589 nm light. The interference color of 1 will, therefore, be different and, if not found in the normal interference color sequence, is an anomalous color.

ing to that color is read from the bottom of the interference color chart. If the birefringence of the mineral is known, it is a simple matter to calculate the thickness.

Thin Section

Quartz is an abundant and easily identified mineral, so it is commonly used to determine the thickness of thin sections. In a thin section, the interference color for different quartz grains ranges from black to a maximum color because birefringence varies from zero up to a maximum of 0.009, depending on how the grains happen to be oriented. It is the highest interference color (i.e., the one furthest to the right on the chart) that is of interest, because the quartz grains showing that color are oriented to have the maximum birefringence of 0.009.

Once the highest interference color for quartz has been recognized, the retardation corresponding to that color can be read from the chart, and a little simple arithmetic yields the thickness (Equation 5.3). The arithmetic can be bypassed by using the three sets of lines shown on the color chart. The vertical lines are retardation lines, the horizontal lines are thickness lines, and the diagonal lines are birefringence lines. An example (Figure 5.8) illustrates their use. Assume that the highest interference color shown by quartz in a thin section is a pale first-order yellow, indicating a retardation of about 315 nm. Follow the 315-nm line straight up to where it intersects the diagonal 0.009 birefringence line. The thickness (0.035 mm) is read on the left side of the diagram opposite the point of intersection.

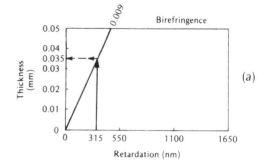

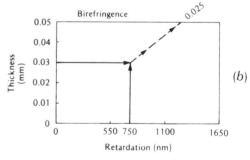

Figure 5.8 Use of the interference color chart (Plate 1). Horizontal lines indicate thickness, vertical lines indicate retardation (interference color), and diagonal lines are for birefringence.   (*a*) Determining thickness. Thickness is indicated by the point where the retardation (interference color) and birefringence lines intersect.   (*b*) Determining birefringence. Birefringence is indicated by the point where the retardation (interference color) and thickness lines intersect.

Any mineral with fairly low birefringence such as plagioclase, K-feldspar, gypsum, etc., can be used to determine thickness, provided that the maximum birefringence is known.

Grain Mount

The thickness of grains in a grain mount can be estimated using the same technique employed for thin sections. However, this method is less precise because it cannot be assumed that all of the grains are the same thickness. Also, minerals with cleavage may not be randomly oriented, because they tend to lie flat on cleavage surfaces. In practice, it is sometimes better to estimate thickness by one of the following methods.

1. Measure the dimensions of a number of grains using a micrometer eyepiece calibrated as described in Chapter 2. Because grains tend to lie flat on the slide, the thickness will be somewhat less than either the width or length measured with the micrometer eyepiece.
2. Determine maximum and minimum dimensions based on the size of sieves used to separate the grains (Table 5.1). If the grains are equant, they must be smaller than the sieve openings they pass through and larger than the openings through which they will not pass. The thickness of elongate or rodlike grains can be estimated in the same manner, although the length may be greater than the size of openings through which they pass. Platy minerals give the most trouble, because very thin grains can be trapped if their width and length are larger than the sieve openings.

Table 5.1. Sieve openings

ASTM[a] Number	Openings (mm)
80	0.177
100	0.149
120	0.125
140	0.105
170	0.088
200	0.074
230	0.062
270	0.053

[a] American Society for Testing and Materials.

Determining Birefringence from the Color Chart

Thin Section

If the thickness of a thin section is known, then the birefringence of an unknown mineral can be determined. Because the maximum birefringence of the mineral usually needs to be determined, grains of the unknown showing the highest-order interference color should be sought. In the example in Figure 5.8b, the interference color is assumed to be second-order green corresponding to a retardation of 750 nm. Follow the retardation line corresponding to that color up to where it intersects with the thickness line (usually 0.03 mm for correctly made thin sections). The diagonal birefringence line that goes through the point of intersection indicates the birefringence (0.025).

Grain Mount

Because thickness is not generally known with certainty, it is difficult to make accurate estimates of birefringence based on identifying interference colors in grain mounts. An additional complication is that grains may be over 0.1 mm thick, so high-order interference colors are commonly encountered even for minerals with moderate birefringence. As a consequence, it is more difficult to identify which interference color is present. Nonetheless, if an estimate of thickness can be made, and the interference color recognized, an estimate of birefringence can be calculated (Equation 5.3) or obtained from the curves on the color chart (p. 327).

The reader should recognize that there is substantial room for error in determining maximum birefringence of a mineral based on recognizing interference colors. If there are relatively few grains of the mineral in the thin section, there is no assurance that any grain will be correctly oriented to display the maximum birefringence. The thin section may not be of uniform thickness and the thickness usually is not precisely known. There also is difficulty in recognizing precisely which interference color is present. For routine mineral identification in thin section, this method is quite adequate. If a precise value for birefringence is needed, then oil immersion methods must be employed to measure refractive indices. The techniques used to measure refractive indices of anisotropic minerals are described in Chapters 6 and 7.

Recognizing the Different Orders of Interference Colors

The first three orders of interference colors are sufficiently similar that they may be difficult to distinguish at a glance. Colored minerals also tend to mask interference colors and add to the difficulty. In most cases, the order of an interference color up to fourth order and sometimes higher may be determined by looking at the edges of grains, particularly those situated along the margin of a thin section. Many grains are thinner at their edges than in the center, so the entire interference color sequence may be present at the edge of the grain. The order of the color in the center of the grain can be determined by "counting" the colors in. If the color bands are closely spaced it may only be possible to pick out the dark bands formed by the red and blue colors which mark the boundaries between the different orders. Persons with color blindness may find this method particularly useful.

First-order white and high-order white sometimes are confused. First-order white is usually a clear white that grades to bluish gray or yellow. High-order white tends to be somewhat creamy colored and may show pale pastel highlights of color due to irregularities on grain surfaces. If in doubt, check by inserting the gypsum plate: if the color is first-order white, it will change to an upper first- or lower second-order color; if the color is a high-order white, little change in color will occur.

Extinction

Unless an optic axis is vertical, anisotropic minerals go dark or extinct between crossed polars once in every 90° of rotation of the microscope stage. Extinction occurs when one vibration direction of the mineral is oriented parallel to the lower polar (Figure 5.9a). No component of the incident light can be resolved into the mineral vibration direction oriented parallel to the upper polar, so all of the light passing through the mineral is absorbed by the upper polar and the mineral appears dark.

If the stage is rotated so that the vibration directions of the mineral are oriented in the 45° positions (Figure 5.9b), a maximum component of both slow and fast ray is available to be resolved into the vibration direction of the upper polar. A maximum

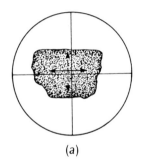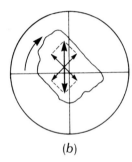

(a)   (b)

Figure 5.9 Extinction. (a) When the vibration directions of the mineral grain are parallel to the lower and upper polar, the mineral is dark or extinct between crossed polars. (b) If the grain is rotated so that its vibration directions are not parallel to the polars, then vector components of both rays pass the upper polar and the mineral appears bright.

amount of light passes and the mineral appears brightest. The interference color does not change with rotation other than to get brighter or dimmer. The phase relation between slow and fast rays and, therefore, the interference color, is unaffected by stage rotation.

Equation 5.4 can be used to predict extinction. If the angle τ between mineral vibration direction and lower polar direction is 0°, percent transmission for all wavelengths is zero. If the angle between mineral and lower polar vibration directions is 45°, then a maximum amount of light is allowed to pass the upper polar.

Many minerals are elongate or have easily recognized cleavage. **The angle between the length or cleavage of a mineral and the mineral's vibration directions is a diagnostic property called the extinction angle.** It is easily measured as follows.

1. Rotate the stage of the microscope until the length or cleavage of the mineral grain is aligned with the north–south crosshair (Figure 5.10a). Record the reading from the stage goniometer.
2. Rotate the stage until the mineral grain goes extinct (Figure 5.10b). It does not matter whether you rotate to the right or left. Record the new reading from the stage goniometer.
3. The extinction angle (EA) is the angle of rotation needed to make the mineral go extinct and is the difference in goniometer readings in 1 and 2. Note that if the extinction angle determined with clockwise rotation is EA, then the extinction angle determined with counterclockwise rotation is 90° − EA. It is usually the smaller of the two angles that needs to be reported, although there are cases where it is necess-

ary to specifically measure the extinction angle to the slow (or fast) vibration direction. The technique to determine which ray is which is discussed in a later section.

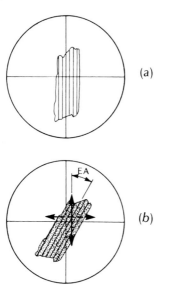

Figure 5.10 Measurement of extinction angle. (*a*) Grain oriented so that the cleavage or length is parallel to one of the crosshairs. (*b*) Stage rotated so that the grain is extinct. The extinction angle (EA) is the angle that the stage was rotated to go from (*a*) to (*b*).

If extinction angles on a number of different grains of a mineral are measured, it usually is found that they vary depending on exactly how the grains are oriented. It is usually the maximum extinction angle that is a diagnostic property.

Categories of Extinction

There are four different categories of extinction: **parallel extinction, inclined extinction, symmetrical extinction**, and **no extinction angle**.

If a mineral displays parallel extinction (Figure 5.11*a*), it is extinct when the cleavage or length is aligned with one or the other of the crosshairs. The extinction angle is 0°.

If a mineral displays inclined extinction (Figure 5.11*b*), it is extinct when the cleavage or length is at some angle to the crosshairs. The extinction angle is measured as previously described.

Minerals that have symmetrical extinction display either two cleavage directions or two distinct crystal faces (Figure 5.11*c*). Because there are two cleavages or faces, two extinction angles can be measured, one from each cleavage or crystal face. If the two extinction angles are the same, the mineral displays symmetrical extinction.

Many minerals do not have an elongated habit or prominent cleavage. While they still go extinct once

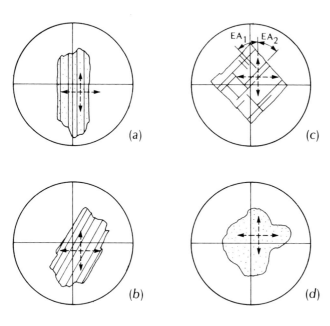

Figure 5.11 Categories of extinction. (*a*) Parallel extinction. The grain is extinct when the cleavage or length is parallel to a crosshair. (*b*) Inclined extinction. The mineral is extinct when the cleavage or length is at an angle to the crosshairs. (*c*) Symmetrical extinction. Extinction angles EA$_1$ and EA$_2$ measured to the two cleavages are the same. (*d*) No extinction angle. Grains without cleavages or a distinctive habit do not have an extinction angle.

in every 90° of stage rotation, there is no cleavage or elongation to measure an extinction angle from. Therefore, these minerals do not have an extinction angle (Figure 5.11*d*). No amount of stage rotation or head scratching will enable an extinction angle to be measured if the mineral does not display a distinct habit or prominent cleavages.

It is common to find that different parts of a single mineral grain may go extinct at different points of stage rotation. There are two causes for this behavior: strain and chemical zonation. In rocks that have been deformed, some of the grains may be bent or strained. As a consequence of the bending, different parts of a single grain are in slightly different orientations and therefore go extinct at different times. If the extinction in a grain follows an irregular or wavy pattern it is called **undulatory extinction**.

A number of minerals, such as plagioclase, which show solid solution, may crystallize so that the center of the grain has a different composition than the outer part. If the optical properties vary as a function of composition, the center of the crystal may go extinct at a different time than the outer part as the stage is rotated. There is no special term for this type of extinction, but grains displaying it are said to be zoned.

Use of the Accessory Plates

It is frequently necessary to determine which of the two rays coming through the mineral is the slow ray and which is the fast ray. This can be done quickly by using one of the accessory plates. The accessory plates consist of pieces of quartz, muscovite, or gypsum mounted in a holder so that their vibration directions are known.

Consider the mineral grain shown in Figure 5.12*a*, which is oriented on the stage so that its vibration directions are in the 45° positions. The light passing through the grain is split into two rays. When these two rays leave the top of the mineral grain, the slow ray is behind the fast ray by a distance equal to the retardation Δ_1. If the accessory plate with retardation Δ_A is superimposed over the mineral so that the slow ray vibration directions are parallel, then the ray that was the slow ray in the mineral is the slow ray in the accessory plate and is further retarded. The result is a higher total retardation ($\Delta_2 = \Delta_1 + \Delta_A$) of the two rays when they reach the upper

polar; therefore, a higher-order interference color is produced.

In Figure 5.12*b*, the mineral is rotated so that its fast ray vibration direction is parallel to the slow ray vibration direction of the accessory plate. The ray that was the slow ray in the mineral becomes the fast ray in the accessory plate. The result is that the accessory plate cancels some of the retardation produced by the mineral. The total retardation is $\Delta_3 = \Delta_1 - \Delta_A$ and the interference color produced at the upper polar is therefore a lower-order color.

Many petrographic microscopes are constructed so that the long dimension of the accessory plate is oriented NW–SE. The fast ray direction of the accessory plate is typically parallel to its length and the slow ray direction is across its width. Hence, when an accessory plate is inserted, the fast ray is NW–SE and the slow ray NE–SW. Note, however, that other conventions may be used and that some manufacturers have constructed gypsum and mica plates so that the optical element can be rotated in its holder. In the discussion that follows, the convention of having the slow ray of the accessory plate oriented NE–SW will be followed.

The gypsum plate produces around 550 nm of retardation. With white light, the interference color is a very distinctive magenta color found right at the boundary between the first and second orders. The mica plate produces 147 nm of retardation, which yields a first-order white interference color. As the name implies, the quartz wedge is wedge shaped and produces a range of retardation. The gypsum plate is most commonly used to determine vibration directions, but the quartz wedge and mica plate also may be used.

To determine which vibration direction in a mineral grain belongs to the slow ray and which to the fast ray, proceed as follows.

1. Rotate the stage of the microscope until the grain is extinct. In this position one of the vibration directions is parallel to the north–south and the other parallel to the east–west crosshair (Figure 5.13*a*).
2. Rotate the stage 45° clockwise (Figure 5.13*b*). The vibration direction that was parallel to the north–south crosshair in step 1 is now oriented NE–SW. Use the stage goniometer to measure the 45° accurately. The grain should be brightly illuminated. Note the interference color, find it on the color chart, and record the retardation that corresponds with that color (Δ_1).
3. Insert the gypsum plate into the accessory slot (Figure

5.13*c*). In most microscopes, the slow ray vibration direction of the plate is oriented NE–SW. Determine whether the interference color now shown by the grain is higher or lower. In many cases, this can be done by inspection. If the color went up, it should be the color on the chart produced by a retardation of Δ_1 + 550 nm. If the color went down, it should be the color on the chart produced by a retardation of Δ_1 − 550 nm. If the latter value is negative, use its absolute value.

4. Interpretation. If the color added (went up) then the slow ray in the accessory plate (NE–SW) is parallel to

the slow ray in the mineral grain (also NE–SW). If the color decreased, then the slow ray of the accessory plate (NE–SW) is parallel to the fast ray in the mineral grain:

> **color increases: slow on slow**
> **color decreases: slow on fast**

If the mineral displays a first-order white or gray interference color (e.g. $\Delta_1 = 200$ nm), color addition with a gypsum plate yields a second-order color ($200 + 550 = 750$ nm) while color subtraction yields a higher first-order color ($|200 − 550| \doteq 350$ nm). This may be confusing because a higher-order color is produced in both cases. The confusion is readily resolved by calculating the net retardation and hence interference color which should be obtained for both addition and subtraction as described in step 3.

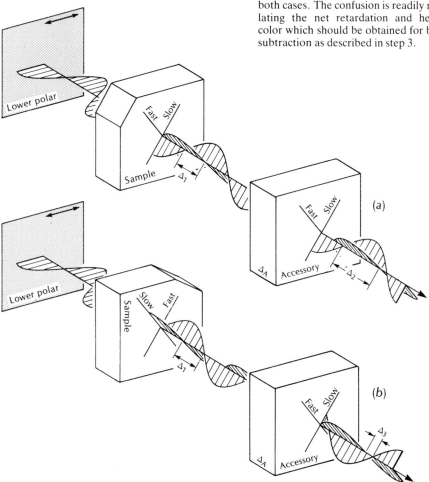

Figure 5.12 Compensation with an accessory plate.   (*a*) The sample is oriented in the 45° position and has produced retardation Δ_1. If the accessory plate with retardation Δ_A is inserted so that slow and fast ray vibration directions in accessory and sample coincide, the total retardation Δ_2 is $\Delta_1 + \Delta_A$. Since the retardation is higher, the interference color is higher.   (*b*) The slow ray vibration direction in the sample is parallel to the fast ray in the accessory. The net retardation Δ_3 is $\Delta_1 − \Delta_A$ and the interference color is lower with the accessory plate.

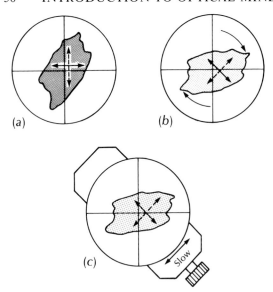

(a)   (b)

(c)   Slow

Figure 5.13 Determining slow and fast ray vibration directions. (*a*) Mineral grain at extinction between crossed polars. Vibration directions in mineral are parallel to the crosshairs. (*b*) Mineral grain rotated 45° clockwise. The vibration direction that was N–S in (*a*) is now NE–SW and the grain is brightly illuminated. (*c*) Accessory plate whose slow ray vibration is NE–SW inserted. If the interference color increases, the NE–SW ray in the mineral is its slow ray. If the interference color decreases, the NE–SW ray in the mineral is its fast ray.

The mica plate also can be used. It causes a smaller change in color as it produces a change in retardation of 147 nm rather than 550 nm.

With grain mounts, it is often useful to employ the quartz wedge instead of the mica or gypsum plate. Grains are often thin on the edges and display a range of interference colors with the lowest-order color along the edges and the highest-order color in the center. As the quartz wedge is inserted thin end first, the bands of interference colors along the edge of the grain either move into or out of the grain. If the lower-order color bands along the edge of the grain move in and displace the higher-order color in the center of the grain, colors are decreasing; the slow ray vibration direction of the wedge is superimposed on the fast ray vibration direction in the mineral. If the color bands move out so that the higher-order colors from the center of the grain displace the lower-order colors along the margin, colors are increasing; the slow ray vibration directions of mineral and quartz wedge are parallel.

Persons who are color blind also may find that the quartz wedge is useful to them. The movement of the color bands along the thin edges of grains, whether in thin section or grain mount, can be recognized even if the individual colors cannot be distinguished.

Sign of Elongation

The terms **length fast** and **length slow** are repeatedly encountered in the mineral descriptions in the latter part of this text. Length fast means that the fast ray vibrates more or less parallel to the length of an elongate mineral. Length slow means that the slow ray vibrates more or less parallel to the length of an elongate mineral. Length slow is called **positive elongation**, and length fast is called **negative elongation**. Sign of elongation is not the same as optic sign. Optic sign is discussed in Chapters 6 and 7.

Not all minerals have a sign of elongation. If a mineral does not have an elongate habit, then the term clearly does not apply. The other case where a sign of elongation cannot be assigned is if the vibration directions are ~45° to the length of a crystal.

Relief

Minerals that display moderate to strong birefringence may display a change of relief as the stage is rotated in plane light. It can be observed both in grain mount and in thin section. The change of relief is a consequence of the fact that the two rays coming through the mineral have different indices of refraction. Consider the grain of calcite shown in Figure 5.14, which is immersed in an oil having an index $n = 1.550$. If the stage is rotated so that the fast ray vibration of the mineral is E–W, all the light coming through the mineral is fast ray with $n \cong 1.57$. Because this is nearly the same as the index of the oil, the relief is low. If the grain is oriented so the slow ray vibration direction is oriented E–W, all the light coming through the mineral is slow ray with $n = 1.658$. Because this is substantially different than the index of the oil, the relief is high. In intermediate orientations, the relief appears to be intermediate. This is not because the index of the grain is intermediate. There are two rays of light coming through; one produces an image of high relief, the other produces a superimposed image of low relief. The eye effectively averages the two images so that we see an image of intermediate relief.

Because birefringence depends on the direction that the light passes through a mineral, not all grains of the mineral in a sample necessarily show the same change of relief. Grains that are oriented to display maximum birefringence show a maximum change of relief as well as highest interference color. Note also that a change of relief is seen only when the index of the mounting medium is close to the index of one of the rays. If birefringence is low, the change of relief usually is not noticed.

There also are two different Becke lines produced, one for each ray. If both rays have indices of refraction either higher or lower than the oil, the two Becke lines are superimposed and are indistinguishable. However, if the oil is selected so that its index is between the indices of the slow and fast ray, one Becke line moves into the grain, and the other moves out as the stage is lowered.

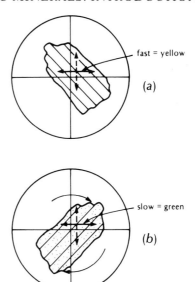

Figure 5.15 Pleochroism is seen in plane light. (*a*) The fast ray is yellow, so when it is parallel to the lower polar (E–W) the mineral is yellow. (*b*) The slow ray is green, so when it is parallel to the lower polar the mineral is green. Intermediate positions will have intermediate colors.

Pleochroism

Many colored anisotropic minerals display a change of color as the stage is rotated in plane light (upper polar removed). This change of color on rotation is called pleochroism (or dichroism). It is produced because the two rays of light are absorbed differently as they pass through the colored mineral and therefore have different colors. In Figure 5.15, the fast ray is yellow and the slow ray is green. When the stage is rotated so that the fast ray vibration direction is parallel to the lower polar, the light coming from the lower polar passes entirely as fast ray and the mineral appears yellow. If the stage is rotated 90° so the slow ray is parallel to the lower polar, the mineral appears green because all the light passes as slow ray. In intermediate positions, both slow and fast rays are present, and the color is intermediate. It should be emphasized that pleochroism is seen with the upper polar removed (i.e., in plane light) and is not related to interference colors that are seen between crossed polars.

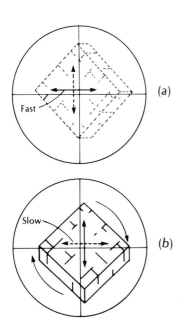

Figure 5.14 Change of relief on rotation of an anistropic mineral. (*a*) Calcite immersed in oil with $n = 1.550$. The calcite is oriented so that only fast ray with $n \cong 1.57$ passes, so the grain shows low relief. (*b*) The stage is rotated so that only slow ray with $n = 1.658$ passes, so the grain shows high relief.

REFERENCES

Ditchburn, R. W., 1976, Light (two volumes): Academic Press, London, 776 p.

Jenkins, F. A., and White, H. E., 1976, Fundamentals of optics: McGraw-Hill, New York, 746 p.

Johannsen, A., 1918, Manual of petrographic methods: McGraw-Hill, New York, 649 p.

Phemister, T. C., 1954, Fletcher's indicatrix and the electromagnetic theory of light: American Mineralogist, v. 39, p. 172–182.

SUGGESTIONS FOR ADDITIONAL READING

Jaffe, H. W., 1988, Crystal chemistry and refractivity; Cambridge University Press, Cambridge, 335 p.

Mandarino, J. A., 1959, Absorption and pleochroism: two much neglected optical properties of crystals: American Mineralogist, v. 44, p. 65–77.

Nye, J. F., 1985, Physical properties of crystals, 2nd Edition: Oxford University Press, Oxford, 329 p.

6

Uniaxial Optics

A cleavage rhomb of calcite can be used to illustrate some of the optical properties of uniaxial minerals. As described in Chapter 5 (Figure 5.1), if the cleavage rhomb is placed on a dot or other image on a piece of paper, two images appear, each composed of plane polarized light vibrating at right angles to the other. If the calcite rhomb is rotated about a vertical axis, one of the dots remains stationary, but the other rotates with the calcite about the stationary image (Figure 6.1). The image that moves behaves in a manner very different from anything found with isotropic materials, so it is called the **extraordinary ray** or ϵ ray. The stationary image is formed of light that behaves as though it were in an isotropic material, so it is called the **ordinary ray** or ω ray. The vibration vector of the ordinary ray always lies in the {0001} plane and is at right angles to the c axis. The extraordinary ray always vibrates perpendicular to the ordinary ray vibration direction in a plane that contains the c axis (Figure 6.1b). If instead of a cleavage rhomb, a slab of calcite cut in a random direction is placed on the dot, two images

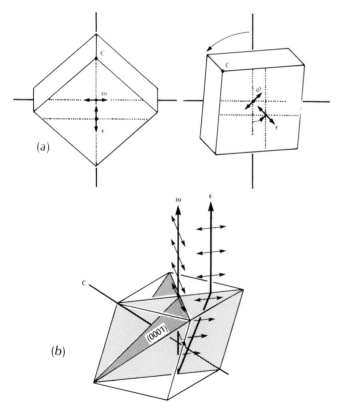

(a)

(b)

Figure 6.1 The double refraction of calcite. (*a*) Two images are produced when a clear cleavage rhomb of calcite is placed on a cross. The vibration directions for the two images indicated by the arrows are mutually perpendicular (cf. Figure 5.1). If the calcite rhomb is rotated, the image formed of ordinary rays (ω) remains stationary, and the image formed of extraordinary rays (ϵ) rotates about the stationary image. The emergence of the c axis is labeled c. (*b*) The ordinary ray (ω) pierces the (001) plane (dark shading), but the vibration vector shown by the transverse arrows lies within the (0001) plane. The extraordinary ray (ϵ) vibrates in a plane (light shading) containing the c axis and the two rays.

53

are produced for all cases except when the slab is cut so the light travels along the *c* axis. When the light travels along the *c* axis, only one image is produced; the light does not become polarized. Therefore, the *c* axis coincides with the optic axis, which is a direction through the mineral along which the light can propagate without being split into two rays.

If the indices of refraction of the two rays in calcite are determined, it will be found that the index of the ordinary ray (n_ω) is a uniform 1.658, regardless of direction through the crystal. The index of the extraordinary ray, however, is variable and ranges from 1.486 to 1.658. If the light is propagating perpendicular to the *c* axis, the extraordinary ray shows the 1.486 index. If the light is propagating almost parallel to the *c* axis, it shows an index of just slightly less than 1.658, and in intermediate directions, it shows intermediate indices. Hence, birefringence is zero for light propagating along the optic axis (*c* axis) and is a maximum for light propagating at right angles to the optic axis.

The optical properties of other uniaxial minerals are similar to those of calcite. The amount that the two images are split depends on the birefringence. For minerals with low birefringence, like quartz, the images show only slight separation even for very thick sections of mineral.

Optic Sign

In calcite the extraordinary ray has a lower index of refraction than the ordinary ray. In other minerals, however, the extraordinary ray may have the higher index. This provides the basis for defining the optic sign. **In optically positive uniaxial minerals, n_ϵ is greater than n_ω. In optically negative uniaxial minerals, n_ϵ is less than n_ω.** Or, alternately, if the extraordinary ray is the slow ray the mineral is optically positive, and if the extraordinary ray is the fast ray the mineral is optically negative.

Because the index of the extraordinary ray is variable, a few words on terminology are in order. The term n_ϵ refers to the maximum or minimum index of the extraordinary ray. This is the value recorded in the mineral descriptions in the latter part of this book. The term n_ϵ' refers to an index for an extraordinary ray that lies between n_ω and n_ϵ.

Crystallographic Considerations

Uniaxial minerals are all either hexagonal or tetragonal. Their common characteristic is a high degree of symmetry about the *c* crystallographic axis. There is uniform chemical bonding in all directions within the (001) or (0001) plane, which is at right angles to the *c* axis, and different strength bonding between these planes. The crystal structure of calcite illustrates this nicely (Figure 6.2) Calcite is constructed of alternating layers of calcium ions and triangular carbonate anion groups parallel to (0001). The chemical bonding and electronic environment is uniform in all directions within the planes of ions and is very different from the bonding between the planes.

Light that travels along the *c* axis is equally free to vibrate in any direction within (001) or (0001) (Figure 6.3). Because there are no preferred vibration directions in this plane, the light is not split into two rays; it passes through the mineral just as it would in an isotropic mineral. If light passes at some angle to the *c* axis, it finds a different electronic

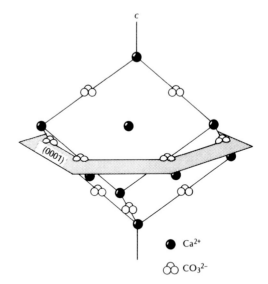

Figure 6.2 The crystal structure of calcite consists of alternating layers of Ca^{2+} and CO_3^{2-} parallel to the (0001) crystal plane. Light whose vibration vector lies in this plane finds the same electronic vibration regardless of propagation direction through the crystal. Light that vibrates across the (0001) plane finds a different electronic environment, depending on the angle made between the vibration vector and the (0001) plane.

environment for different vibration directions and the light is split into two rays with different velocities. The vibration vector of the ordinary ray is parallel to the (001) or (0001) plane and the extraordinary ray vibrates across these planes. The ordinary ray has the same velocity regardless of path because it always vibrates in the same electronic environment. The velocity of the extraordinary ray varies depending on direction. For light traveling almost parallel to the c axis, the extraordinary ray vibrates almost parallel to (001) or (0001). It encounters almost the same electronic environment as the ordinary ray, so its index n_ϵ' is nearly the same as n_ω. For light traveling at right angles to the c axis, the extraordinary ray vibrates directly across the (001) or (0001) planes and, therefore, has an index n_ϵ most different from n_ω. For intermediate directions, the extraordinary ray vibrates at an angle to (001) or (0001), so it has an intermediate index n_ϵ'. Whether the extraordinary ray has a higher or lower index than the ordinary ray depends on the chemical bonding and crystal structure of the mineral.

Uniaxial Indicatrix

As seen in the previous chapter, it is important to know the indices of refraction and vibration directions of the two rays coming through an anisotropic mineral. To provide that information, a geometric figure called an **indicatrix** is used. **The indicatrix is constructed so that the indices of refraction are plotted as radii that are parallel to the vibration direction of the light.** Figure 6.4a shows a unixial positive indicatrix. All light traveling along the Z axis, which is the optic axis, has index n_ω whether it vibrates parallel to the X or Y axis or any other direction in the X–Y plane. The X–Y plane of the indicatrix must therefore be a circle whose radius is n_ω. Light traveling along the X axis is split into two rays. The ordinary ray vibrates parallel to Y, so n_ω is again plotted along the Y axis. The extraordinary ray vibrates parallel to Z, so n_ϵ is plotted along the Z axis. The X–Z and Y–Z sections through the indicatrix are identical ellipses whose axes are n_ω and n_ϵ. If the indices for light traveling in all directions are plotted, the result is an ellipsoid of revolution whose axis is the optic axis. If the mineral is optically positive $(n_\epsilon > n_\omega)$, the ellipsoid is prolate (i.e., stretched out along the optic axis). If the mineral is optically negative $(n_\omega > n_\epsilon)$ (Figure 6.4b), the ellipsoid is oblate (i.e., flattened along the optic axis). In each case, the circular section of the indicatrix is perpendicular to the optic axis and has a radius equal to n_ω. The radius of the indicatrix along the optic axis is always n_ϵ.

A section through the indicatrix that includes the optic axis is called a **principal section**, which is an ellipse whose axes are n_ω and n_ϵ. A section through the indicatrix perpendicular to the optic axis is the **circular section** whose radius is n_ω. Any random cut through the indicatrix produces an ellipse whose axes are n_ω and n_ϵ' where n_ϵ' is between n_ω and n_ϵ.

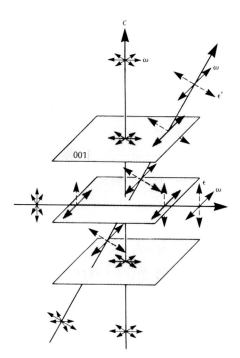

Figure 6.3 Light propagating along the c or optic axis vibrates in the (001) plane and passes entirely as ordinary ray (ω), because the electronic environment is uniform for all vibration directions with (001). Light propagating at right angles to the c axis is split into an ordinary ray whose vibration vector lies in (001) and an extraordinary ray (ϵ) that vibrates parallel to the c axis. The index for the extraordinary ray is most different from n_ω, so birefringence is maximum. Light traveling in a random direction is composed of an ordinary ray whose vibration vector lies in (001) and an extraordinary ray that vibrates at some angle to (001). The extraordinary ray has an intermediate index n_ϵ', so birefringence is intermediate.

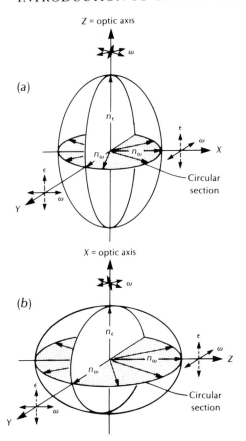

Figure 6.4 Uniaxial indicatrix. Indices of refraction are plotted parallel to *vibration* directions. (*a*) Uniaxial positive. The indicatrix is a prolate spheroid whose axes are n_ω and n_ϵ. The circular section and a principal section containing Y and Z are shown. (*b*) Uniaxial negative. The indicatrix is an oblate spheroid. The circular section and a principal section containing the X and Y axes are shown. The axes are oriented to be consistent with the convention used in biaxial minerals (Chapter 7).

Use of the Indicatrix

In practice, the uniaxial indicatrix is constructed so the circular section has radius n_ω, and the radius along the optic axis is equal to n_ϵ. It is oriented within the mineral so that the optic axis is parallel to the c crystallographic axis. The light is considered to pass directly through the center of the indicatrix. Determining the indices of refraction and vibration directions of wave normals passing in a random direction through the mineral requires the following steps.

1. Construct the wave normal direction through the center of the indicatrix (Figure 6.5*a*). Recall that the wave normal is the direction that the wave fronts are traveling.
2. Construct a slice through the center of the indicatrix perpendicular to the wave normal (Figure 6.5*b*). Unless it is perpendicular to the optic axis, the section is an ellipse whose axes are n_ϵ' and n_ω. The section is parallel to the wave fronts for both rays.
3. Interpretation. The vibration directions are parallel to the axes of the elliptical section and the indices are the lengths of the axes. In Figure 6.5*b*, the ordinary ray vibrates parallel to AB and has index n_ω. The extraordinary ray vibrates parallel to AC and has index n_ϵ'. If the angle θ between the ϵ wave normal and the optic axis is known, the value of n_ϵ' can be calculated from the equation

$$n_\epsilon' = \frac{n_\omega}{[1 + (\frac{n_\omega^2}{n_\epsilon^2} - 1) \sin^2 \theta]^{\frac{1}{2}}} \qquad 6.1$$

which is the equation of the ellipse expressed in polar coordinates. The extraordinary ray direction can be determined by constructing a tangent to the indicatrix from the wave normal and parallel to AC (Figure 6.5*c*). The extraordinary ray goes through point D where the tangent touches the indicatrix. The angle ψ between the optic axis and the extraordinary ray is given by the equation (Wahlstrom, 1979; Appendix B):

$$\tan \psi = \frac{n_\omega^2}{n_\epsilon^2} \cot (90 - \theta) \qquad 6.2$$

If the ray direction for ordinary and extraordinary rays is specified instead of the wave normal direction, the following steps allow the wave normal directions and indices of refraction to be determined.

1. Construct the ray direction through the center of the indicatrix (Figure 6.6*a*) at angle ψ from the optic axis.
2. Construct a surface tangent to the indicatrix at the point where the rays pierce it (Figure 6.6*b*). The angle ϕ between the tangent surface and the optic axis is given by the equation

$$\cot \phi = \frac{n_\epsilon^2}{n_\omega^2} \tan \psi$$

which is derived from Equation 6.2 by substituting $\phi = 90 - \theta$ (cf. Figures 6.5*d* and 6.6*d*).
3. Construct a section through the center of the indicatrix parallel to the tangent surface (Figure 6.6*b*). This section is an ellipse whose axes are n_ω and n_ϵ'.

4. Interpretation. The indices of refraction and vibration directions for the ordinary and extraordinary rays are given by the axes of the elliptical section. The ordinary ray vibrates parallel to AB and has index n_ω. The extraordinary ray vibrates parallel to AC and has index n_ϵ'. The wave front for the extraordinary ray is parallel to the section through the indicatrix. The wave front for the ordinary ray is perpendicular to the ordinary ray direction (Figure 6.6c).

Birefringence and Interference Colors

The birefringence, and therefore the interference color of uniaxial minerals depends on the direction that the light passes through the mineral. Four different cases will be examined. Three involve light that is normally incident to the surface of a mineral cut in different orientations and the fourth involves inclined incidence. The three cases of normal incidence apply when the microscope is set up for nor-

mal viewing (orthoscopic illumination). Inclined incidence applies when the auxiliary condensor is inserted to provide strongly convergent light (conoscopic illumination).

Case 1. Normal incidence on a sample cut perpendicular to the optic axis. In Figure 6.7a, a section of a mineral is cut so that the top and bottom surfaces are perpendicular to the optic axis. This could be a grain of an unknown in a thin section or in a grain mount. Because the light entering the mineral has an angle of incidence of 0°, the wave fronts are not refracted and remain parallel to the surface of the mineral section. A cut through the indicatrix parallel to the bottom of the mineral gives the indices and vibration directions of the light. In this case, the slice through the indicatrix is a circular section with radius n_ω. There is no preferred vibration direction, so the light passes along the optic axis as an ordinary ray and retains whatever vibration direction it had

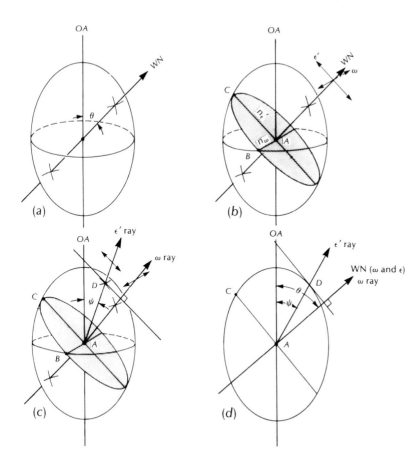

Figure 6.5 Determining indices and vibration directions for a random wave normal (WN). (a) The wave normal at angle θ to the optic axis (OA). (b) Elliptical section through the indicatrix constructed at right angles to the wave normal. The section is parallel to the wave front for both waves, and the axes of the ellipse indicate the vibration directions and indices of refraction n_ω and n_ϵ'. (c) The ϵ ray pierces the indicatrix at D, where a line parallel to the ϵ vibration direction comes tangent to the indicatrix. (d) Principal section through the indicatrix showing the relations between wave normals, rays, and the section through the indicatrix.

before entering the mineral. When placed between crossed polars, the light is entirely absorbed by the upper polar, so the mineral should appear dark on rotation just like an isotropic mineral. In an actual microscope, however, the light is somewhat converging. Even though the optic axis is vertical, some of the light passes at an angle to the optic axis and, therefore, experiences double refraction and develops retardation. The grain may not appear entirely black, particularly if the mineral has high birefringence. If the aperture diaphragm is stopped down to restrict the cone of light, the sample is more likely to appear entirely black.

Case 2. In Figure 6.7b the mineral is cut parallel to the optic axis. The indicatrix section is a principal section with axes equal to n_ω and n_ϵ. The incident light is split into two rays. The ordinary ray has index n_ω and vibrates perpendicular to the optic axis. The extraordinary ray has index n_ϵ and

vibrates parallel to the optic axis. The birefringence is $n_\epsilon - n_\omega$, which is a maximum. In a thin section or grain mount where all grains of the mineral are the same thickness, this grain would show the highest interference color. Note that this is the only case where the extraordinary ray and its wave normal are parallel.

Case 3. In Figure 6.7c, the mineral is cut in a random direction with normally incident light. As always, the ordinary ray has index n_ω and vibrates perpendicular to the optic axis. The extraordinary ray has index n_ϵ' and vibrates in a plane containing the optic axis. Because n_ϵ' is intermediate, the birefringence is intermediate. In a sample where all the grains of the mineral are the same thickness, this grain would show an intermediate interference color. If the extraordinary ray path is determined as described earlier, it will be found that it diverges from the ordinary ray path.

This is the same geometry found with the calcite

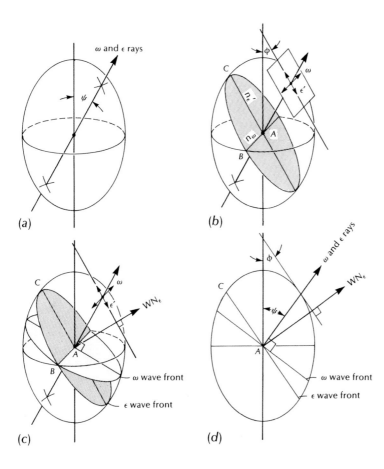

(a)   (b)   (c)   (d)

Figure 6.6 Determining indices of refraction and vibration directions for a random ray direction through an indicatrix. (*a*) Ray direction for the ω and ε rays constructed through an indicatrix. (*b*) Tangent surface constructed at the point where the rays pierce the indicatrix. The elliptical section through the center of the indicatrix is parallel to the tangent surface. The axes of the elliptical section indicate the indices and vibration directions for the two rays, and is parallel to the wave front for the extraordinary ray. (*c*) Wave front for the ω ray (light stipple) constructed at right angles to the ω ray. The ε wave normal (WN$_\epsilon$) is perpendicular to the tangent surface. (*d*) Principal section through the indicatrix showing the relation between wave normals, rays, and the sections through the indicatrix.

experiment (Figure 6.1). In Figure 6.8, the indicatrix (n_ω=1.658, n_ϵ=1.486) is constructed in the calcite rhomb at a point where normally incident light enters the rhomb from below. The section through the indicatrix parallel to the wave fronts has axes n_ω=1.658, n_ϵ'=1.566. The value of n_ϵ'=1.566 can either be determined by measuring the length of the section parallel to the ϵ vibration direction or by calculation from Equation 6.1. The inclination ψ=50.8° of the extraordinary ray from the optic axis can be calculated from Equation 6.2 or can be determined by construction as shown. From this, it can be seen that the extraordinary ray diverges from the ordinary ray by an angle of 6.2°. Because the image follows the ray, the ordinary image emerges at point O and the extraordinary image emerges at point E.

Case 4. Inclined incidence. In Figure 6.9, the mineral is cut in a random direction and the incident and refracted light lie in a principal section of the indicatrix. The refraction of the ordinary ray can be determined by applying Snell's law (Equation 1.3).

The angle of refraction (θ_ϵ) of the extraordinary wave normal is more difficult to determine since the index of the extraordinary wave (n_ϵ') varies depending on how much the light is refracted. Hence, we must solve simultaneously for both the index n_ϵ' and the angle of refraction θ_ϵ. This requires two equations that contain both terms. One equation that

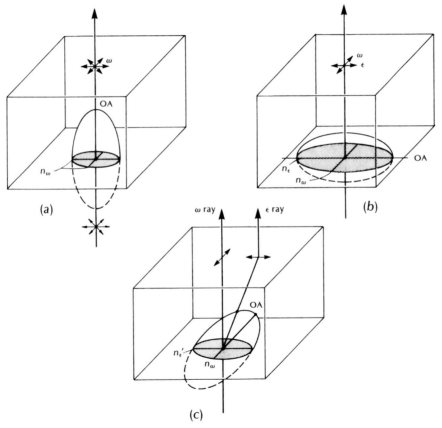

Figure 6.7 Normally incident light. The optic axis is labeled OA.   (a) Crystal cut perpendicular to the optic axis. The section through the indicatrix is the circular section, so all light passes as ordinary ray and preserves whatever vibration direction it had before entering the mineral.   (b) Mineral cut parallel to the optic axis. The section through the indicatrix is a principal section with axes n_ω and n_ϵ, so the mineral displays maximum birefringence. This is the only case where the extraordinary ray and wave normal coincide.   (c) Mineral cut in a random direction. The section through the indicatrix is an ellipse with axes n_ω and n_ϵ', so birefringence is intermediate and the ω and ϵ rays diverge.

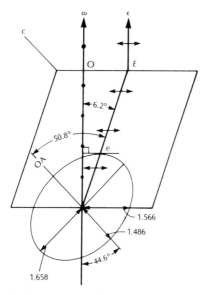

Figure 6.8 Use of the indicatrix to explain the double refraction of calcite. The section through the indicatrix parallel to the wave fronts has axes $n_\omega = 1.658$ and $n_\epsilon' = 1.566$. The ϵ-ray path is found by constructing a tangent to the indicatrix parallel to the wave front at point e. The extraordinary image, which follows the ray, must pass through e and emerge from the top of the rhomb at E. The ordinary ray emerges at O. The dimensions of the indicatrix and the angle between ω and ϵ rays are not drawn to scale.

must be satisfied is Snell's law, which for the present geometry gives

$$n_\epsilon' = \frac{\sin \theta_1}{\sin \theta_\epsilon} \qquad 6.3$$

The second equation that must be satisfied is Equation 6.1 for the radius of the indicatrix (n_ϵ'), which is at right angles to the wave normal. It can be rewritten to suit the geometry in Figure 6.9a as:

$$n_\epsilon' = \frac{n_\omega}{\left[1 + \left(\dfrac{n_\omega^2}{n_\epsilon^2} - 1\right) \sin^2 (A - \theta_\epsilon)\right]^{\frac{1}{2}}} \qquad 6.4$$

where A is the angle between the normal to the surface and the optic axis. If values of θ_ϵ are selected, the corresponding values of n_ϵ' can be calculated from Equations 6.3 and 6.4 and are plotted in Figure 6.9b. The point where the two curves cross corresponds to the only values of n_ϵ' and θ_ϵ that satisfy both equations. The ϵ ray direction can be deter-

mined by constructing a tangent to the indicatrix that is at 90° to the wave normal or by using Equation 6.2.

The birefringence in this case must be intermediate since n_ϵ' is intermediate. In general, the interference color will be higher than if the same birefringence were experienced with normal incidence because the inclined path through the mineral is longer (cf. Equation 5.3).

The most general case is where the incident and refracted rays do not lie in a principal section. The index of the extraordinary ray and its angle of refraction can be determined by a method similar to that described in case 4. However, the procedure is more complicated, because it involves dealing mathematically with angles and planes in three dimensions rather than just two. Because this type of numerical calculation will not be needed, the derivation will not be persued here.

Extinction

Recall from the discussion in Chapter 5 that anisotropic minerals go extinct between crossed polars when the vibration direction of the two rays in the mineral coincide with the vibration direction of the lower and upper polars. The previous section examined how the vibration directions in the mineral can be determined. The object in this section is to look at the relation between the vibration directions and both cleavages and crystal outlines for a variety of different cuts through both tetragonal and hexagonal minerals.

In a thin section, cleavages typically appear as thin parallel cracks in a mineral grain. If relief is high, cleavages usually can be recognized, but if relief is low, the cleavages may be difficult to see. Typically, only those cleavages that are at a substantial angle to the plane of the section are likely to be seen. However, in Figures 6.10, 6.11, and 6.12, all cleavages are shown so that the overall geometry can be seen. When examining these figures, bear in mind that cleavages not at a substantial angle to the plane of the slice through the crystal would probably not be seen if the slice were part of a real thin section.

Note also that Figures 6.10–6.12 contain information about pleochroism and interference figures, which are discussed in the following sections.

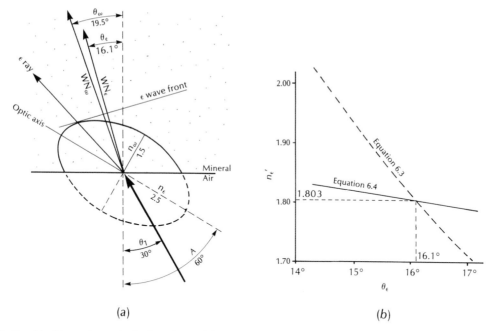

(a)   (b)

Figure 6.9 Inclined incidence in a principal section. (a) The principal section of the indicatrix is shown where an inclined ray of light enters the mineral. Refraction of the ω wave normal (WN$_\omega$) is computed from Snell's law. The ω wave front is not shown but is perpendicular to the ω wave normal. The angle of refraction (θ_ϵ) of the ε wave normal (WN$_\epsilon$) is determined in (b). The ε-ray path is determined graphically (cf. Figure 6.8). (b) Simultaneous solution of Equations 6.3 (dashed) and 6.4 (solid). The intersection of the curves gives $n_\epsilon' = 1.803$ and $\theta_\epsilon = 16.1°$.

Tetragonal Minerals

Tetragonal minerals are typically prismatic and either elongate or stubby parallel to the c axis. The faces are commonly combinations of prisms parallel to the c axis, pinacoids perpendicular to c, and pyramids, although other forms are possible. The usual cleavages are prismatic and pinacoidal. To illustrate the extinction possible with common tetragonal crystals, five sections through a typical elongate prismatic crystal displaying both prismatic and pinacoidal cleavage (Figure 6.10a) are described as follows.

Case 1. Figure 6.10b. The crystal is cut perpendicular to the optic axis, so the light traveling through the crystal follows the optic axis. The section through the indicatrix is the circular section, so all the light passes with index n_ω. The two directions of the prismatic cleavage are visible at right angles to each other. The pinacoidal cleavage is parallel to the plane of the section, so it is not seen. If the polars

are crossed, the section should be uniformly dark on rotation, so no extinction angle can be measured.

Case 2. Figure 6.10c. The crystal is cut parallel to the optic axis. Light travels through the crystal perpendicular to the optic axis. The pinacoidal cleavage shows as cracks across the width of the crystal and the prismatic cleavages show as parallel cracks along the length. The section through the indicatrix is a principal section with the ε axis parallel to the length of the crystal and the ω axis perpendicular to the length. The extinction must be parallel because the vibration directions for ω and ε are parallel to the length and across the width of the crystal. If the mineral is optically positive ($n_\omega < n_\epsilon$) it is length slow, and if it is optically negative ($n_\omega > n_\epsilon$) it is length fast.

Case 3. Figure 6.10d, e, f. The crystal is cut on an angle. As described earlier, the birefringence is intermediate. All three cleavages may be visible. The section through the indicatrix shows that

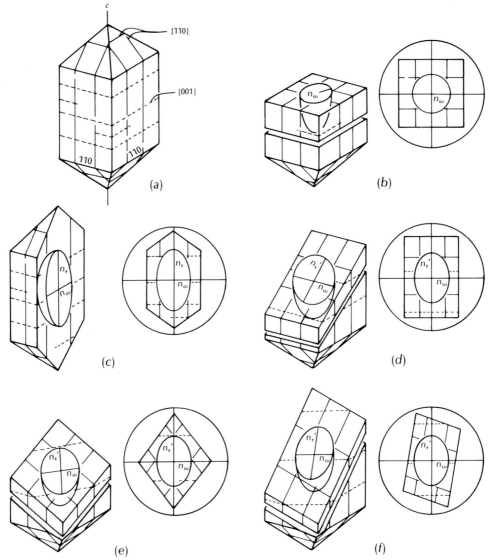

Figure 6.10 Extinction in a typical tetragonal mineral.   (*a*) Mineral showing trace of pinacoidal {001} (dashed) and prismatic {110} (solid) cleavage.   (*b*) Section cut perpendicular to optic axis (*c* axis). The indicatrix section is circular, so the grain behaves like an isotropic material. In plane light this grain would display the ω color. The interference figure would be an optic axis figure.   (*c*) Section cut parallel to the *c* axis. The indicatrix section is a principal section, so birefringence is maximum and extinction is parallel. In plane light the grain would display the ω color as shown (lower polar E–W), and if rotated 90° would display the ε color. The interference figure would be a flash figure.   (*d*) Miter cut through the mineral. Birefringence is intermediate. Extinction is parallel, as the axes of the indicatrix section are parallel to the trace of the cleavages.   (*e*) A diagonal cut produces a diamond-shaped section with intermediate birefringence. Vibration directions are parallel to the diagonals of the diamond, so extinction is parallel to the pinacoidal cleavage and symmetrical to the prismatic cleavage.   (*f*) A random cut produces a parallelogram-shaped section. Extinction is parallel to the trace of the pincoidal cleavage and asymmetric with respect to the prismatic cleavage. The grains shown in (*d*), (*e*), and (*f*) would all show the ω color in plane light as drawn. If rotated 90°, all would display a color intermediate between ω and ε. All three grains would also yield off-center interference figures. Mineral color (pleochroism) and interference figures are discussed beginning on page 65.

the index for the extraordinary ray is intermediate. Depending on how the cut through the crystal is made, the extinction to the prismatic cleavage may be parallel (Figure 6.10*d*), symmetrical (Figure 6.10*e*), or anything in between (Figure 6.10*f*). The ϵ vibration direction lies within the acute angle made between the two prismatic cleavages. The ω ray vibrates parallel to the trace of the pinacoidal cleavage, so extinction is always parallel to that cleavage.

Hexagonal Minerals

The forms commonly found in hexagonal minerals are prisms, pinacoids, pyramids, and rhombohedrons, although a number of other forms are possible. The common cleavages are prismatic, pinacoidal, and rhombohedral.

Rhombohedral cleavage

With rhombohedral cleavage, there are three cleavage planes that intersect at angles other than 90° (Figure 6.11*a*).

Case 1. The mineral is cut perpendicular to the optic axis (Figure 6.11*b*). This cut through the crystal produces either a three- or six-sided outline with cleavage parallel to the edges. Because the optic axis is vertical, the crystal displays uniform dark color on rotation between crossed polars.

Case 2. The mineral is cut parallel to the optic axis (Figure 6.11*c*). All three cleavage directions may be visible, although the angles between them depend on exactly where the crystal is cut. If it is cut perpendicular to one of the planes of symmetry, as shown in Figure 6.11*c*, then the extinction is symmetrical to two cleavage traces and parallel to the third. If the

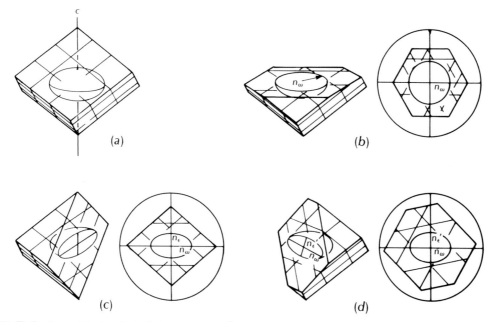

Figure 6.11 Extinction with rhombohedral cleavage {10$\bar{1}$1}. (*a*) Rhombohedral mineral showing the trace of the three cleavage directions. (*b*) Section cut perpendicular to the optic axis. The indicatrix section is circular, so the mineral should remain dark on rotation between crossed polars. All three cleavage directions should be visible. In plane light the color would be the ω color. The interference figure would be an optic axis figure. (*c*) Section cut parallel to the optic axis and perpendicular to a symmetry plane (not shown). Extinction is parallel to one cleavage and symmetrical to the others, and birefringence is maximum. If the section were cut in a random direction parallel to the *c* axis, extinction would be neither symmetrical nor parallel. In plane light the grain would display the color of the ω ray (lower polar E–W) as shown, and the color of the ϵ ray if rotated 90° to place the ϵ vibration direction parallel to the lower polar. The interference figure would be a flash figure. (*d*) Cut in a random direction. All three cleavage directions are visible, but extinction is not parallel or symmetrical to any of them. Birefringence is intermediate. In plane light the grain would display the color of the ω ray as drawn, and a color intermediate between ω and ϵ if rotated 90°. The interference figure would be an off-center figure. Mineral color (pleochroism) and interference figures are discussed beginning on page 65.

cut is in any other orientation parallel to the c axis, then the extinction will typically not be symmetrical or parallel to any of the cleavages.

Case 3. The mineral is cut in a random direction (Figure 6.11d). This is the most commonly found situation in thin sections. As can be seen from the section through the indicatrix, the vibration directions are not generally parallel or symmetrical to any of the cleavages. If, by coincidence, the section hap-

pens to be perpendicular to one of the symmetry planes, the extinction will be essentially the same as illustrated in Figure 6.11c.

Prismatic and Pinacoidal Cleavage

Figure 6.12a shows a crystal with three prismatic $\{10\bar{1}0\}$ cleavages which intersect at angles of 60° and 120°, and a $\{0001\}$ pinacoidal cleavage at right angles to the c axis.

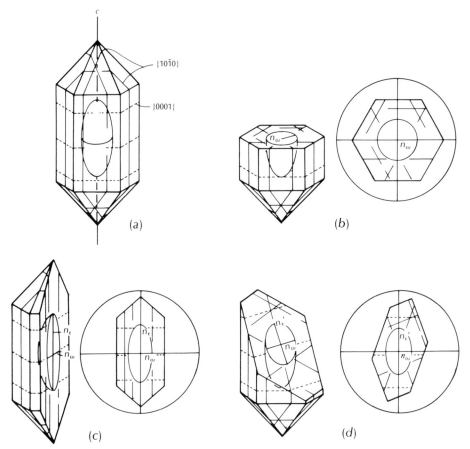

Figure 6.12 Extinction in hexagonal crystals.   (*a*) Prismatic cleavage $\{10\bar{1}0\}$ is shown with solid lines and pinacoidal cleavage $\{0001\}$ is shown with dashed lines.   (*b*) Section cut perpendicular to the optic axis. The indicatrix section is circular, so the grain should remain dark on rotation between crossed polars. In plane light this grain would display the ω color. The interference figure would be an optic axis figure.   (*c*) Section cut parallel to the optic axis. The indicatrix section is a principal section whose axes are parallel to the length and width of the grain. Extinction is parallel and birefringence is a maximum. In plane light this grain would display the ω color as shown (lower polar E–W), and the ϵ color if rotated 90°. The interference figure would be a flash figure.   (*d*) Cut in a random direction. Birefringence is intermediate and extinction is parallel to the trace of the pinacoidal cleavage. Extinction to the traces of the prismatic cleavages will generally not be parallel or symmetric unless the crystal is cut at right angles to a symmetry plane. In plane light this grain would display the ω color as shown, and a color intermediate between ω and ϵ if rotated 90°. The interference figure would be an off-center figure. Mineral color (pleochroism) and interference figures are discussed beginning on page 65.

Case 1. The mineral is cut perpendicular to the optic axis (Figure 6.12*b*). The three prismatic cleavage planes may be visible but the pinacoidal cleavage will not. The angles between the cleavages are 60° or 120°. The grain remains uniformly dark on rotation between crossed polars because the section through the indicatrix is the circular section.

Case 2. The mineral is cut parallel to the optic axis (Figure 6.12*c*). This is essentially the same as for the tetragonal mineral (Figure 6.10*c*). Extinction is parallel to both prismatic and pinacoidal cleavages.

Case 3. Random cut through the crystal (Figure 6.12*d*). It may be possible to see all of the cleavage directions. The vibration direction of the ordinary ray is always parallel to the trace of the pinacoidal cleavage. The extraordinary ray vibration direction usually is not parallel or symmetrical to the other cleavages unless the section happens to be cut perpendicular to one of the symmetry planes.

The reader should be cautioned that the examples described here are representative of the habit and cleavage found with many tetragonal and hexagonal minerals, but by no means include all possibilities. However, by studying these examples, and sketching in the same manner used here, it should be possible to get a good idea of the vibration directions and cleavages for any section through a crystal.

Pleochroism

Pleochroism, or change of color on rotation in plane light, occurs when the extraordinary and ordinary rays are absorbed differently on passing through a mineral and, therefore, have different colors. Colored uniaxial minerals are usually pleochroic. To describe the pleochroism, it is necessary to specify the color of both the ω and ϵ rays. For example, in common tourmaline (schorl), the pleochroism could be described as ω = dark green, ϵ = pale green. An alternative is to specify which ray is darker colored: $\omega > \epsilon$. If the change of color is substantial, it is described as strong pleochroism, and if there is relatively little color change it is weak pleochroism. The reader is cautioned that the description of colors is somewhat subjective and that two observers may describe the same color somewhat differently.

The following procedure can be used to determine the color of each ray for uniaxial minerals.

1. Cross the polars and search the sample for a grain that shows the lowest-order interference color. The optic axis is vertical, so all the light passes as ordinary ray. These are grains such as those shown in Figures 6.10*b*, 6.11*b*, and 6.12*b*.

2 Uncross the polar and note the color of the grain. This color is the color of the ordinary ray. Because there is no extraordinary ray present, the grain should remain the same color on rotation of the stage.

3. Cross the polars and search the sample for a grain that shows the highest-order interference color. The optic axis is horizontal, and both ordinary and extraordinary rays are present. These are grains such as those shown in Figures 6.10*c*, 6.11*c*, and 6.12*c*.

4. Uncross the polars. This grain should show the maximum change of color as the stage is rotated. The extraordinary ray color differs from the ordinary ray color, and is seen when the grain is in one of its extinction positions (check by crossing the polars). In Figures 6.10*c*, 6.11*c*, and 6.12*c*, the grains are oriented to display the color of the ω ray in plane light (lower polar E–W). If rotated 90° to place the ϵ vibration direction E–W, the grains would display the color of the ϵ ray.

Grains cut in a random direction such as those shown in Figures 6.10*d*, *e*, and *f*, 6.11*d*, and 6.12*d* display the color of the ω ray in plane light when the ω vibration is parallel to the lower polar as shown. If rotated 90° to place the ϵ' vibration direction parallel to the lower polar, they will display colors intermediate between ω and ϵ.

Interference Figure

The interference figure provides the basis for determining whether an anisotropic mineral is uniaxial or biaxial and also for determining the optic sign. The following procedure is used to obtain an interference figure.

1. Focus on a mineral grain with the high-power objective.
2. Flip in the auxiliary condensor; refocus if needed and open the aperture diaphragm.
3. Cross the polars.
4. Insert the Bertrand lens *or* remove the ocular and look down the microscope tube. The image seen with the ocular removed is smaller but somewhat crisper. Instead of an image of the grain, the interference figure consisting of a pattern of interference colors and dark bands appears near the top surface of the objective lens. The nature of the pattern depends on the orientation of the mineral grain.

Optic Axis Interference Figure

If the optic axis of the mineral grain is perpendicular to the stage, the interference figure looks like the image in Figure 6.13. It consists of a black cross superimposed on circular bands of interference colors. The cross is formed of black bars called **isogyres**. The point in the center where the two isogyres cross is called the **melatope**, and it marks the emergence of the optic axis. The interference colors increase in order outward from the melatope. Those nearest the melatope are low first order. Each band of color is called an **isochrome**. If the optic axis of the mineral is vertical, the interference figure does not move or change as the stage is rotated. The grains shown in Figures 6.10b, 6.11b, and 6.12b are oriented to produce optic axis interference figures.

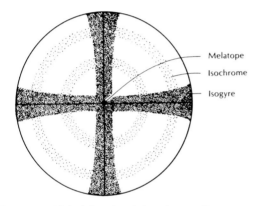

Figure 6.13 Uniaxial optic axis interference figure.

Formation of Isochromes

The formation of the isochromes is illustrated in Figure 6.14. The auxiliary condensor of the substage provides strongly convergent light that passes through the mineral and is collected by the objective lens. Light following path 1 parallel to the optic axis is not split into two rays and exits the mineral with zero retardation to form the melatope. Light following path 2 experiences moderate retardation because the value of n_ϵ' is close to n_ω. Light following path 3 at a greater angle to the optic axis encounters higher birefringence and must traverse a longer path through the mineral, so its retardation is proportionately greater. Because the optic axis is vertical and the optical properties are symmetric about the optic

axis, rings of equal retardation are produced about the melatope.

The number of isochromes visible within the field of view depends on the birefringence and thickness of the sample. Samples that are thick or that have high birefringence show more isochromes than thin or low birefringence samples because retardation is a function of both thickness and birefringence. The number of isochromes in the microscope image also depends on the numerical aperture of the objective lens. Lenses of high numerical aperture can accept a larger cone of light, so more of the higher-order isochromes are visible.

The paths of light rays as they pass through the mineral plate are somewhat simplified here. In fact, each ray that enters the mineral is split into two rays that are refracted differently and, therefore, leave the top of the mineral plate at different points. As seen earlier, in most minerals the divergence between the rays is too small to be effectively illustrated and, even if the divergence is large, to show it would unduly clutter the diagram.

Formation of Isogyres

Isogyres are formed where the vibration directions in the interference figure correspond to the vibration directions of the lower and upper polars. They are areas of extinction.

Figure 6.15a schematically shows the vibration directions for light that pierces the uniaxial positive indicatrix of the mineral shown in Figure 6.14. The vibration directions are determined using the procedure described in an earlier section. Ordinary rays vibrate parallel to lines analogous to lines of latitude. Extraordinary rays vibrate parallel to lines analogous to lines of longitude.

Vibration directions for the strongly convergent light passing through the mineral plate in Figure 6.15b can be projected into the interference figure. The extraordinary rays vibrate along radial lines like the spokes on a wheel and the ordinary rays vibrate tangent to the circular isochromes. If the polars are crossed, isogyres form where the vibration directions in the figure are parallel to the vibration directions of the polars.

Determining Optic Sign

The optic sign can be identified if it can be determined whether the ordinary ray is the fast ray or the slow ray. If it is the fast ray, the mineral is optically positive; if it is the slow ray, the mineral is optically negative. Because the vibration directions of ordinary and extraordinary rays in the interference figure are known, one of the accessory plates can be used to determine which is fast and which is slow. The procedure is as follows (Figure 6.16).

1. Obtain an optic axis interference figure.
2. Insert the accessory plate into the slot above the objective lens.
3. Observe the interference colors. In two quadrants, the interference colors increase; in the other two quadrants, the interference colors decrease.
4. Interpretation. Consider the southeast quadrant. In

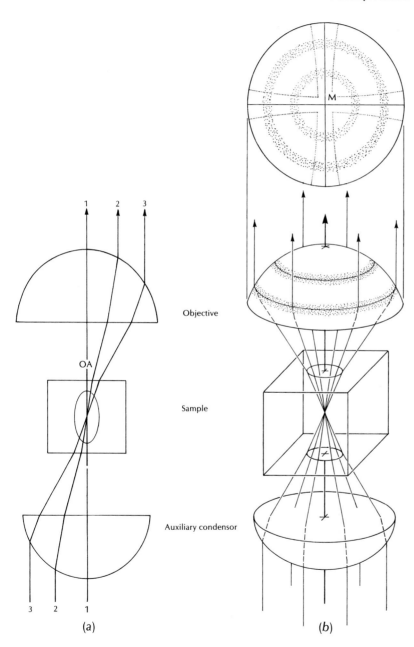

Objective

Sample

Auxiliary condensor

(a)   (b)

Figure 6.14 Formation of isochromes. (a) Light following path 1 experiences zero retardation because it follows the optic axis (OA). Paths 2 and 3 produce progressively higher retardation because both birefringence and distance through the sample increase as the inclination of the light path to the optic axis increases. (b) Optical properties are symmetric about the optic axis, so rings of equal retardation are produced around the melatope (M).

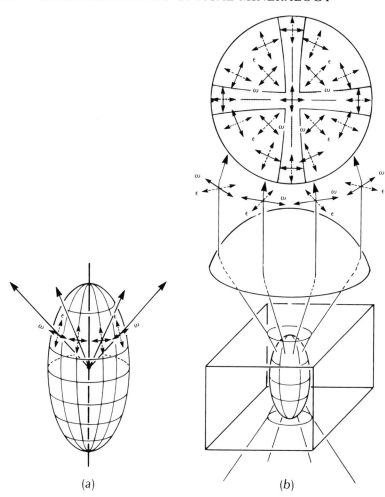

Figure 6.15 Formation of isogyres. (*a*) Vibration directions on a uniaxial positive indicatrix. Ordinary rays (ω) vibrate parallel to lines of latitude and extraordinary rays (ε) vibrate parallel to lines of longitude. (*b*) Strongly convergent light passing through the mineral exits with a vibration pattern that is symmetric about the melatope. Extraordinary rays (dashed) vibrate along radial lines and ordinary rays (solid) vibrate tangent to the circular isochromes. Isogyres are formed between crossed polars in areas where the vibration directions in the figure are parallel to the vibration directions of the lower and upper polars.

(a)   (b)

this quadrant, the ordinary ray vibrates NE–SW parallel to the isochromes and the extraordinary ray vibrates NW–SE. If the slow ray vibration direction of the accessory plate is oriented NE–SW, it is parallel to the ordinary ray vibration direction. If the colors increase in the southeast quadrant, the ordinary ray must be the slow ray and the mineral is optically negative. If the colors decrease in the southeast quadrant, the ordinary ray must be the fast ray and the mineral is optically positive. Because the vibration directions of ω and ε are the same in the northwest quadrant, its colors will increase or decrease the same as the southeast quadrant. In the northeast and southwest quadrants, the extraordinary ray vibration direction is parallel to the slow ray vibration of the accessory plate, so if the colors added in the northeast and

southwest quadrants they will decrease in the northwest and southeast quadrants and vice versa.

Either the gypsum or mica plate may be used to determine the optic sign, provided there are not too many isochromes. With the gypsum plate, first-order gray with a retardation of 200 nm either increases to second-order green (200 nm + 550 nm = 750 nm) or decreases to first-order yellow (550 nm − 200 nm = 350 nm). With the mica plate, first-order gray would increase to first-order white or yellow and decrease to low first-order gray or black as it adds or subtracts only 147 nm of retardation. If there are numerous isochromes in the interfer-

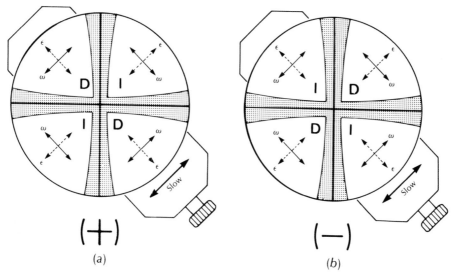

Figure 6.16 Determining optic sign. I indicates that colors increase and D indicates that colors decrease when the accessory plate is inserted.   (a) Optically positive.   (b) Optically negative.

ence figure, it may be desirable to use the quartz wedge to determine the optic sign. As the wedge is inserted thin end first, the isochromes move (Figure 6.17). In the quadrants where the colors subtract, the isochromes move outward as lower-order colors from near the melatope displace higher-order col-

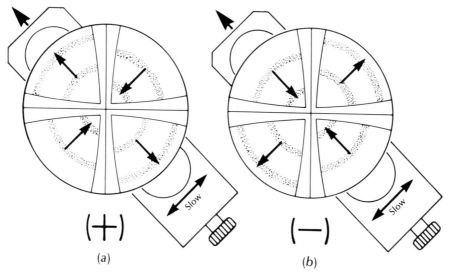

Figure 6.17 Movement of isochromes with insertion of the quartz wedge.   (a) Optically positive. Isochromes move inward in the NE and SW quadrants and outward in the NW and SE quadrants.   (b) Optically negative. Isochromes move inward in the NW and SE quadrants and outward in the NE and SW quadrants.

ors. In the quadrants where the colors add, the isochromes move in towards the melatope as higher-order colors displace lower-order colors.

With the accessory plate inserted, the isogyres adopt the interference color corresponding to the retardation of the accessory plate. Because the isogyres are areas of extinction, only one ray vibrating parallel to the lower polar direction is present. This ray is split on entering the accessory plate and develops the retardation of the accessory plate.

Off-center Optic Axis Figure

If the optic axis is inclined somewhat from the vertical, the interference figure will no longer be centered in the field of view. The isogyres still form a cross centered on the melatope, but the melatope swings in an arc around the center of the field as the stage is rotated (Figure 6.18). The isogyres retain their essentially N–S, E–W orientations and sweep across the field of view centered on the melatope. If the melatope is in the field of view, the optic sign can be determined just as it was for the centered optic axis figure. If the melatope is substantially out of the field of view (Figure 6.19), the isogyres sweep across the field in sequence as the stage is rotated. By noting the direction of isogyre movement as the stage is rotated, it is generally possible to identify which quadrant is present in the field at any particular time. If the quadrant is identified, the vibration directions of the ω and ϵ rays can be recognized and the optic sign determined by inserting an accessory plate. However, there is a hazard in working with interference figures that do not contain a melatope, because you cannot be entirely certain whether the mineral is uniaxial or biaxial. The grains shown in

Figures 6.10d, e, and f, 6.11d, and 6.12d are oriented to produce off-center figures.

Flash Figure

If the mineral grain is oriented with the optic axis horizontal, a flash figure is produced. As can be seen in Figure 6.20, the vibration directions throughout the field of view are nearly parallel. The ϵ rays vibrate approximately parallel to the trace of the optic axis, and the ω rays vibrate at approximately 90° to the optic axis. If the grain is oriented so that the optic axis is either N–S or E–W most of the field of view will be occupied by broad fuzzy isogyres. If the stage is rotated a small amount (e.g., $< 5°$) (view II), the isogyres quickly split and move out of the field of view in opposite quadrants corresponding to the quadrants into which the optic axis is being moved.

When the optic axis is placed in a 45° position (NW and SE in view III, Figure 6.20a), the entire field of view is occupied with interference colors and the isochromes are concave outward. The color in the center of the figure is the color normally displayed by the mineral between crossed polars. In the quadrants that contain the optic axis, the colors decrease away from the center because the birefringence for inclined rays is less than for rays passing at right angles to the optic axis. The decreased birefringence more than compensates for the slightly longer path length for the inclined rays (Figure 6.20b). In the remaining quadrants, the interference colors increase away from the center of the figure.

The number of isochromes depends on the thickness and birefringence of the mineral. Thick or high birefringence minerals display more isochromes

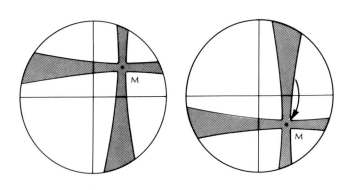

Figure 6.18 Off-center optic axis figure. As the stage is rotated, the melatope (M) sweeps around the field of view. The isogyres stay centered on the melatope and retain their essentially N–S and E–W orientation.

than thin or low birefringence minerals. If the central part of the figure is first-order white, the quadrants containing the optic axis will be first-order gray and the other quadrants will be pale first-order yellow.

The grains shown in Figures 6.10c, 6.11c, and 6.12c are oriented to produce flash figures.

The optic sign can be determined from a flash figure as follows.

1. Rotate the stage until the isogyre fills the field of view. The optic axis is now either parallel to the N–S or to the E–W crosshair.
2. Slowly rotate the stage and note the quadrants from which the isogyres leave the field of view. The optic axis will be entering the quadrants from which the isogyres exited.
3. If the isogyres left the field from the NW and SE quadrants, continue rotating the stage to make a total of 45°. This places the optic axis NW–SE. If the isogyres left the field from the NE and SW quadrants, reverse the direction of rotation so that they leave from the

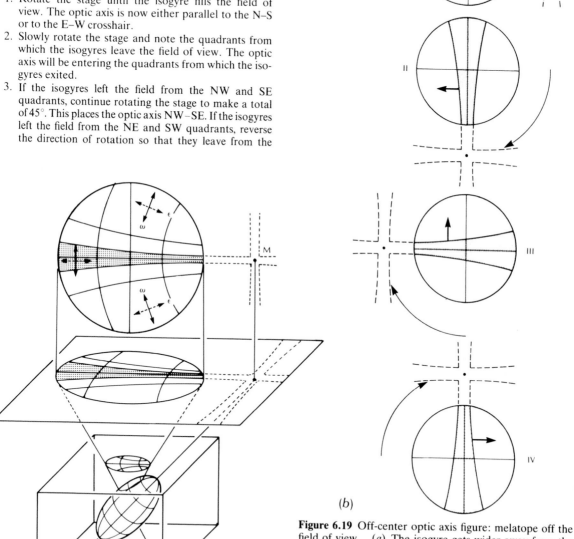

(a)

(b)

Figure 6.19 Off-center optic axis figure: melatope off the field of view. (a) The isogyre gets wider away from the position of the melatope (M). (b) Clockwise rotation of the stage causes the isogyres to sweep across the field of view in the sequence I to IV. The isogyres remain roughly parallel to the crosshairs and move in the direction indicated by the arrows.

NW and SE quadrants, then continue rotating in the new direction to make a total of 45°. Near the center of the field of view, the extraordinary ray will now be vibrating NW–SE parallel to the optic axis, and the ordinary ray will vibrate NE–SW.

4. Insert the accessory plate with slow vibration direction NE–SW and note the change in interference color.

5. Interpretation. With the optic axis oriented NW–SE, a shown in view III of Figure 6.20a, in the center of the field of view the extraordinary ray vibrates parallel to the optic axis and the ordinary ray vibrates NE–SW parallel to the slow ray vibration direction of the accessory plate. If the colors added, then the ordinary ray is slow and the mineral is negative. If the colors decreased, then the ordinary ray is fast and the mineral is positive.

Using the flash figure to determine optic sign is usually not a good practice. It is not possible to

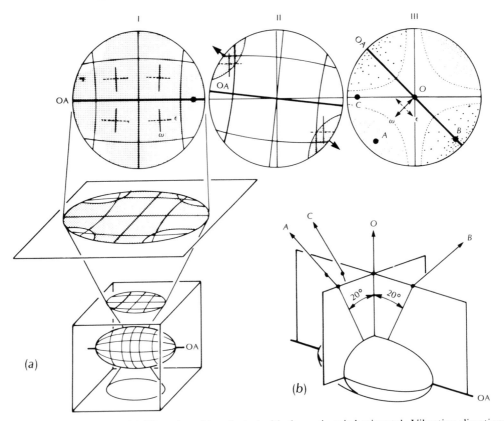

Figure 6.20 Uniaxial flash figure.   (a) The mineral is oriented with the optic axis horizontal. Vibration directions in the figure are derived in the same manner as in Figure 6.15. In view I, the optic axis (OA) is oriented exactly east–west. The isogyre is a broad, fuzzy cross because vibration directions in all but the outer parts of the four quadrants are essentially parallel to the vibration directions of the polars. In view II, the stage has been rotated a few degrees clockwise, and the isogyres quickly split and leave the field of view (the isochromes are not shown). The isogyres exit from the quadrants that the optic axis is being rotated into because only in the outer parts of those quadrants are the vibration directions oriented N–S and E–W. View III shows the figure with the optic axis NW–SE. The isochromes are concave outward. Interference colors decrease outward in the quadrants containing the optic axis and increase outward in the other quadrants.   (b) If the mineral is a 0.03-mm-thick piece of quartz with $n_\omega = 1.544$, $n_\epsilon = 1.553$, and $\delta = 0.009$, ray O at the center of the figure has a retardation of 270 nm (upper first white). Paths such as A, B, and C, which are inclined 20° from the normal, are 0.032 mm long. Path A experiences $\delta = 0.009$ because it is at right angles to the optic axis and yields a retardation of 289 nm (pale first yellow) (Equation 5.3). Path B has $n_\epsilon' = 1.552$ (Equation 6.1), birefringence of 0.008, and retardation of 253 nm (middle first white). Path C has birefringence intermediate between A and B, so its interference color is lower than A and higher than B, hence the isochromes must be concave outward. Points O, A, B, and C are shown in view III above.

determine whether a mineral is actually uniaxial with a flash figure because biaxial minerals also produce very similar flash figures. The flash figure does have some value, because it confirms that the optic axis is nearly parallel to the microscope stage. It is necessary to select grains in this orientation when measuring n_ϵ in grain mount.

Selecting Grains to Give Interference Figures

Optic Axis Figure

The optic axis figure, whether centered or slightly off center, is normally used to determine whether a mineral is uniaxial, and if so, its optic sign. The melatope should be in the field of view to be certain that the mineral is uniaxial. The optic axis of suitably oriented grains must be nearly vertical, so these grains display zero or low birefringence. They are recognized as follows.

1. Arrange the microscope for orthoscopic illumination (auxiliary condensor removed), use the low- or medium-power objective and cross the polars.
2. Scan the slide for a grain of the mineral with the lowest interference color, or ideally, for one which remains dark on rotation. Grains shown in Figures 6.10*b*, 6.11*b*, and 6.12*b* are appropriately oriented.
3. Obtain an interference figure (high-power objective, auxiliary condensor inserted, and Bertrand lens inserted). If the grain is properly oriented, the melatope will be in the field of view. If not, look for another grain.

Flash Figure

A grain that produces a flash figure has its optic axis parallel to the microscope stage. It experiences maximum birefringence and displays the highest-order interference color of all grains of the mineral in the sample. The grains shown in Figure 6.10*c*, 6.11*c*, and 6.12*c* are oriented to yield flash figures.

Determining Indices of Refraction

Grain Mount

The procedures used to measure indices of refraction in grain mounts are essentially the same as those described for isotropic minerals. The indices of oil

and mineral are compared using relief, Becke line, and oblique illumination methods. However, two different indices of refraction must be measured, n_ω and n_ϵ, so it is necessary to select grains that are oriented so that the light passes as only one or the other of the two rays.

Determining n_ω

Finding the index of the ordinary ray is easier because one of the rays passing through every grain is an ordinary ray with index n_ω. There are two ways of selecting and orienting grains so that all of the light passing through is ordinary ray.

1. Cross the polars and search for a grain whose optic axis is vertical. As has been described earlier, this grain has the lowest-order interference color and remains uniformly dark with stage rotation. All of the light passing through the mineral is ordinary ray with index n_ω. To confirm that the optic axis is actually vertical, obtain an interference figure (high-power objective, auxiliary condensor, and Bertrand lens). Determine the optic sign if that has not already been done. If the optic axis is vertical, the melatope will be in the center of the field of view. Uncross the polars, return to the orthoscopic arrangement, and compare the index n_ω to the index of the oil. Repeat with new grain mounts, using the bracketing technique described in Chapter 3, until a match between oil and n_ω is obtained. The disadvantage of this method is that it requires that grains be fortuitously oriented so that the optic axis is vertical. Grains in this orientation are sometimes difficult to find.
2. Every grain in the sample has one ray with index n_ω. The trick is to rotate the stage so that all the light coming through is ordinary ray. The index can then be compared with the oil. There are three methods that can be used alone or in conjunction with each other to place the ω vibration direction parallel to the lower polar vibration direction.
   a. Identify which of the two rays is the ω ray with one of the accessory plates. If the mineral is positive ω is the fast ray, and if it is negative ω is the slow ray. Identify the slow and fast ray vibration directions using the procedure described in Chapter 5 and place the ray corresponding to the ordinary ray parallel to the lower polar vibration direction.
   b. Scan the mount for a grain with low interference colors so that the optic axis is reasonably close to vertical. Obtain an interference figure. In most cases, it will be an off-center optic axis figure. Rotate the stage until an isogyre is superimposed on the crosshair that is perpendicular to the lower polar (Figure 6.21). Return to orthoscopic illumination (remove Bertrand lens and auxiliary condensor). The ordinary ray vibration direction is now oriented parallel to the lower

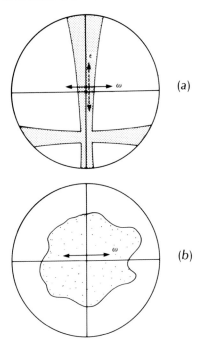

(a)

(b)

Figure 6.21 Orienting a grain using an interference figure so that only an ordinary ray passes (lower polar E–W). (*a*) The stage is rotated so that one arm of the isogyre is bisected by the crosshair that is perpendicular to the lower polar vibration direction. The vibration directions under the crosshair correspond to the vibration directions in the grain with orthoscopic illumination. (*b*) In orthoscopic illumination, the grain is now oriented so that the ordinary ray vibration direction in the grain is parallel to the lower polar vibration directions. All the light now passes as ordinary ray.

polar. As seen in the discussion of the interference figure, the vibration directions of the light in the center of the field of view are the same as the vibration directions found in the mineral in conventional orthoscopic illumination.

    c. If the mineral is pleochroic, the ω ray vibration direction can be identified by observing the color. In plane light, rotate the stage until the mineral has the color of the ω ray determined as described in an earlier section. The ω ray can be placed exactly parallel to the lower polar by crossing the polars and rotating slightly to get complete extinction.

Determining n_ϵ

Accurate measurement of n_ϵ requires grains oriented so that the optic axis is horizontal. If grains

are in any other orientation, only a value of n_ϵ', intermediate between n_ϵ and n_ω, can be measured. The technique is to select grains with the optic axis horizontal and then rotate the stage so the ϵ ray vibration direction is parallel to the lower polar. The procedure to do this is as follows.

1. Scan the grain mount for grains that show the highest interference colors with polars crossed. These are grains with the optic axis close to horizontal. To confirm that the optic axis is horizontal, obtain an interference figure. If the optic axis is horizontal, a symmetrical flash figure is produced.
2. Rotate the stage so that the ϵ ray vibration direction is parallel to the lower polar. There are two ways to do this.
    a. Use the accessory plate. If the mineral is optically positive the extraordinary ray is the slow ray, if negative it is the fast ray. From the extinction position, rotate the grain 45° and insert an accessory plate to determine which of the two rays is fast and which is slow, as described in Chapter 5. Then rotate the slow (for positive) or fast (for negative) vibration direction so it is parallel to the lower polar.
    b. Use the flash figure. Starting with the broad fuzzy isogyres occupying much of the field of view, rotate the stage slightly clockwise. If the isogyres leave the field of view from the NE and SW quadrants, the ϵ ray vibration direction was oriented N–S before rotation. If the isogyres leave from the NW and SE quadrants, then the ϵ ray vibration direction was oriented E–W. Place the ϵ ray vibration direction parallel to the lower polar, return to orthoscopic illumination, and compare n_ϵ to the index of the oil.

    Time sometimes can be saved by not being overly concerned with grain orientation in the first few oils used to identify n_ϵ. For example, if the mineral is optically positive, n_ϵ has the higher index. If even a few grains show indices higher than the oil, then it is clear that n_ϵ is higher than the oil and there is little to be gained by searching out grains with the optic axis horizontal. To enable a large number of grains to be quickly scanned, the oblique illumination method usually is more convenient than the Becke line method.

    Time also can be saved by determining the birefringence $(n_\omega - n_\epsilon)$ of the mineral, as described earlier, and computing an approximate value for n_ϵ based on knowledge of n_ω and the optic sign. Oils can then be selected accordingly. A further expedient is to make comparisons for both n_ω and n_ϵ from the same grain mounts, particularly if the birefringence is relatively low.

It is sometimes found that minerals such as the carbonates with good cleavages lie flat on their cleavage surfaces so that very few grains are oriented with their optic axis horizontal. Hence, it becomes more difficult to obtain reliable values for n_ϵ directly. Several methods can be used to get around this problem. The first is just to prepare the grain mounts with more grains to increase the odds of having some correctly oriented. The second is to prepare slides to hold the grains in random orientation. Finely ground glass may be added to the mount to prop up some of the recalcitrant grains, or gelatin-covered slides may be used. The third method is to calculate the value of n_ϵ from values of n_ϵ' (Bloss, 1961 after Tobi, 1956; Loupekine, 1947). The calculation method is the least desirable but may be useful in identifying carbonates.

Thin Section

It is not practical to accurately measure indices of refraction in thin section. However, estimates can be made by using the Becke line to compare the indices of an unknown mineral with the index of the cement or other known minerals in the thin section, and by examining the relief. It is generally more successful to compare the indices of the unknown mineral with the cement, rather than another anisotropic mineral whose indices vary depending on orientation. To make the comparison, grains need to be oriented so that only the ordinary ray or only the extraordinary ray is present. Grains in suitable orientations can be selected using the same techniques employed for grain mounts.

Spindle Stage

The spindle stage (Appendix A) allows a single grain of the unknown mineral to be rotated about a horizontal axis. This makes it possible to place the optic axis of a uniaxial mineral horizontal so that an accurate value of n_ϵ can be measured according to the following procedure.

1. With orthoscopic illumination, rotate the spindle until the grain shows maximum birefringence and the highest interference color. The optic axis should now be horizontal.
2. Change to conoscopic illumination (high-power objective, auxiliary condensor, Bertrand lens, crossed polars) and obtain an interference figure. When properly oriented, a symmetrical flash figure is observed. Rotate the microscope stage to place the optic axis at right angles to the lower polar using the technique described in an earlier section on measuring indices in grain mount.
3. Return to orthoscopic illumination. The grain should be in an extinction position. If it is not, rotate the few degrees needed to make it extinct. Uncross the polars and compare the index n_ω to the index of the immersion oil using the Becke line method.
4. Rotate the stage 90° so that the optic axis is parallel to the lower polar and all the light passes as ϵ ray. Compare the index n_ϵ to the index of the immersion oil.
5. Remove the index oil from the spindle stage using a piece of blotting paper and repeat these procedures using new index oils until matches are obtained for n_ω and n_ϵ. Use the bracketing techniques described in Chapter 3.

This procedure presumes that it is already known that the mineral is uniaxial. If this is not known, then it will be necessary to obtain an interference figure with the melatope in the field of view to determine whether the mineral is uniaxial or biaxial. By chance a grain may be mounted on the spindle in an orientation that will allow a useable interference figure to be obtained. A more reliable procedure, however, would be to examine the unknown mineral in a conventional grain mount or a thin section to determine whether it is uniaxial or biaxial, then use the spindle stage to accurately measure the indices of refraction. Bloss (1981) describes a more precise method of determining the orientations from which the indices of refraction can be measured. It involves analysis on a stereographic net of extinction angles measured for different settings of the spindle stage.

REFERENCES

Bloss, F. D., 1961, An introduction to the methods of optical crystallography: Holt, Rinehart & Winston, New York, 294 p.

Bloss, F. D., 1981, The spindle stage: principles and practice: Cambridge University Press, Cambridge, 340 p.

Loupekine, I. S., 1947, Graphical derivation of refractive index for the trigonal carbonates: American Mineralogist, v. 32, p. 502–507.

Tobi, A. C., 1956, A chart for measurement of optic axial angles: American Mineralogist, v. 41, p. 516–519.

Wahlstrom, E. E., 1979, Optical crystallography, 5th Edition: Wiley, New York, 488 p.

SUGGESTIONS FOR ADDITIONAL READING

Fairbairn, H. W., 1943, Gelatin coated slides for refractive index immersion mounts: American Mineralogist, v. 28, p. 396.

Fletcher, L., 1891, The optical indicatrix and the transmission of light in crystals: Mineralogical Magazine, v. 9, p. 278–388.

Johannsen, A., 1918, Manual of petrographic methods: McGraw-Hill, New York, 649 p.

Kamb, W. B., 1958, Isogyres in interference figures: American Mineralogist, v. 43, p. 1029–1067.

Phemister, T. C., 1954, Fletcher's indicatrix and the electromagnetic theory of light: American Mineralogist, v. 39, p. 172–192.

7

Biaxial Optics

It is useful at this point to compare crystallographic and optical properties in isotropic, uniaxial, and biaxial minerals. Isotropic minerals, which belong in the isometric system, are all highly symmetric and display the same chemical bonding and crystal structure in all directions. The unit cell has the same dimension along all three crystallographic axes, and only one index of refraction is necessary to describe the optical properties for monochromatic light.

Uniaxial minerals, which belong in the tetragonal and hexagonal systems, are less symmetric and display the same chemical bonding and crystal structure in all directions at right angles to the c crystallographic axis but different structure and bonding in other directions. To describe their crystallographic properties, it is necessary to specify two unit cell dimensions: one along the c axis, the other at right angles. To describe the optical properties, it is similarly necessary to specify two indices of refraction, n_ϵ and n_ω.

Biaxial minerals include the orthorhombic, monoclinic, and triclinic crystal systems. They are less symmetric than uniaxial minerals and vary in crystal structure and chemical bonding in all directions. To describe their crystallographic properties, it is necessary to specify the lengths of the unit cell along all three crystallographic axes. Similarly, it is necessary to specify three different indices of refraction. Various conventions have been used to identify the three principal indices. One of the more popular is α, β, and γ. While there is precedent for using this nomenclature, it is ignored here so that all indices of refraction are identified with the symbol n and so that the symbols for the indices are not confused with the angles between the crystal axes, which are usually identified α, β, and γ. **The convention in this book is to identify the three indices n_α, n_β, and n_γ, where $n_\alpha < n_\beta < n_\gamma$.** Other conventions that may be encountered include n_x, n_y, n_z; N_x, N_y, N_z; nX, nY,

nZ; n_a, n_b, n_c; X, Y, Z; n_1, n_2, n_3; and n_p, n_m, n_g. **The maximum birefringence of a mineral is always $n_\gamma - n_\alpha$.**

Two points need to be made to avoid confusion. The first is that while it takes three indices of refraction to describe the optical properties of biaxial minerals, the light that enters biaxial minerals is still broken into two rays: one is the fast ray, the other the slow ray. The second point is that the ordinary/extraordinary terminology is abandoned in biaxial minerals. Both rays behave as the extraordinary ray did in uniaxial minerals. Because both rays are extraordinary, the terminology presented in Chapter 5 is used; the two rays are called the slow ray and the fast ray. The index of the slow ray is identified n_γ', which is always between n_γ and n_β ($n_\gamma \geq n_\gamma' \geq n_\beta$). The index of the fast ray is identified n_α', which is always between n_α and n_β ($n_\alpha \leq n_\alpha' \leq n_\beta$).

Biaxial Indicatrix

The biaxial indicatrix is constructed and used in much the same manner as the uniaxial indicatrix; the major difference is that there are three principal indices to be dealt with instead of two. To construct the indicatrix, the three principal indices are plotted along three mutually perpendicular axes (Figure 7.1): n_α is plotted along the X axis, n_β is plotted along the Y axis, and n_γ is plotted along the Z axis. In every case, n_α is the smallest index and n_γ is the largest index. The indicatrix is therefore a triaxial ellipsoid elongate along the Z axis and flattened along the X axis.

There are three **principal sections** through the indicatrix: the X–Y, X–Z, and Y–Z planes (Figure 7.1a). The X–Y section is an ellipse with axes n_α and n_β. The X–Z section is an ellipse with axes n_α and n_γ.

The Y–Z section is an ellipse with axes n_β and n_γ. Random sections through the indicatrix also are ellipses.

The indicatrix has two circular sections with radius n_β that intersect in the Y axis. Consider the X–Z plane (Figure 7.1a). It is an ellipse whose radii vary

in length from n_α to n_γ, so there must be radii with lengths equal to n_β. Radii shorter than n_β are n_α' and those longer are n_γ'. The length of the indicatrix along the Y axis also is n_β, so the Y axis and the n_β radii in the X–Z plane define the two circular sections (Figure 7.1b). **Directions perpendicular to the**

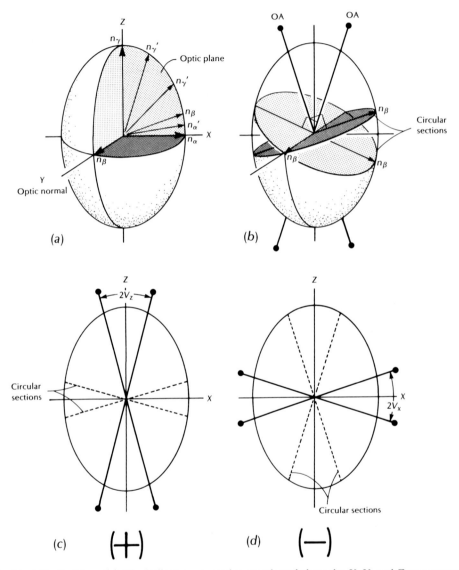

Figure 7.1 The biaxial indicatrix.   (a) The indices n_α, n_β, and n_γ are plotted along the X, Y, and Z axes, respectively. Principal sections are the X–Y, X–Z, and Y–Z planes. Within the X–Z plane, the radii of the ellipse vary from n_α to n_γ with certain radii equal to n_β.   (b) The circular sections have radius n_β. The primary optic axes (OA) are perpendicular to the circular sections and lie in the X–Z, or optic, plane.   (c) An optic plane of biaxial positive indicatrix.   (d) An optic plane of biaxial negative indicatrix.

two circular sections are the **primary optic axes**[1] and they both lie in the X–Z plane.

The X–Z plane that contains the optic axes is called the **optic plane**. The acute angle between the optic axes is the **optic angle** or **2V angle**. The axis (either X or Z) that bisects the 2V angle is the **acute bisectrix** or **Bxa**. The axis (either Z or X) that bisects the obtuse angle between the optic axes is the **obtuse bisectrix** or **Bxo**. The Y axis, which is perpendicular to the optic plane, is called the **optic normal**.

The optic sign is defined based on whether the X or Z axis is the acute bisectrix (Figure 7.1c, d). **If the acute bisectrix is X, the mineral is optically negative. If the acute bisectrix is Z, the mineral is optically positive.** In the special case where 2V is exactly 90°, the mineral is optically neutral.

A slightly different convention is to identify the angle between the optic axes bisected by the X axis as $2V_x$ and the angle between the optic axes bisected by the Z axis as $2V_z$. The angles then can vary between 0 and 180° with the restriction that $2V_x + 2V_z = 180°$. If $2V_z$ is less than 90°, the mineral is positive, and if $2V_z$ is greater than 90°, the mineral is negative. This nomenclature is particularly useful in describing optical data for minerals that change from positive to negative as the chemical composition changes.

The uniaxial indicatrixes may be considered special cases of the biaxial indicatrix. If $n_\alpha = n_\beta$, the Z axis is the optic axis, the X–Y plane is the circular section, and the optic sign is uniaxial positive (cf. Figure 6.4a). If $n_\beta = n_\gamma$, the X axis is the optic axis, the Y–Z plane is the circular section, and the optic sign is uniaxial negative (cf. Figure 6.4b).

Mathematical Relationships

The equation for the indicatrix is conveniently expressed as

$$1 = \frac{x^2}{n_\alpha^2} + \frac{y^2}{n_\beta^2} + \frac{z^2}{n_\gamma^2} \qquad 7.1$$

where x, y, and z are the coordinates for some point on the indicatrix surface. The length of any radius (n′) is the index of refraction for light vibrating par-

allel to the radius and whose wave normal is perpendicular to the radius. It is given by the equation

$$n' = \frac{1}{\left[\dfrac{\sin^2\rho\cos^2\delta}{n_\alpha^2} + \dfrac{\sin^2\rho\sin^2\delta}{n_\beta^2} + \dfrac{\cos^2\rho}{n_\gamma^2}\right]^{\frac{1}{2}}} \qquad 7.2$$

where ρ and δ are the angles shown in Figure 7.2. This is based on the equation of a triaxial ellipsoid expressed in polar coordinates. Note that if $n_\alpha = n_\beta$, or $n_\beta = n_\gamma$, the indicatrix is uniaxial and Equation 7.2 can be rewritten to give Equation 6.1, if allowance is made for the different manner in which the angles are defined in the two equations.

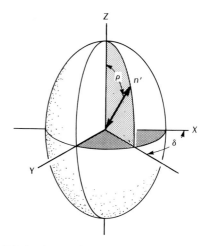

Figure 7.2 Angles used to describe the orientation of a random radius (n′) of the indicatrix.

The relationship between the optic angle and the principal indices of refraction is given by the equation (Wright, 1951)

$$\cos^2 V_z = \frac{n_\alpha^2\,(n_\gamma^2 - n_\beta^2)}{n_\beta^2\,(n_\gamma^2 - n_\alpha^2)} \qquad 7.3$$

where V_z is half the angle $2V_z$.

A simple nomogram, based on Equation 7.3, devised by Mertie (1942) (Figure 7.3) also can be used to determine 2V from the principal indices. Indices of refraction are plotted along the vertical axis and 2V is plotted along the horizontal axis. The value for n_α is plotted on the left side of the diagram and n_γ is plotted on the right side. A line is drawn between these two points and the value of n_β pro-

---

[1] The primary optic axes are hereafter usually referred to as the optic axes. Secondary optic axes are defined with reference to the ray velocity surfaces, which are described in Appendix B.

jected to where it intersects the line at point A. The value of $2V$ is read from the bottom of the diagram directly below A. To avoid marking the chart, a straightedge can be used instead of drawing lines. For convenience, values of $2V_x$ and $2V_z$ also are shown. Note that if any three of the variables in Equation 7.3 are known, the fourth can be determined either by calculation or by using Figure 7.3.

Use of the Indicatrix

The biaxial indicatrix is used in the same manner as the uniaxial indicatrix but, because of lower symmetry, involves a few additional considerations.

In Figure 7.4, a wave normal WN is shown in a random direction through a biaxial indicatrix. To find the indices of refraction, vibration directions for the slow and fast waves, and the ray paths the following procedure is used:

1. Construct a section through the indicatrix perpendicular to the wave normal (Figure 7.4a). This section through the indicatrix is parallel to the wave fronts and is an ellipse, unless the wave normal is parallel to an optic axis. The axes of the elliptical section indicate the vibration directions and indices of refraction n_α' and n_γ' for the fast and slow rays, respectively.
2. To find the point of emergence of the slow ray, construct a line through the wave normal that is parallel to the slow vibration direction and tangent to the indi-

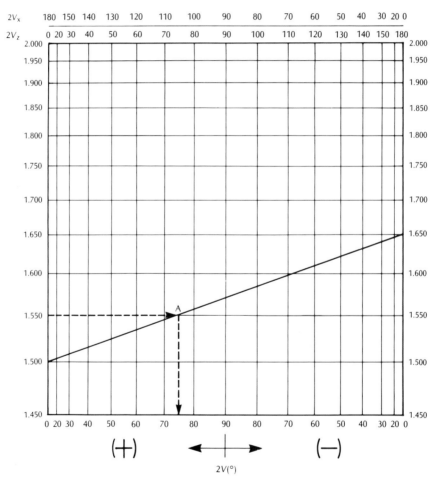

Figure 7.3 The Mertie diagram. If three of the values n_α, n_β, n_γ, and $2V$ are known, the fourth can be determined. In the example shown here, $n_\alpha = 1.500$, $n_\beta = 1.550$, and $n_\gamma = 1.650$. The $2V$ angle must, therefore, be 75°, and the mineral is optically positive.

catrix as shown in Figure 7.4*b*. The slow ray emerges at point *a* where the tangent line touches the indicatrix. The point of emergence of the fast ray (*b*) is found in a similar manner. Because both tangent lines are parallel to the vibration directions they are also perpendicular to the wave normal. Note that both rays behave as extraordinary rays.

The Biot–Fresnel rule also may be used to determine the vibration directions associated with a wave normal (Figure 7.5). Two planes are constructed through the indicatrix: one plane contains the wave normal and one of the optic axes, the other contains the wave normal and the second optic axis. The vibration directions for light waves traveling along the wave normal bisect the angles between the two planes.

To find the vibration directions, indices of refraction, and wave normal directions for slow and fast rays emerging at a particular point on the indicatrix the following procedure is used:

1. Construct the ray path through the center of the indicatrix and construct a section through the indicatrix parallel to a surface (*T*) tangent to the indicatrix at the point where the rays pierce the indicatrix. The indices of refraction and vibration directions for the two rays are given by the lengths and orientations of the axes of the elliptical section (Figure 7.6*a*).
2. The wave normals for the two rays are lines from the origin that are perpendicular to the vibration directions (Figure 7.6*b*). The wave fronts for the two rays

are surfaces perpendicular to the wave normals (Figure 7.6*c*).

While it is possible to calculate values of n_α' and n_γ' for a random wave normal direction through the indicatrix, that undertaking will not be pursued here.

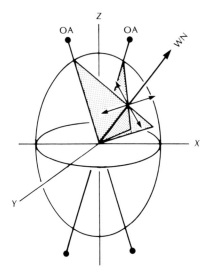

Figure 7.5 Biot–Fresnel construction. The vibration directions associated with the wave normal (WN) bisect the angles between planes constructed from the optic axes (OA) to the wave normal.

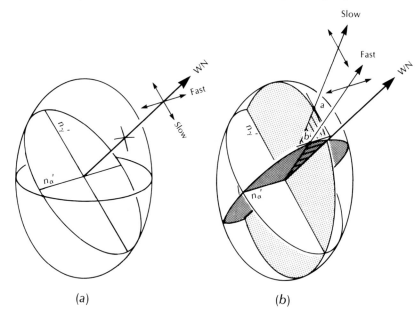

(a)   (b)

Figure 7.4 Determining indices of refraction and vibration directions given the wave normal direction. (*a*) Indicatrix with wave normal (WN) and elliptical section perpendicular to the wave normal. The axes of the elliptical section (n_γ' and n_α') indicate the vibration directions and indices of refraction of the slow and fast rays. (*b*) Ray directions associated with the wave normal. Tangents to the indicatrix are constructed through the wave normal and parallel to the axes of the elliptical section. The points of tangency *a* and *b* indicate where the slow and fast rays emerge from the indicatrix.

The graphical procedures that illustrate the principles involved serve current purposes quite adequately.

We will examine the behavior of light passing through a mineral plate in four cases: normal incidence parallel to one of the principal axes of the indicatrix, normal incidence parallel to an optic axis, normal incidence in a random direction, and inclined incidence. With normal incidence, the wave normals are not refracted, so the wave normals for slow and fast rays are parallel in the mineral. With inclined incidence, the wave normals for the two rays are refracted but by different amounts, because the slow and fast rays have different indices of refraction.

Normal Incidence Parallel to an Indicatrix Axis

Figure 7.7 illustrates the case where the incident light passes parallel to the X indicatrix axis. The indicatrix is situated so that it is cut in half by the bottom surface of the mineral. Because this elliptical section through the indicatrix is perpendicular to the wave normal, its axes, which are n_γ and n_β along the Z and Y axes, respectively, give the indices of refrac-

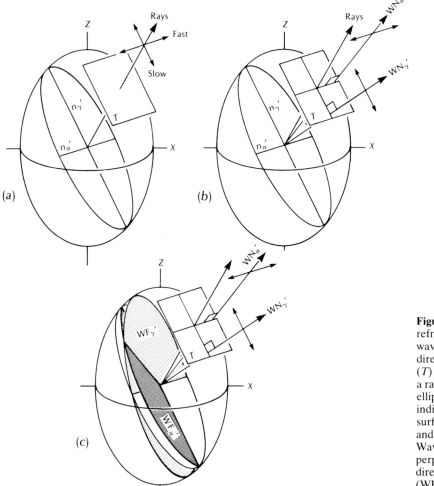

(a)

(b)

(c)

Figure 7.6 Determining indices of refraction, vibration directions, and wave normal directions, given a ray direction. (a) Tangent surface (T) constructed at the point where a ray pierces the indicatrix. An elliptical section through the indicatrix parallel to the tangent surface gives indices of refraction and vibration directions. (b) Wave normals WN$_\alpha$' and WN$_\gamma$' are perpendicular to the vibration directions. (c) Wave fronts (WF$_\alpha$' and WF$_\gamma$') are perpendicular to the wave normals.

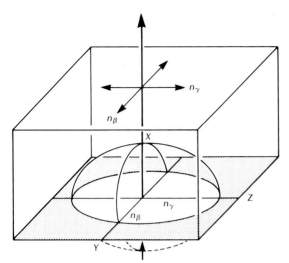

Figure 7.7 Normal incidence parallel to the X indicatrix axis. The two rays vibrate parallel to the Y and Z indicatrix axes and have indices n_β and n_γ.

tion and vibration directions for the two rays. If the construction is done to find the ray paths, it will be found that they coincide with the wave normals.

Normal Incidence Parallel to an Optic Axis

Figure 7.8 illustrates the case where the mineral has been cut so that one of the optic axes is perpendicular to the top and bottom surfaces of the mineral. The section through the indicatrix is one of the circular sections, indicating that light waves propagate along the optic axis with equal velocity regardless of vibration direction and have index n_β, which is the radius of the circular section.

A unique feature about a biaxial optic axis is that unless the light is vibrating parallel to the optic normal (Y axis), the ray and wave normal do not coincide. If a beam of unpolarized light enters the mineral, it forms into a hollow cone of light in the mineral and a cylinder of light after exiting the top of the mineral plate. The explanation of this phenomenon, called **interior conical refraction**, can be found by determining the ray path associated with each different vibration direction of the incident

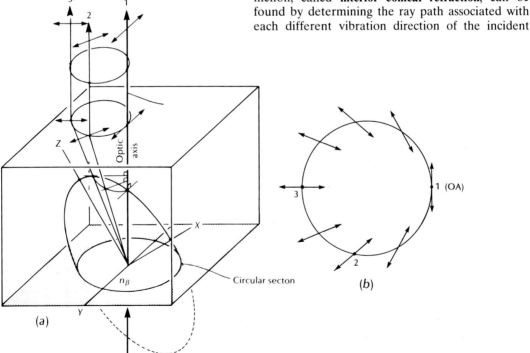

Figure 7.8 Normal incidence parallel to an optic axis: interior conical refraction. (*a*) The section through the indicatrix is a circular section, so all light has index n_β with no preferred vibration directions. The wave normal for all light follows the optic axis. Ray paths such as 1, 2, and 3 form a hollow cone of light within the mineral and a hollow cylinder above the mineral. (*b*) Vibration directions for light in the cone.

light. As described earlier, this is done by constructing a tangent to the indicatrix, which goes through the wave normal and is parallel to the vibration direction of the light. The ray paths for three different vibration directions are shown in Figure 7.8a. In each case, the wave normal is parallel to the optic axis, but the associated rays are deflected (except for ray 1 which vibrates parallel to the optic normal). The pattern of vibration directions as seen from above is shown in Figure 7.8b. The example shown in Figure 7.8 is greatly exaggerated for illustration purposes. With the thicknesses and birefringence usually encountered, the radius of the cone is actually quite small. For practical purposes, the light can be considered to behave the same as light following the optic axis of uniaxial minerals which cannot display conical refraction.

Normal Incidence in a Random Direction

Normal incidence into a crystal plate cut in a random direction is shown in Figure 7.9. Because the light is normally incident to the bottom surface of the crystal plate, the wave normals are not refracted. The section through the indicatrix is an ellipse with axes equal to n_α' and n_γ', where $n_\alpha < n_\alpha' < n_\beta$ and $n_\beta < n_\gamma' < n_\gamma$. The vibration directions of the slow and fast rays are parallel to the axes of the elliptical section. The construction technique described in Figure 7.5 allows determination of the ray paths.

Inclined Incidence

As was seen in the uniaxial case, inclined incidence is somewhat more complicated because the slow and

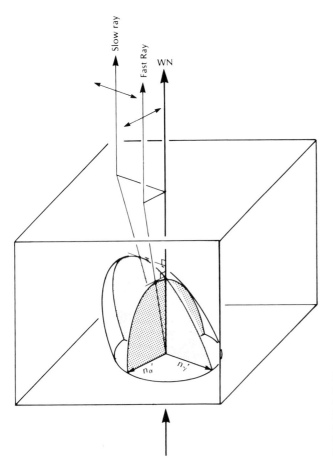

Figure 7.9 Normal incidence to a crystal plate cut in a random direction. The section through the indicatrix is an ellipse with axes n_α' and n_γ'. The ray paths both diverge from the wave normal (WN) as shown.

fast wave normals are refracted different amounts depending on their respective indices of refraction. Only the restricted case where the incident and refracted wave normals lie in one of the principal planes of the indicatrix is described here. The most general case where the refracted wave normals do not lie in a principal plane of the indicatrix is solvable but is rather complex. Because little will be gained for present purposes by going through the procedure, it will not be pursued.

Figure 7.10 shows the case of inclined incidence where the incident and refracted wave normals lie in the X–Z plane of the indicatrix. Every section through the indicatrix for possible refracted wave normal directions must have one axis along the Y axis. Hence, one of the wave normals must have index n_β and vibrate parallel to the Y axis. The refraction of this wave normal can be calculated directly from Snell's law. If the ray path associated with this wave normal is determined, it will be found

that it is coincident with the wave normal. By analogy to the uniaxial case, this ray behaves like the extraordinary ray when propagating perpendicular to the optic axis.

The angle of refraction of the second wave normal, which vibrates within the X–Z plane, can be determined using the same method employed to determine the refraction of the extraordinary ray in uniaxial minerals (Figure 6.9). For the case described here, Equation 6.1 may be rewritten:

$$n_s = \frac{n_\alpha}{[1 + \left(\frac{n_\alpha^2}{n_\gamma^2} - 1\right) \sin^2 (A + \theta_s)]^{\frac{1}{2}}} \qquad 7.4$$

where θ_s and n_s are the angle and index of refraction for the second wave normal, and A is the angle between the Z axis and the normal to the surface. The index is identified n_s because it will not be known whether it is n_α' or n_γ' until the problem is solved.

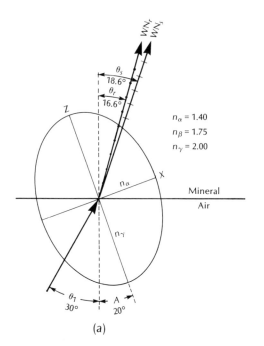

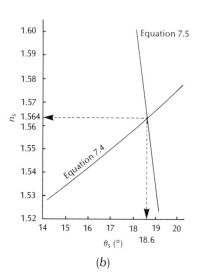

(a)   (b)

Figure 7.10 Inclined incidence in the X–Z principal section of the indicatrix. (a) The light in wave normal r (WN$_r$) vibrates parallel to the Y indicatrix axis (in and out of the page) and has index of refraction n_β. Its angle of refraction (16.6°) is calculated from Snell's law. The light in wave normal s (WN$_s$) vibrates within the X–Z indicatrix plane. (b) Simultaneous solution of Equations 7.4 and 7.5 for $n_s = 1.564$ and $\theta_s = 18.6°$.

Snell's law also must be satisfied; for the geometry shown in Figure 7.10 it can be written:

$$n_s = \frac{\sin \theta_1}{\sin \theta_s} \qquad 7.5$$

Simultaneous solution of Equations 7.4 and 7.5 is shown graphically in Figure 7.10b.

Crystallographic Orientation of Indicatrix Axes

Optically biaxial minerals include those in the orthorhombic, monoclinic, and triclinic crystal systems. Because the optical properties of minerals reflect their crystal structure and symmetry, we can expect that the optic orientation must be consistent with mineral symmetry.

Orthorhombic Minerals

Orthorhombic minerals have three mutually perpendicular crystallographic axes of unequal length. These axes must coincide with the three indicatrix axes (Figure 7.11a), and the three mutually perpendicular symmetry planes in the mineral must coincide with the principal sections of the indicatrix. However, it is not necessary for the X, Y, and Z indicatrix axes to correspond with the a, b, and c crystallographic axes, respectively. Any indicatrix axis may correspond with any crystallographic axis. For example, in aragonite, $X = c$, $Y = a$, and $Z = b$, whereas in anthophyllite, $X = a$, $Y = b$, and $Z = c$.

Monoclinic Minerals

In monoclinic minerals, the b crystallographic axis coincides with the single twofold rotation axis, and is perpendicular to the single mirror plane. The a and c axes are perpendicular to b, lie within the mirror plane, and intersect in an obtuse angle. One of the indicatrix axes always coincides with the b crystallographic axis, and the other two lie within the (010) mirror plane but are not parallel to either a or c (Figure 7.11b), except by chance. The b crystallographic axis may be parallel to either the X, Y, or Z indicatrix axis.

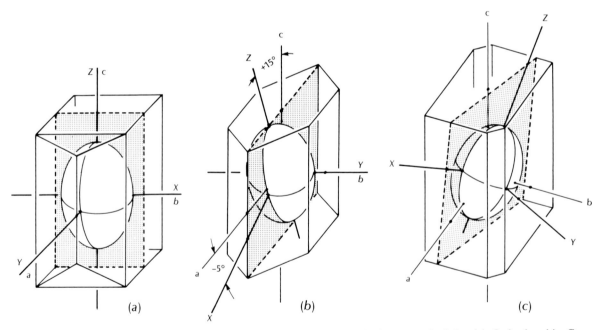

(a)   (b)   (c)

Figure 7.11 Relationship between crystal axes and indicatrix axes. The optic planes are shaded.   (a) Orthorhombic. Crystal axes and indicatrix axes coincide.   (b) Monoclinic. The b crystal axis coincides with one of the indicatrix axes. The other axes do not coincide except by chance. In this case $b = Y$, so the optic plane is parallel to (010). If $b = X$ or Z, then the optic plane is at right angles to (010).   (c) Triclinic. None of the crystal axes coincide with indicatrix axes except by chance.

The angle between the a or c crystal axis and an indicatrix axis is a positive angle if the indicatrix axis falls in the obtuse angle between the a and c axes and a negative angle if it falls in the acute angle. Using this convention, the β angle between the a and c crystal axes is

$$\beta = 90° + a \wedge (X, Y, \text{or } Z) + c \wedge (X, Y, \text{or } Z).$$

The optic orientation of the sample in Figure 7.11b is $X \wedge a = -5°$, $Y = b$, $Z \wedge c = +15°$. The angle β between a and c must therefore be $90° - 5° + 15° = 100°$. The reader is cautioned, however, that other conventions have been used. The most common is that, when viewed down the b axis, angles measured clockwise from the a and c crystal axes are positive, and negative if measured counterclockwise.

Triclinic Minerals

Triclinic minerals have three crystallographic axes of different lengths, none of which are at right angles to the others. Because there can be no symmetry elements other than center, the indicatrix axes are not parallel to the crystal axes (Figure 7.11c), except by chance.

Biaxial Interference Figure

Biaxial interference figures are obtained in the same manner as uniaxial figures, with the microscope arranged for conoscopic illumination. The high-power objective is used, polars are crossed, and the auxiliary condensor and Bertrand lens inserted. The Bertrand lens can be omitted if the ocular is removed. Biaxial interference figures are distinctly different from their uniaxial counterparts for most mineral orientations. This provides the basis for distinguishing between uniaxial and biaxial minerals. The nature of the interference figure depends on the orientation of the mineral grain. Five cases will be examined: acute bisectrix vertical, optic axis vertical, obtuse bisectrix vertical, optic normal vertical, and random figures. In each case, the interference figure consists of isochromes and isogyres.

Acute Bisectrix Figure

The acute bisectrix interference figure is obtained when the acute bisectrix is oriented perpendicular to the stage of the microscope (Figure 7.12). If $2V$ is small enough, the melatopes marking the points of

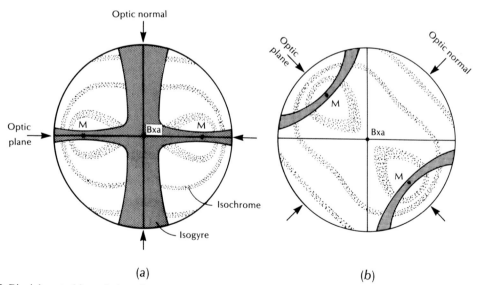

(a)   (b)

Figure 7.12 Biaxial acute bisectrix interference figure. The melatopes (M) mark the points of emergence of the optic axes. The acute bisectrix (Bxa) is in the center of the field of view. (a) Trace of the optic plane oriented east–west. (b) Optic plane placed in a 45° position. With rotation from (a) to (b), the isogyres split and move across the field of view to form hyperbolae centered on the melatopes.

emergence of the optic axes are in the field of view. The isochromes form an oval or figure-eight pattern about the melatopes. The pattern of isogyres changes as the stage is rotated.

Formation of Isochromes

A strongly convergent cone of light enters the mineral from below. Only light that happens to follow one of the optic axes emerges from the top of the mineral plate with zero retardation (Figure 7.13). Light following any other path experiences varying amounts of retardation depending on the length of the path followed through the mineral and the birefringence experienced. Retardation increases outward from the optic axes. Light following paths 1, 2, 3, and 4 has developed 600 nm of retardation when it exits the top of the mineral and defines the 600-nm isochrome. Closer to the melatopes, the light experiences lower birefringence and

has a shorter path through the mineral, so the retardation is less. The retardation increases relatively slowly from the melatopes towards the acute bisectrix because the increased birefringence is partially compensated by the shorter paths followed by light emerging near the acute bisectrix. Consequently, the isochromes tend to be stretched out towards the acute bisectrix. If white light is used, the isochromes show the normal interference color sequence. If monochromatic light is used, the isochromes consist of dark bands where the retardation is an integer number of wavelengths and bright illumination where the retardation is $\frac{1}{2}\lambda$, $1\frac{1}{2}\lambda$, etc. The melatopes in both cases are black because the retardation there is zero.

The number of isochromes visible depends on the partial birefringence between n_α and n_β (optically positive) or n_β and n_γ (optically negative) and the thickness of the crystal. High birefringence and thick crystal plates give rise to more isochromes. The

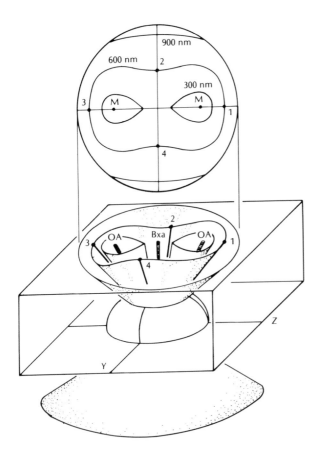

Figure 7.13 Formation of isochromes. Light that propagates along the optic axes (OA) emerges at the melatopes (M) with zero retardation. Light following paths within the flattened cone outlined by paths 1, 2, 3, and 4 develops 600 nm of retardation by the time it exits the top of the mineral. Light inclined further from the optic axes develops greater retardation, and light inclined at a lower angle from the optic axes develops less retardation. The isochromes are developed along the bands of equal retardation.

numerical aperture of the objective lens also controls the number of isochromes. High numerical aperture lenses intercept a larger cone of light, so more isochromes are visible than with a lower numerical aperture lens.

Vibration Directions and Formation of Isogyres
The vibration directions in the acute bisectrix figure can be derived in the same manner employed for the uniaxial figure. Figure 7.14 shows a biaxial negative mineral. Vibration directions for a number of wave paths determined from the indicatrix are projected onto the top surface of the mineral and into the interference figure. The approximate vibration direction at any point in the interference figure can be graphically derived from the Biot–Fresnel law, as shown in Figure 7.15. To determine the vibration directions at a point in the interference figure, lines are constructed from the two melatopes to the point.

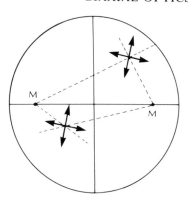

Figure 7.15 Biot–Fresnel construction. The vibration directions at selected points in an acute bisectrix figure bisect lines drawn from the melatopes to the points.

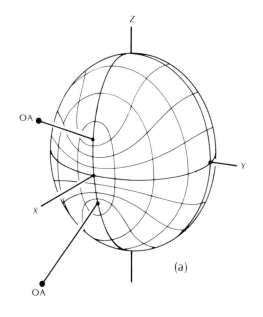

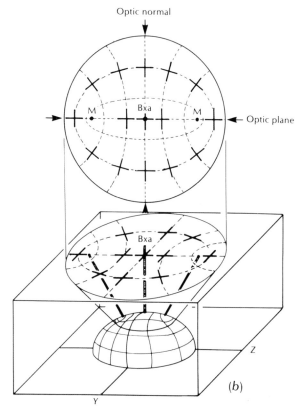

Figure 7.14 Vibration directions in the acute bisectrix interference figure.   (*a*) Vibration directions for all paths emerging from the biaxial indicatrix are determined as described in Figure 7.4 and have been projected onto the indicatrix. (*b*) Vibration directions for a number of wave paths have been projected onto the upper surface of the mineral and are shown as they appear in the interference figure.

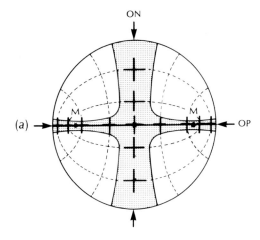

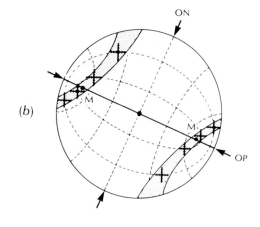

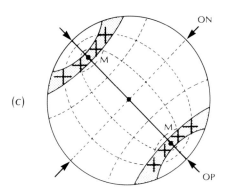

Figure 7.16 Change of isogyres with rotation of the stage; melatopes (M) in the field of view. The isogyres are located where the vibration directions in the figure are parallel to the N–S and E–W vibration directions of the polars. (*a*) With the optic plane (OP) oriented E–W, the isogyres form a cross that is wider along the trace of the optic normal (ON). (*b*) With a few degrees of stage rotation, the cross splits into two segments that pivot about the position of the melatopes. (*c*) With the optic plane in a 45° position, the isogyres form hyperbolae centered on the melatopes.

The vibration directions bisect the angles made between the construction lines. Kamb (1958) provided a more rigorous mathematical approach to determine the vibration directions within the interference figure. Within the field of the interference figure, isogyres are formed where the vibration directions correspond with the vibration directions of the lower and upper polars. The isogyres are areas of extinction. If the lower polar passes E–W vibrating light, only E–W vibrating light emerges in the areas occupied by the isogyres, and this light is absorbed by the upper polar.

The location and shape of the isogyres change as the stage is rotated. If the optic plane is oriented N–S or E–W (Figure 7.16*a*), the isogyres form a cross. The isogyre parallel to the trace of the optic normal is somewhat wider than the isogyre parallel to the trace of the optic plane. The positions of the melatopes are marked by a narrowing of the isogyres. The narrowing in the vicinity of the melatope

may be obscured in minerals with low birefringence because the low first-order gray in the vicinity of the melatope may be indistinguishable from the extinction area of the isogyre. If the optic plane is rotated away from N–S or E–W, the cross-shaped isogyre splits into two separate isogyre segments that appear to pivot about the positions of the melatopes (Figure 7.16*b*). When the trace of the optic plane is placed in a 45° position, the isogyres form two hyperbolic arcs whose vertices are the melatopes (Figure 7.16*c*). These isogyres are typically narrowest at the melatopes and fan out somewhat toward the edge of the field of view. If the melatopes are outside the field of view, no isogyres are visible in the 45° positions (Figure 7.17). The pattern defined by the isochromes remains fixed relative to the melatopes and rotates as the stage is rotated. The pattern of extinction forming the isogyres is superimposed on the isochromes. When the optic plane is oriented N–S or E–W, the mineral is in an extinction position.

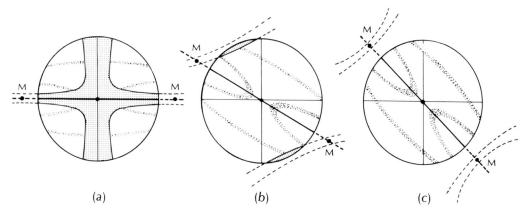

Figure 7.17 Change of isogyres in acute bisectrix figure with rotation of the stage; melatopes out of field of view (2V greater than roughly 60°). (a) Optic plane on the E–W, melatopes off the field of view. (b) With rotation of 30°, the center of the isogyres are tangent to the edge of the field of view. If 2V were larger, the isogyres would leave the field of view with less rotation. (c) Optic plane in a 45° position. The isogyres are off the field of view, which is entirely occupied by isochromes.

Centered Optic Axis Figure

A centered optic axis interference figure is produced when one of the optic axes is vertical. The melatope corresponding to this optic axis is positioned immediately beneath the crosshair. The other melatope will be visible in the field of view if 2V is less than roughly 30°, otherwise it will be outside the field of view. If 2V is relatively small, the interference figure looks like an off-center acute bisectrix figure (Figure 7.18).

If 2V is greater than roughly 60°, the pattern shown in Figure 7.19 is seen. When the optic plane is oriented along the N–S or the E–W, only a single arm of the cross pattern is within the field of view and it narrows somewhat at the position of the melatope. As the stage is rotated clockwise, the isogyre pivots around the melatope counterclockwise, and vice versa. When the trace of the optic plane is on the N–S or E–W, the isogyre is straight. When the trace of the optic plane is in a 45° position, the isogyre shows a maximum curvature and the position of the acute bisectrix lies on the convex side of the isogyre.

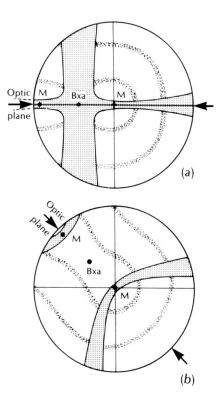

Figure 7.18 Centered optic axis figure with 2V less than about 30°. Both melatopes are in the field of view. (a) Optic plane E–W. (b) Optic plane in a 45° position.

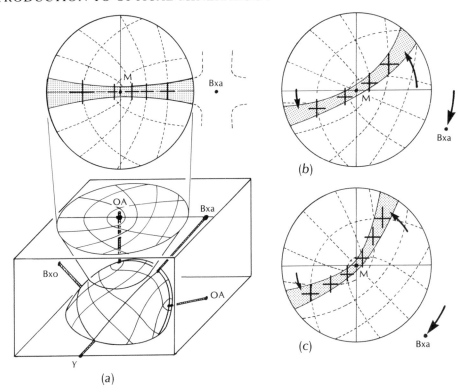

Figure 7.19 Centered optic axis figure with large $2V$. Only one melatope is in the field of view. The isochromes are not shown but form a pattern similar to that shown in Figure 7.18. (*a*) Pattern of vibration directions; optic plane E–W. (*b*) Stage rotated $22\frac{1}{2}°$ clockwise. The isogyre pivots counterclockwise about the melatope (M). (*c*) Optic plane in a 45° position. The acute bisectrix is on the convex side of the isogyre.

Obtuse Bisectrix Figure

Obtuse bisectrix interference figures (Figure 7.20) are produced when the obtuse bisectrix is oriented perpendicular to the microscope stage. Because the angle between the obtuse bisectrix and the optic axes must be greater than 45°, the melatopes are outside the field of view. The pattern of isochromes and geometry of vibration directions are essentially the same as those shown in Figures 7.13 and 7.14, except that the melatopes are well out of the field of view. The isogyres form a cross pattern when the trace of the optic plane is N–S or E–W. Usually between 5° and about 15° of stage rotation is needed to cause the isogyres to split and leave the field of view. Note that for $2V$ of near 90°, the acute and obtuse bisectrix figures are very similar, and if $2V = 90°$, there is no distinction between them. If $2V$ is

small, the obtuse bisectrix figure looks much like an optic normal figure.

Optic Normal Figure

An optic normal or flash figure is produced when the optic normal is vertical. The pattern of vibration directions within the figure is nearly rectilinear (Figure 7.21*a*) and is very similar to that seen with the uniaxial flash figure. When the X and Z indicatrix axes are parallel to the lower and upper polars, the field of view is occupied by a broad, fuzzy isogyre because vibration directions at all but the outer edges of the quadrants are N–S and E–W. If the stage is rotated a few degrees (Figure 7.21*b*), the isogyres split into two curved segments that leave the field of view from the quadrants into which the acute bisectrix is being rotated. For minerals with $2V$ of

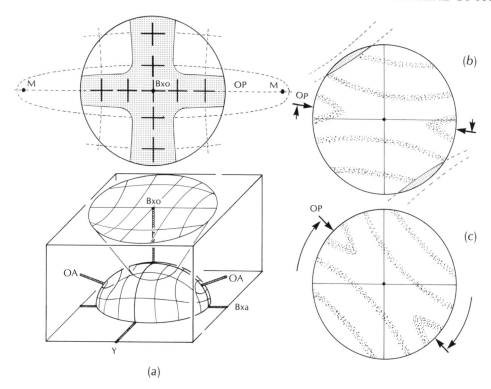

Figure 7.20 Obtuse bisectrix interference figure.   (*a*) Pattern of vibration directions in the interference figure. With the optic plane (OP) E–W, the isogyre forms a broad cross. The positions of the melatopes well out of the field of view are schematically shown at M. Compare with Figure 7.14 for the acute bisectrix figure.   (*b*) Less than ∼15° of rotation causes the isogyres to split and leave the field. The pattern of isochromes is similar to that seen with the acute bisectrix figure (Figure 7.13).   (*c*) In a 45° position, the field is occupied by isochromes only.

nearly 90°, the diffuse cross-shaped isogyre does not split into two curved segments; it simply dissolves as the stage is rotated. The amount of rotation required to cause the isogyres to completely leave the field of view typically is less than 5°.

The pattern of isochromes (Figure 7.21*c*) also is similar to the uniaxial flash figure. If 2*V* is moderate to small, the interference colors increase less rapidly outwards along the trace of the acute bisectrix than along the trace of the obtuse bisectrix. These variations often are difficult to see, however. If 2*V* is large, all four quadrants of the figure show essentially the same pattern of interference colors.

Off-center Figures

When none of the indicatrix or optic axes is vertical, an off-center interference figure is produced. It is easiest to interpret off-center figures when one of the indicatrix axes is parallel to the stage of the microscope and one of the principal sections of the indicatrix is vertical. If the optic normal (*Y* axis) is horizontal, the interference figure will be an off-center acute bisectrix, optic axis, or obtuse bisectrix figure, depending on which of these three axes happens to be closest to vertical. Figure 7.22*a* shows the case where the acute bisectrix is close to vertical. As long as the optic normal is horizontal, the figure is symmetrical and is bisected by the optic plane.

If the acute bisectrix is horizontal, then the interference figure can vary anywhere between an obtuse bisectrix figure and an optic normal figure. Figure 7.22*b* shows the case where the obtuse bisectrix is within the field of view. The figure is symmetric about the trace of the plane containing the optic normal and the obtuse bisectrix.

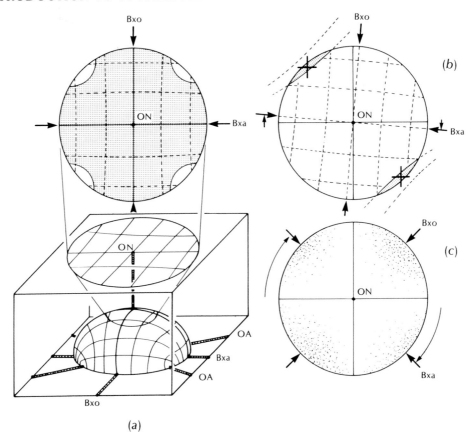

Figure 7.21 Optic normal (flash) figure. The optic normal (ON) emerges at the center of the figure.   (*a*) Vibration directions within the figure. If the bisectrixes are N–S and E–W, the field is occupied by a broad fuzzy cross because all but the outer edges of the four quadrants are extinct.   (*b*) Isogyres split and leave the field with very few degrees of rotation. The isogyres exit the field from the quadrants into which the acute bisectrix is being rotated.   (*c*) Isochrome pattern. In a 45° position no isogyres are present. The quadrants containing the trace of the obtuse bisectrix have slightly higher interference colors than the quadrants containing the acute bisectrix, but the difference may be difficult to see.

If the obtuse bisectrix is horizontal, the field of view is bisected by the trace of the plane containing the acute bisectrix and the optic normal. Figure 7.22*c* shows the case where the trace of the optic plane is just outside the field of view. A single isogyre marks the trace of the optic normal when it is parallel to the E–W or N–S. If the trace of the optic normal is placed in a 45° position, the isogyres may be visible near the edges of the field of view, provided that 2*V* is not too large.

The most general case is where none of the indicatrix axes are either parallel or perpendicular to the stage. Figure 7.22*d* shows the case where the acute bisectrix and one optic axis are reasonably close to the edge of the field of view. As the stage is rotated, isogyres sweep across the field of view. The thinner end of the isogyre roughly points towards the position of the melatope. The wide end of the isogyre sweeps across the field more rapidly than the thin end, and the sense of rotation of the isogyre is opposite the rotation direction of the stage. Unless the acute bisectrix or one of the optic axes is in the field of view, off-center figures are not of any particular value in interpreting 2*V* or the optic sign.

Determining Optic Sign

Acute Bisectrix Figure

Of the two rays of light that propagate along the acute bisectrix and emerge in the center of the acute

bisectrix figure, one vibrates parallel to the Y axis and has index n_β (Figure 7.23). The other vibrates parallel to the obtuse bisectrix and has index n_{Bxo}. If the mineral is optically positive, the obtuse bisectrix is the X axis and $n_{Bxo} = n_\alpha$. If the mineral is optically negative, the obtuse bisectrix is the Z axis and n_{Bxo}

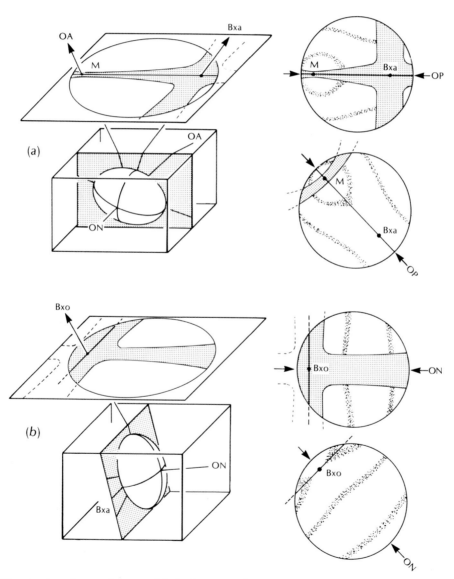

(a)

(b)

Figure 7.22 Off-center interference figures.   (a) Optic normal (ON) horizontal, and the acute bisectrix (Bxa) and one melatope (M) within the field of view.   (b) Acute bisectrix horizontal and obtuse bisectrix (Bxo) in the field of view.

$= n_\gamma$. Our task is to determine whether the ray vibrating parallel to the obtuse bisectrix is the fast or slow ray. The steps required to do this are as follows.

1. Obtain an acute bisectrix interference figure. Rotate the stage so that the trace of the optic plane is

NE–SW. If the melatopes are out of the field of view, it may be difficult to recognize the trace of the optic plane, which is located in the quadrants from which the isogyres leave the field of view (cf. Figure 7.17) as the stage is rotated. Near the center of the figure, the ray with index n_{Bxo} vibrates NE–SW, parallel to the trace of the optic plane.

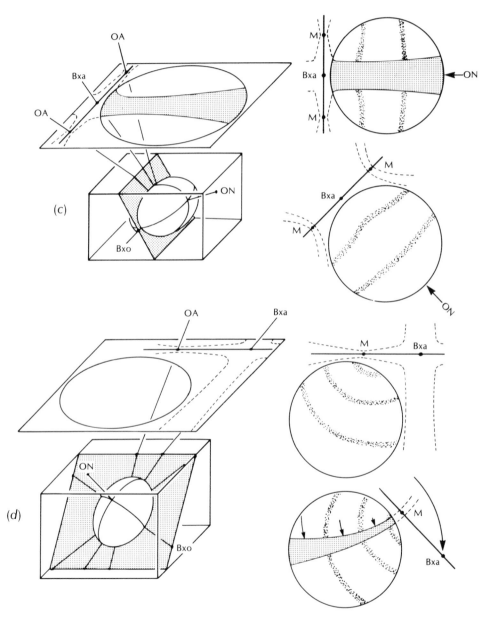

Figure 7.22 (continued) (c) Obtuse bisectrix horizontal and no melatope or indicatrix axis in the field of view. (d) Random section. As the stage is rotated, the isogyres sweep across the field of view but are not parallel to the cross hairs.

2. Insert an accessory plate with its slow ray oriented NE–SW and note the change of color in the central part of the figure.
3. Interpretation. If the interference colors between the melatopes decrease, the ray vibrating parallel to the Bxo must be the fast ray, so the Bxo is the X axis, the Bxa must be the Z axis, and the mineral is optically positive (Figure 7.23a). If the interference colors between the melatopes increase, then the ray vibrating parallel to the Bxo must be the slow ray, so the Bxo is the Z axis, the Bxa is the X axis, and the

mineral is optically negative (Figure 7.23b). If the trace of the optic plane is placed NW–SE, the areas of addition and subtraction will be reversed.

If the interference figure has relatively few isochromes, the gypsum and mica plates give color changes that are easy to interpret.

If there are many isochromes, the color changes produced by the quartz wedge are easiest to interpret, because the isochromes move as the wedge is

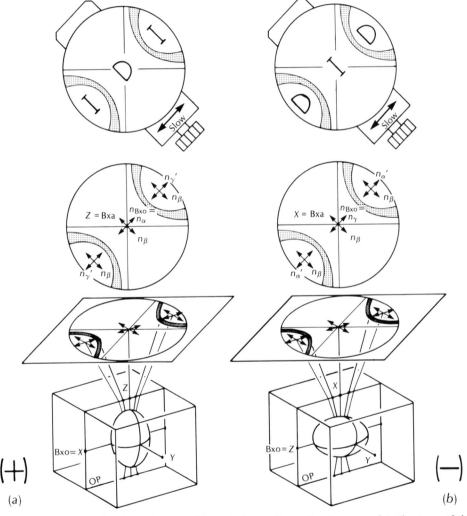

Figure 7.23 Determining optic sign from the acute bisectrix figure. Rays vibrating normal to the trace of the optic plane (OP) have index n_β. The accessory plate is inserted with its slow direction parallel to the trace of the optic plane (NE–SW). (a) Optically positive. Rays vibrating parallel to the trace of the optic plane have index n_α' (fast) between the melatopes and n_γ' (slow) outside the melatopes, so colors decrease (D) between the melatopes and increase (I) outside. (b) Optically negative. Rays vibrating parallel to the trace of the optic plane have index n_γ' (slow) between the melatopes and n_α' (fast) outside the melatopes, so colors increase between the melatopes and decrease outside. If the optic plane is NW–SE, the color changes are reversed.

inserted (Figure 7.24). In areas where the colors increase, the isochromes move into the figure towards the melatopes as higher-order colors from the edge of the figure displace lower-order colors near the melatope. In areas where the colors decrease, the isochromes move outward from the melatopes toward the edge of the figure.

Obtuse Bisectrix Figure

An obtuse bisectrix figure can be interpreted in a similar manner. The only difference is that the ray vibrating parallel to the trace of the optic plane has index n_{Bxa} instead of n_{Bxo}. If this ray proves to be the fast ray, then the acute bisectrix is the X axis and the mineral is negative. If this ray is the slow ray, the acute bisectrix is the Z axis and the mineral is positive. As was the case with the acute bisectrix figure, the trace of the optic plane is in the quadrants from

which the isogyres leave the field of view as the stage is rotated (Figure 7.20).

If $2V$ is large, it may be difficult to distinguish acute from obtuse bisectrix figures. Sign determination with these figures is impractical and an optic axis figure should be obtained instead.

Optic Axis Figure

If both melatopes are in the field of view, an optic axis figure can be treated the same as an acute bisectrix figure. If only one melatope is visible, then it can be considered to be half of an acute bisectrix figure with the acute bisectrix on the convex side of the isogyre when the trace of the optic plane is oriented in a 45° position (Figure 7.25). If $2V$ is near 90°, it may be difficult to interpret the curvature of the isogyre unless the figure is accurately centered. In this case, the sign should be recorded as either positive or negative with $2V$ of around 90°.

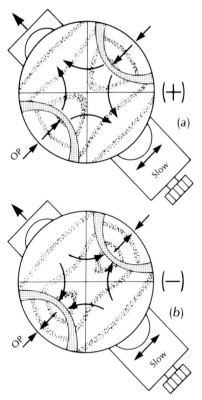

Figure 7.24 Movement of isochromes in acute bisectrix figures with insertion of the quartz wedge. The trace of the optic plane (OP) is parallel to the slow ray direction in the quartz wedge. If the optic plane is placed NW–SE, the directions of movement are reversed. (a) Optically positive. (b) Optically negative.

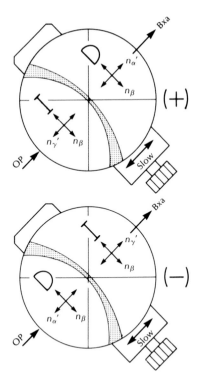

Figure 7.25 Determining optic sign with an optic axis figure. The optic plane is parallel to the slow direction in the accessory plate. Interference colors decrease (D) on the convex side of the isogyre for optically positive minerals and increase (I) for negative minerals. If the optic plane is oriented NW–SE, the color changes are reversed.

Flash Figure

A flash figure also may be used to determine optic sign because the vibration directions parallel to the obtuse and acute bisectrixes can be identified (cf. Figure 7.21). The stage is rotated so that the isogyres exit from the NW and SE quadrants in order to place the trace of the obtuse bisectrix NE–SW and the acute bisectrix NW–SE. The accessory plate can then be used to determine if n_{Bxo} is associated with the fast ray or the slow ray. Using the flash figure is not recommended, because $2V$ cannot be measured from it, nor can it be used to distinguish uniaxial from biaxial minerals.

Determining 2V

2V Versus 2E

Figure 7.26 shows the paths followed by light waves that propagate parallel to the optic axes in a crystal

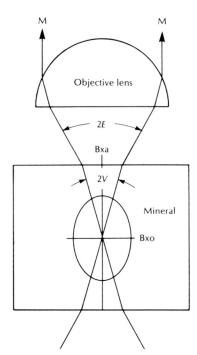

Figure 7.26 $2E$ versus $2V$. Within the mineral, the angle between the optic axes is $2V$. When light that is following the optic axes exits the top of the mineral, it is refracted to form the apparent optic angle $2E$. The position of the melatopes (M) within the interference figure depends on the value of $2E$.

cut so that the acute bisectrix is vertical. The light that follows the optic axes has index n_β and is refracted when it enters the air above the mineral plate. The angle between the refracted optic axes is a larger angle known as $2E$. The size of $2E$ can be calculated from Snell's law, if $2V$ and n_β are known:

$$\frac{\sin E}{\sin V} = \frac{n_\beta}{n_{air}}$$

Since $n_{air} \cong 1$, this becomes

$$\sin E = n_\beta \sin V \qquad 7.6$$

Mallard's Method

Mallard's method for determining $2V$ uses an acute bisectrix figure. It depends on the observation that the melatopes must be further apart for larger values of $2V$ (and $2E$). If $2V$ is small, the melatopes are close together, and if $2V$ is larger, the melatopes are further apart. Values of $2V$ can be calculated using Mallard's equation:

$$\sin V = \frac{D}{Kn_\beta} \qquad 7.7$$

where D is half the distance between the melatopes measured with a micrometer ocular and V is half of the $2V$ angle (Figure 7.27). K is Mallard's constant whose value depends on the numerical aperture of the objective lens, the magnification of the ocular, and the micrometer scale used. Values of K can be determined by examining the acute bisectrix figure of a mineral for which $2V$ and n_β are known. Figure 7.27 shows an acute bisectrix figure with $2V = 45°$ and $n_\beta = 1.50$. The distance $2D$ between the melatopes measured with the micrometer eyepiece is 40

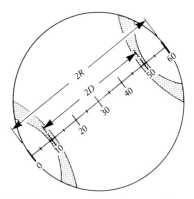

Figure 7.27 Measurements of the acute bisectrix figure used in the Mallard and Tobi methods of determining $2V$. Measurements are made with a micrometer ocular.

units, so Mallard's constant K, calculated from Equation 7.7, is 34.84 in this example.

Tobi's Method

Tobi's (1956) method is much the same as Mallard's method except that a chart (Figure 7.28a) is used to obtain values for $2E$ and $2V$. The value

$$\frac{2D \ (NA)}{2R \ (0.85)}$$

is plotted on the left side of the diagram. NA is the numerical aperture of the objective lens, $2D$ is the distance between the melatopes, and $2R$ is the diameter of the field of view (Figure 7.27). If $2D$, $2R$, and n_β are known, then $2V$ can be read from the chart, as shown in Figure 7.28b. If only $2D$ and $2R$ are known, values for $2E$ can be read from the chart, as shown in Figure 7.28c.

It should be evident that visual estimates based on the Mallard and Tobi methods can be made without using either calculations or charts. Figure 7.29 shows a number of interference figures for lenses with numerical apertures of 0.85, 0.75, and 0.65, and minerals with n_β equal to 1.60. These figure provide the bases for making visual estimates to within 10° or 15°, which is adequate for routine work.

For a given objective lens and value of n_β, the largest value of $2V$ ($2V_{max}$) (Table 7.1), which still allows the melatopes to be within the field of view of

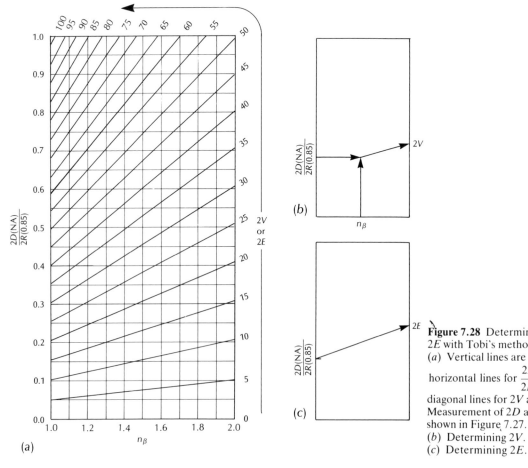

Figure 7.28 Determining $2V$ and $2E$ with Tobi's method.
(a) Vertical lines are for n_β, horizontal lines for $\dfrac{2D \ (NA)}{2R \ (0.85)}$, and diagonal lines for $2V$ and $2E$. Measurement of $2D$ and $2R$ is shown in Figure 7.27.
(b) Determining $2V$.
(c) Determining $2E$.

an acute bisectrix figure, can be determined by combining Equation 2.1 and Equation 7.6:

$$NA = n_\beta \sin V_{max} \qquad 7.8$$

Note that if the melatopes are at the edge of the field of view, then $2E$ must equal the angular aperture (AA) of Equation 2.1.

Kamb's Method

Kamb's (1958) method is an updated version of the Michel–Lévy method and is used when $2V$ is large and the melatopes are outside of the field of view. It involves measuring the angle of stage rotation needed to cause the cross-shaped isogyre in the acute bisectrix figure to split apart and leave the field of view. It also may be used with moderate accuracy for obtuse bisectrix figures and is the basis for distinguishing between obtuse and acute bisectrix figures when $2V$ is large. The following steps are needed to employ this method.

1. Obtain an acute bisectrix figure. In general, it will not be possible to find one that is exactly centered. A slightly off-center figure can be used, but the accuracy becomes progressively worse the more off-center the figure is. Rotate the stage until the isogyres form the cross pattern (Figure 7.30).
2. Rotate the stage clockwise until the center of the isogyres are tangent to the edge of the field of view and record the amount of rotation required. In general, one isogyre leaves the field before the other, so two angles of rotation, r_1 and r_2, must be measured.
3. Starting again with the isogyres in the cross position, rotate the stage counterclockwise to measure r_3 and r_4, which are the angles of rotation needed to bring

first one, and then the other, isogyre tangent to the edge of the field of view.
4. Compute an average value of r_1, r_2, r_3, and r_4. This value $\bar{r}$ is plotted along the vertical axis of Figure 7.31.
5. Interpretation. The value of $2V$ is read from the bot-

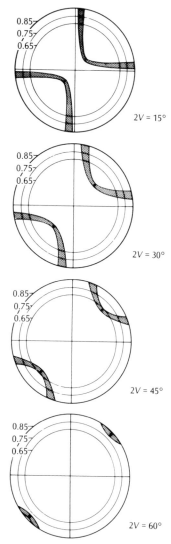

Figure 7.29 Rapid estimate of $2V$ based on separation of isogyres in acute bisectrix figures. The positions of the isogyres are constructed for $n_\beta = 1.60$ and an objective lens with a numerical aperture of 0.85. The two circles in each diagram show the field of view with 0.75 and 0.65 numerical aperture lenses.

Table 7.1. $2V$ angle ($2V_{max}$) for melatopes at the edge of the field of view for different numerical aperture lenses and values of n_β

n_β	$2V_{max}$		
	NA = 0.85	NA = 0.75	NA = 0.65
1.40	75°	65°	55°
1.50	69°	60°	51°
1.60	64°	56°	48°
1.70	60°	52°	45°
1.80	56°	49°	42°

tom of the graph, provided that the value of n_β is known or can be estimated or guessed at. Note that a relatively large error in the estimate of n_β leads to only a modest error in the value of $2V$. Two sets of curves are provided: one for objectives with a numerical aperture of 0.85, the other for objectives with a numerical aperture of 0.65. For other numerical apertures, interpolate between the curves. Note that the curves work for both acute and obtuse bisectrix figures. The size of $2V$ is less than 90° in acute bisectrix figures and greater than 90° in obtuse bisectrix figures. The angle of rotation $\bar{r}$ may be used to distinguish between acute and obtuse bisectrix interference figures. If $\bar{r}$ is greater than 30°, then the figure has the acute bisectrix vertical. If $\bar{r}$ is between approximately 5 and 15°, then the figure has the obtuse bisectrix vertical. Figures with intermediate values of $\bar{r}$ indicate a large $2V$ and could be either acute or obtuse bisectrix figures, depending on the value of n_β.

Wright Method

The Wright (1905) method is used with optic axis interference figures and depends on the observation that curvature of the isogyre in a 45° position is a function of $2V$ (Figure 7.32). If $2V$ is 90°, the isogyre forms a straight line in a 45° position. For smaller values of $2V$, the isogyres are progressively more curved. If $2V$ is less than about 30°, both melatopes are usually in the field of view. For $2V$ of less than 5°, the distance between the melatopes is very small and the two isogyres look almost like the uniaxial cross except for the small gap right in the middle.

To accurately orient the figure in a 45° position, proceed as follows.

1. With an optic axis figure obtained, rotate the stage so that the isogyre is oriented due N–S. This places the trace of the optic plane N–S. Unless the melatope is in the center of the field of view, the isogyre will probably be to one side of the N–S crosshair.
2. Rotate the stage 45° clockwise using the stage goniometer to measure the angle. The trace of the optic plane will now be oriented NE–SW and the curvature of the isogyre can be compared with Figure 7.32. Note that this procedure orients the figure so that the optic sign can also be determined (cf. Figure 7.25).

The Wright method is the most useful for estimating $2V$, because grains oriented to give optic axis

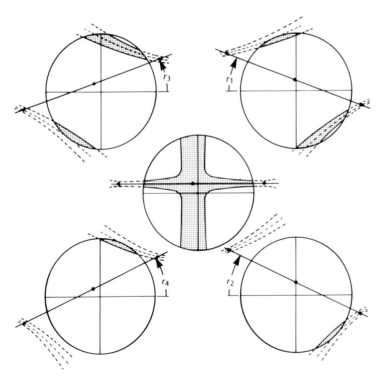

Figure 7.30 Measurements in the acute bisectrix figure for Kamb's method of determining $2V$. The angle $\bar{r}$ is the average of r_1, r_2, r_2 and r_4.

figures are much easier to find than for acute bisectrix figures and the method is relatively insensitive to values of n_β. Accuracy is not as great as with the more rigorous Mallard, Kamb, and Tobi methods, but great accuracy often is not needed for routine work.

Selecting Grains to Produce Interference Figures

Optic axis figures are produced by grains that display the lowest interference color in the sample, because an optic axis is vertical. If the birefringence of the mineral is low, appropriately oriented grains will remain extinct with rotation of the stage. Minerals with higher birefringence may show interference colors, because some of the light passes through the mineral at an angle to the optic axis and, therefore, must experience some retardation. It is best to use optic axis figures for routine work, because grains that produce optic axis figures are easy to recognize, and both $2V$ and the optic sign can be determined from them.

Optic normal (flash) figures are produced if the optic normal (Y) is vertical. Because the X and Z axes are horizontal, birefringence is a maximum, and appropriately oriented grains show the highest interference color of any grain in the sample. Although the optic normal figure is not useful in determining $2V$ or the optic sign, it can be used to confirm that the optic normal is nearly vertical.

Identifying grains that give acute bisectrix figures is a trial-and-error proposition. It is often necessary to obtain interference figures on many grains before stumbling on one with the acute bisectrix nearly vertical. Note, however, that the interference color dis-

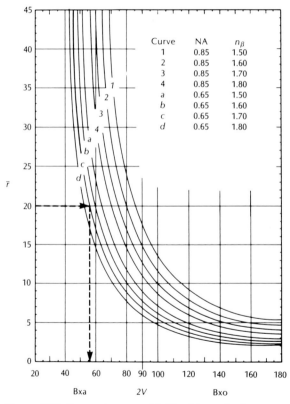

Curve	NA	n_β
1	0.85	1.50
2	0.85	1.60
3	0.85	1.70
4	0.85	1.80
a	0.65	1.50
b	0.65	1.60
c	0.65	1.70
d	0.65	1.80

Figure 7.31 Curves for use with Kamb's method of estimating $2V$. Measurement of $\bar{r}$ is shown in Figure 7.30. Curves are shown for different values of n_β, and 0.85 and 0.65 numerical aperture (NA) lenses. If $\bar{r} = 20°$, $n_\beta = 1.70$, and NA = 0.65, then $2V = 56°$. Assume that $n_\beta = 1.60$ or 1.70 if it is not known.

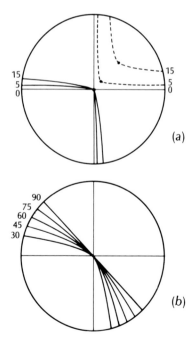

Figure 7.32 Isogyre curvature in optic axis interference figures. (a) Both melatopes in the field of view. The position of the second isogyre depends on the value of n_β and the NA of the objective lens. When $2V$ is $0°$, the figure is a uniaxial cross. (b) One melatope in the field of view.

played by grains with the acute bisectrix vertical is somewhere in the lower part of the range displayed by the mineral in the sample. This is because the birefringence must be less than about $(n_\gamma - n_\alpha)/2$. Although it is sometimes time consuming to search for suitably oriented grains, it may be desirable to do so, because these figures provide a more accurate estimate of $2V$ than do optic axis figures.

There is no systematic way of recognizing grains with the obtuse bisectrix vertical in orthoscopic illumination. These grains must also be located by trial and error. The birefringence displayed by minerals with the obtuse bisectrix vertical is greater than about $(n_\gamma - n_\alpha)/2$, so we can expect these grains to show interference color somewhere in the upper half of the range displayed by the mineral. Unfortunately, many other orientations also may give the same birefringence.

The acute bisectrix, obtuse bisectrix, and optic normal (flash) figure all form a cross pattern that splits as the stage is rotated. If $2V$ is large, the acute and obtuse bisectrix figures may look very much alike because the melatopes are well outside the field of view in both (cf. Figures 7.17 and 7.20). Similarly, the flash figure may look like an obtuse bisectrix figure. In some cases, these figures may be distinguished by using Kamb's method of determining $2V$, which is described above. As a rule of thumb, however, if less than 5° of stage rotation is required to cause the isogyres to split and leave the field of view, the figure is a flash figure. If between 5 and 15° is required, the figure is probably an obtuse bisectrix figure, and if more than 30° is required, the figure is probably an acute bisectrix figure. If between 15 and 30° is required, it is uncertain whether it is an obtuse or acute bisectrix figure, although it is probable that $2V$ is large. Recall that if $2V = 90°$ there is no distinction between the acute and obtuse bisectrixes. If there is doubt as to the identity of the figure, an unambiguous figure such as an optic axis figure should be used to determine the optic sign and $2V$.

Other Optical Properties of Biaxial Minerals

Pleochroism

To describe the pleochroism of biaxial minerals completely it is necessary to specify three colors.

One color is for light vibrating parallel to the X indicatrix axis, the second is for light vibrating parallel to the Y axis, and the third is for light vibrating parallel to the Z axis. For example, the pleochroism of some hornblende can be described as X = yellow, Y = pale green, and Z = dark green. An alternate nomenclature is n_α = yellow, n_β = pale green, and n_α = dark green. Because in this case the light vibrating parallel to X is the least strongly absorbed and the light vibrating parallel to Z is the most strongly absorbed, it also can be described as $Z > Y > X$.

One procedure that can be used to identify the three different colors and their associated vibration directions is as follows.

1. Identify the color associated with Y.
   a. Cross the polars with orthoscopic illumination and search for a grain with the lowest-order interference color.
   b. Convert to conoscopic illumination and obtain an interference figure. If the optic axis is vertical, an optic axis figure will be produced.
   c. Rotate the stage until the isogyre is parallel to the N–S crosshair (for E–W lower polar). This places the optic normal (Y) parallel to the E–W.
   d. Return to orthoscopic illumination without rotating the stage and uncross the polars. The color of the mineral is the color associated with Y, because only light vibrating parallel to the optic normal passes through the mineral.

2. Identify the colors associated with X and Z.
   a. Cross the polars again with orthoscopic illumination and search for a grain that shows the highest-order interference colors. As long as all the grains are the same thickness, the grain with the highest-order color has the X and Z indicatrix axes horizontal because birefringence is a maximum.
   b. Convert to conoscopic illumination and obtain an interference figure. A flash figure confirms that the X and Z axes are horizontal. Return to orthoscopic illumination.
   c. With one of the accessory plates, determine which of the two rays passing through the mineral is the slow ray and which is the fast ray (Chapter 5). Rotate the fast ray direction parallel to the lower polar.
   d. Uncross the polars. The color of the mineral is the color associated with X.
   e. Rotate the stage 90°. The slow ray (n_γ) must now be parallel to the lower polar, so the color of the mineral is the color associated with Z.

The colors also may be identified in the course of determining other optical properties of the mineral. For example, grains oriented to give an acute bisectrix figure can be used to determine the color associated with Y, and either X or Z, depending on optic sign. Similarly, the search for principal indices in grain mounts requires that grains be oriented in such a way that the X, Y, and Z colors can be recognized.

Extinction

There are too many different cleavages and crystal habits to allow a complete description of all the different extinction characteristics possible with biaxial minerals. Our approach here is to describe the extinction for relatively simple examples and let the reader work out the optics for specific cases. Note that all cleavages are shown in the microscopic views in Figures 7.33, 7.34, and 7.35, so that the geometry can be more easily visualized. However, in most cases, only those cleavages which are inclined at a substantial angle to the plane of the section are likely to be seen in an actual thin section.

Orthorhombic Minerals

The usual cleavages in orthorhombic minerals are pinacoids and prisms parallel to the three crystallographic axes. Figure 7.33 shows a mineral that has $\{110\}$ prismatic and $\{001\}$ pinacoidal cleavage. For convenience of illustration, the c axis is placed parallel to the long axis of the mineral. Sections cut parallel to the long axis of the mineral show parallel extinction. Cross sections cut parallel to (001) show symmetrical extinction. Sections cut in a random direction typically show inclined extinction to all the cleavages unless the section happens to be perpendicular to one of the principal sections of the indicatrix, in which case the extinction is parallel to the pinacoidal cleavages and symmetrical to the prismatic cleavages.

Monoclinic Minerals

Monoclinic minerals offer substantially more variability than orthorhombic minerals because the symmetry is lower and the only constraint on the

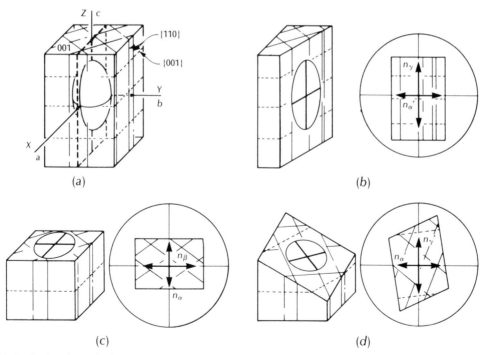

Figure 7.33 Extinction in orthorhombic minerals.   (a) Crystal habit showing trace of prismatic $\{110\}$ (solid lines) and pinacoidal $\{001\}$ (dashed lines) cleavage.   (b) Section cut parallel to the c axis: parallel extinction.   (c) Section cut parallel to (001): symmetrical extinction.   (d) Random section: neither symmetrical nor parallel extinction.

orientation of the indicatrix is that one indicatrix axis is parallel to the b crystallographic axis. The usual cleavages are pinacoids (front, side, and basal) and prisms. Figure 7.34 shows a mineral with $\{110\}$ prismatic and $\{001\}$ pinacoidal cleavage. The indicatrix is oriented so that $b = Y$, $c \wedge Z = +25°$, and $\beta = 100°$. Sections cut parallel to (001) show symmetrical extinction and intermediate birefringence. If the mineral is cut parallel to (100), the optic plane is vertical, extinction is parallel, and birefringence is intermediate. If the material is cut parallel to (010), the extinction angle measured to the prismatic cleavage is 25° and the angle measured to the pinacoidal cleavage is 15°. For the optic orientation used here, the birefringence is a maximum. Random sections typically show inclined extinction to all the cleavage directions. In general, monoclinic minerals show either parallel or symmetrical extinction if (010) is vertical, and inclined extinction in all other orientations.

The extinction angle measured when (010) is horizontal is a diagnostic property in minerals such as the amphiboles and pyroxenes. It has commonly been assumed that the extinction angle to prismatic $\{hk0\}$ cleavage decreases systematically from a maximum with (010) horizontal to zero with (010) vertical for sections cut with the c axis horizontal. These sections are easily identified because the traces of the prismatic cleavages are parallel (e.g., Figure 7.34b, c). The usual technique has been to find a number of grains with parallel prismatic cleavage, measure the extinction angle, and report the maximum angle as the angle between c and its nearest indicatrix axis. However, if the optic plane is (010) and the obtuse

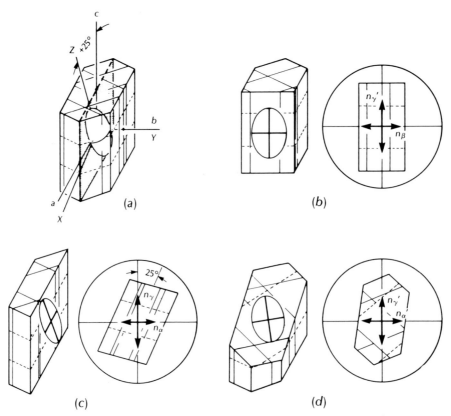

Figure 7.34 Extinction in monoclinic minerals.   (a) Crystal habit showing trace of prismatic $\{110\}$ (solid lines) and pinacoidal $\{001\}$ (dashed lines) cleavage.   (b) Section cut parallel to (100): parallel extinction.   (c) Section cut parallel to (010): inclined extinction.   (d) Random section: inclined extinction.

bisectrix is less than 45° from the c axis, extinction angles larger than c ∧ Bxo may be measured in intermediate orientation (Su and Bloss, 1984). Fortunately, with this optic orientation, grains with (010) horizontal are easily recognized because they must display the maximum birefringence (e.g., Figure 7.34c). Grains that yield the maximum extinction angle are typically oriented with an optic axis nearly vertical and yield off-center optic axis figures and low birefringence.

Triclinic Minerals

There are no restrictions placed on the orientation of the indicatrix axes in triclinic minerals, so neither parallel nor symmetrical extinction should be expected except by chance.

Sign of Elongation

The sign of elongation depends on which indicatrix axis is close to the long dimension of elongate mineral grains or cleavage fragments (Figure 7.35a). If the X axis is parallel to the length, the mineral is length fast. If the Z axis is parallel to the length, the mineral is length slow. If the Y axis is parallel to the length, the mineral may be either length fast or length slow, depending on whether the Z or the X axis happens to be closer to horizontal.

Sections through platy minerals such as the micas (Figure 7.35b) are length slow if the X axis is oriented roughly at right angles to the plates and length fast if the Z axis is roughly at right angles. If the Y axis is at right angles to the plates, sections through them may be either length fast or length slow, depending on whether the X or Z axis happens to be closer to horizontal.

Indices of Refraction

It is necessary to measure three different indices (n_α, n_β, and n_γ) in biaxial minerals. This can be done in grain mount or with a spindle stage. It is generally not possible to measure indices in thin section.

Grain Mount

The procedure used to measure indices of refraction of biaxial minerals in grain mount is as follows.

1. Determine n_β.
   a. With orthoscopic illumination and crossed polars, scan the sample for a grain that shows the lowest interference color. This grain should have an optic

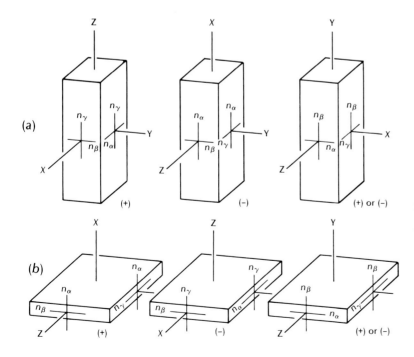

Figure 7.35 Sign of elongation. (a) Elongate minerals or cleavage fragments are length slow (+) if Z is parallel to the length, length fast (−) if X is is parallel to the length, and either if Y is parallel to the length. (b) Platy minerals are length slow if X is normal to the plates, length fast if Z is normal to the plates, and either if Y is normal to the plates.

axis vertical so that all light passes with index n_β. Confirm that the optic axis is vertical with an interference figure, determine the optic sign and an approximate value of $2V$, then return to orthoscopic illumination.

    b. Uncross the polars and compare the index n_β with the immersion oil, using the Becke line or oblique illumination method.

    c. Repeat until a match is obtained.

2. Determine n_α.

    a. With orthoscopic illumination and crossed polars, scan the sample for a grain that shows the highest interference color. This grain should have the optic plane horizontal so that the two rays passing through the mineral have indices n_α and n_γ. Correctly oriented grains yield optic normal (flash) figures.

    b. In orthoscopic illumination, determine which vibration direction is the fast ray with index n_α by using an accessory plate (Chapter 5). Rotate the fast ray vibration direction parallel to the lower polar so that all light passes with index n_α.

    c. Remove the upper polar and compare the index n_α with the index of the immersion oil, using the Becke line or oblique illumination method.

    d. Repeat with new oils until a match is obtained.

3. Determine n_γ.

    a. With orthoscopic illumination and crossed polars, scan the sample for a grain that shows the highest interference color. This grain should have the optic plane horizontal, so that the two rays passing through the mineral have indices n_α and n_γ. Correctly oriented grains yield optic normal (flash) figures.

    b. In orthoscopic illumination, determine which vibration direction is the slow ray with index n_γ by using an accessory plate (Chapter 5). Rotate the slow ray vibration direction parallel to the lower polar so all light passes with index n_γ.

    c. Remove the upper polar and compare the index n_γ with the index of the immersion oil, using the Becke line or oblique illumination method.

    d. Repeat with new oils until a match is obtained.

While n_β is usually measured first, time can be saved by comparing all three indices with each immersion oil. Once n_β is identified, oils needed to match n_α and n_γ can be selected based on knowledge of the birefringence, $2V$, and comparisons made in earlier oils. Note that the grain used to compare n_α can be rotated 90° to compare n_γ.

Spindle Stage
The spindle stage allows all three indices of refraction to be measured on a single grain, because the grain can be rotated to place each indicatrix axis parallel to the stage of the microscope. The following procedure is used to identify the indices.

1. Mount a grain on the spindle stage (Appendix A) and obtain an interference figure.

2. Find the three orientations which place indicatrix axes E–W[2] parallel to the stage of the microscope. These are recognized when an isogyre is orientated symmetrically along the N–S crosshair. Beginning with the spindle rotated fully counterclockwise, rotate the stage to bring an isogyre directly through the center of the field of view. Next, rotate the spindle a few degrees clockwise and then rotate the stage to again bring the isogyre through the center of the field. Repeat these small adjustments until the isogyre is oriented symmetrically along the N–S crosshair. This places one of the three indicatrix axes E–W, parallel to the stage of the microscope. Record the spindle and microscope stage settings to make it easy to return to this orientation. Based on the geometry of the interference figure, determine which indicatrix axis (X, Y, or Z) is oriented E–W (cf. Figure 7.24).

    Continue the process of rotating the spindle in small increments clockwise and adjusting the microscope stage to find the other two orientations where an isogyre is oriented along the N–S crosshair. Record these orientations and, based on the geometry of the interference figures, determine which indicatrix axis is aligned E–W in each case.

3. Return to orthoscopic illumination with plane light and compare the index of refraction for each of the three settings (n_α, n_β, n_γ), to the index of refraction of the immersion oil using the Becke line or oblique illumination method. Repeat with new oils as needed to systematically bracket each of the indices of refraction.

This presumes that it is known that the mineral is biaxial. If that is not known, then a diagnostic interference figure, either on the grain mounted on the spindle, or in a conventional grain mount or thin section, should be obtained to confirm the optical character. A more precise method of finding the orientations which allow measurements of the three indices of refraction is described by Bloss (1981). It involves analysis on a stereographic net, of extinction angles measured for different settings of the spindle stage.

Dispersion in Biaxial Minerals

The indices of refraction n_α, n_β, and n_γ vary for different wavelengths of light just as the index of isotropic materials varies. Consequently, the value of $2V$ and orientation of the indicatrix can be expected

---

[2] Provided that the lower polar is E–W.

to vary for different wavelengths of light. Variation in the size of $2V$ is called **optic axis dispersion** and variation in the orientation of the indicatrix is called **bisectrix dispersion** or **indicatrix dispersion**. If the dispersion is pronounced, color fringes may be visible along the isogyres in interference figures; their disposition may provide information regarding which crystal system the mineral belongs to. The amount of dispersion is termed **weak, moderate,** or **strong** depending on how noticeable the color fringes are.

Orthorhombic Minerals

Orthorhombic minerals can display only optic axis dispersion because the orientations of the indicatrix axes are fixed parallel to the crystallographic axes. Figure 7.36a shows a hypothetical mineral in which $2V$ increases for longer wavelengths of light. Because $2V$ for red light is larger than that for violet light, this dispersion is described as $r > v$. If $2V$ for violet light is larger than that for red light, it is described as $v > r$ (Figure 7.36b). If $r > v$, a red fringe of color may be seen on the sides of the isogyres facing the acute bisectrix (optic plane in a 45° position), and

a blue fringe may be seen on the outside of the isogyres. If $v > r$, the position of the fringes is reversed. Note that the red fringes are formed where only the isogyres associated with violet light are located (and vice versa), because violet light in these areas is not allowed to pass the upper polar but red light is.

An unusual form of dispersion called **crossed axial plane dispersion** is found in some orthorhombic minerals, such as brookite. As shown in Figure 7.37, the optic plane for light with wavelengths longer than 555 nm is the b–c crystal plane (100) and the optic plane for light with wavelengths shorter than 555 nm is the a–b crystal plane (001). The mineral is uniaxial for light with a wavelength of 555 nm. With white light, the two optic planes at right angles to each other are visible in the interference figure. Because the locations of isogyres for one wavelength are areas of illumination for others, the figure is rather complex and does not show conventional black isogyres. If monochromatic light is used, the figure looks like a normal acute bisectrix figure, with the position of the melatopes dictated by the wavelength of light used. The positions of the isogyres for several wavelengths of light are shown in Figure 7.37.

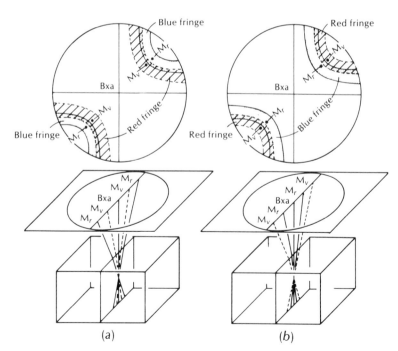

Figure 7.36 Optic axis dispersion. M_v and M_r are the melatopes for violet and red light, respectively. The ruled and stipple patterns show the locations of the isogyres for violet light and red light, respectively. The red fringes are produced where the isogyres and therefore areas of extinction for violet light are located and vice versa. The area of overlap is black. The widths of the color fringes are shown greatly exaggerated. (a) The $2V$ for red is greater than the $2V$ for violet ($r > v$). (b) The $2V$ for red is less than the $2V$ for violet ($r < v$).

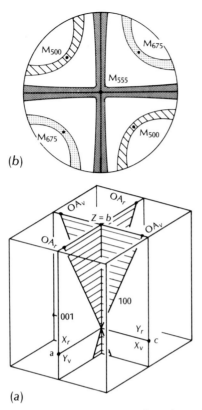

(b)

(a)

Figure 7.37 Crossed axial plane dispersion. (a) The optic plane for light with wavelengths longer than 555 nm in brookite is the (001) crystal plane. The optic plane for light with wavelengths shorter than 555 nm is the (100) plane. The mineral is uniaxial for 555 nm light and the optic axis is the b crystal axis. The subscripts v and r indicate the indicatrix axes and optic axes for violet and red light, respectively. (b) Acute bisectrix interference figure. Location of isogyres and melatopes (M) for light with 500-, 555-, and 675-nm wavelengths are shown.

Monoclinic Minerals

In addition to optic axis dispersion, monoclinic minerals also may display bisectrix dispersion because the indicatrix is free to rotate about the b crystal axis for different wavelengths of light. There are three cases, depending on which indicatrix axis coincides with the b axis.

Inclined dispersion is produced when the optic normal (Y) is parallel to the b axis (Figure 7.38a). The orientation of Y is fixed for all wavelengths, but the orientation of X and Z are free to vary within the (010) plane, which is the optic plane for all wavelengths of light. Color fringes are symmetrical across the trace of the optic plane.

Horizontal or parallel dispersion is produced when the obtuse bisectrix is parallel to the b axis. The optic normal (Y) and acute bisectrix must lie within the (010) symmetry plane and are free to assume different orientations within (010) for different wavelengths of light. The optic plane pivots on the obtuse bisectrix, which is parallel to the b axis. Color fringes like those shown in Figure 7.38b may be seen. The color fringes are symmetrical about the trace of the optic normal in the figure.

Crossed bisectrix dispersion is produced when the acute bisectrix is parallel to the b axis. The optic normal and obtuse bisectrix lie in the (010) symmetry plane. The optic plane pivots on the acute bisectrix, which is parallel to the b axis (Figure 7.38c). The pattern of color fringes does not show a plane of symmetry but is symmetric by twofold rotation about the acute bisectrix.

Triclinic Minerals

Because there are no constraints due to symmetry on the orientation of the indicatrix, the orientation of the optic axes for red light may be rotated in any direction relative to the position of the corresponding axes for violet light. The color fringes that may be found in triclinic interference figures usually are not arrayed with the symmetry found in monoclinic or orthorhombic minerals.

In practice, color fringes may be difficult to see and their symmetry is even more difficult to recognize. If the fringes are recognized, they can be useful in determining which crystal system a mineral belongs to. If the pattern of colors shows two mirrors through the acute bisectrix, the mineral is orthorhombic. If there is only one plane of symmetry to the color fringes, or the acute bisectrix is a twofold axis of rotation, the mineral is monoclinic. If the color fringes show no symmetry, the mineral is triclinic. Caution is needed, however, because the amount of dispersion is often small, and the color fringes in triclinic and monoclinic minerals may look like the fringes produced by orthorhombic minerals.

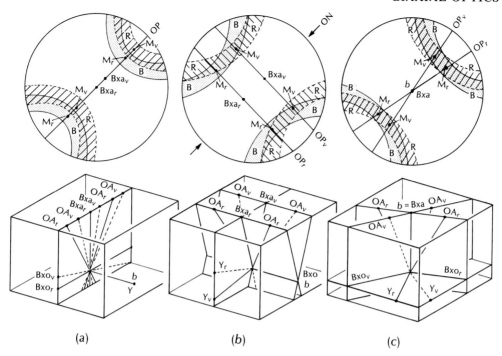

Figure 7.38 Bisectrix dispersion in monoclinic minerals. The subscripts *r* and *v* indicate the indicatrix and optic axes (OA) for red and violet light, respectively. The red color fringes (R) are found where only the isogyres for violet light (ruled) are located, and the blue fringes (B) are found where only the isogyres for red light (stippled) are located. The area of overlap is entirely extinct. The size of the fringes is greatly exaggerated for purposes of illustration. (*a*) Inclined dispersion. The *b* axis is the *Y* indicatrix axis. The figure is symmetric across the trace of the optic plane (OP). (*b*) Parallel dispersion. The *b* axis is the obtuse bisectrix (Bxo). The figure is symmetric across the trace of the optic normal (ON). (*c*) Crossed bisectrix dispersion. The *b* axis is the acute bisectrix (Bxa). The figure is symmetric by twofold rotation about the Bxa.

REFERENCES

Bloss, F. D., 1981, The spindle stage: principles and practice: Cambridge University Press, Cambridge, 340 p.

Kamb, W. E., 1958, Isogyres in interference figures: American Mineralogist, v. 43, p. 1029–1067.

Mertie, J. B., Jr., 1942, Nomograms of optic angle formulae: American Mineralogist, v. 27, p. 538–551.

Su, S. C., and Bloss, F. D., 1984, Extinction angles for monoclinic amphiboles and pyroxenes: a cautionary note: American Mineralogist, v. 69, p. 399–403.

Tobi, A. C., 1956, A chart for measurement of optic axial angles: American Mineralogist, v. 41, p. 516–519.

Wright, F. E., 1905, The determination of the optical character of birefractant minerals: American Journal of Science, v. 170, p. 285–296.

Wright, F. E., 1951, Computation of the optic axial angle from the three principal refractive indices: American Mineralogist, v. 36, p. 543–556.

SUGGESTIONS FOR ADDITIONAL READING

Bloss, F. D., 1978, The spindle stage: a turning point for optical crystallography: American Mineralogist, v. 63, p. 433–447.

McAndrew, J., 1963, Relationship of optical axial angle with the three principal refractive indices: American Mineralogist, v. 48, p. 1277–1285.

Olcott, G. W., 1960, Preparation and use of a gelatin mounting medium for repeated oil immersion of minerals: American Mineralogist, v. 45, p. 1099–1101.

Wahlstrom, E. E., Optical crystallography, 5th Edition: Wiley, New York, 488 p.

8

Identification of Minerals

Rapid identification of minerals in thin section or grain mount requires a systematic approach tempered with common sense and familiarity with a variety of common rocks and minerals. This chapter provides some general information to aid in identification. Identification tables and charts are located in Appendix C.

Chapters 9 through 15 provide descriptions of a variety of common minerals, and experience has shown that this selection covers the large majority of minerals found in most rocks. The reader is cautioned, however, that there are over 3000 different minerals and there is no assurance that an uncommon mineral may not be present in any given rock. More complete compilations of optical data are provided by Troger (1979), Fleischer and others (1984), and Phillips and Griffen (1981).

Descriptive Features

A prodigious amount of terminology has been developed to describe minerals. Some commonly used terms are as follows.

CRYSTAL SHAPE

acicular	elongate needlelike grains
anhedral	without regular crystal faces
bladed	elongate, slender
columnar	shaped like a column; moderately elongate grains with equidimensional cross section
equant	equidimensional grains
euhedral	has well-formed crystal faces
fibrous	individual grains are long slender fibers
lathlike	flat elongate grains
prismatic	the dominant faces are those of a prism
subhedral	has crystal faces but they are poorly formed or irregular
tabular	shaped like a book

MODE OF AGGREGATION

columnar	parallel arrangement of columnar grains
foliated	more or less parallel tabular or platy grains
granular	equant grains, all about the same size
matted	elongate grains in a random pattern
radiating	elongate grains that radiate out from a center

Cleavage

In grain mount, the planar sides to the individual grains indicate the presence of cleavage, and, with careful observation, the number and approximate angular relation between the cleavages can be determined. In thin section, cleavage may be difficult to recognize in minerals with low relief, but careful examination with the aperture diaphragm set to enhance the relief may reveal the presence of fine parallel cracks. The angle between the traces of different cleavages in thin section depends on how the mineral has been cut, and examples are shown in Chapters 6 and 7.

Twinning

Many minerals are twinned and the nature of the twinning is often easily seen. Simple twins consist of two segments that usually go extinct at different points with stage rotation. Contact twins are joined by a smooth twin plane separating the segments, while penetration twins are generally joined on irregular contacts. Polysynthetic twinning consists of numerous twin segments joined on parallel twin planes. If the successive twin planes are not parallel, a cyclic twin may result.

Alteration

Most minerals are subject to alteration from weathering, hydrothermal processes, or other causes. All too often, the alteration obscures the identity of the original mineral, but, in some cases—such as pinite after cordierite, or iddingsite after olivine—the alteration is a useful diagnostic property. If the alteration is severe, the optical properties of the remnants of the original mineral may be significantly changed.

Association

Some minerals are commonly associated in certain rocks or mineral deposits and other minerals are rarely found together. A knowledge of common mineral associations in a variety of rock types can aid in making educated guesses as to the possible identity of an unknown mineral. In addition, a knowledge of associations may suggest the presence of a mineral that might otherwise be overlooked.

While the use of association is valuable, it contains a subtle trap, because it tends to blind the observer to the possibility of an unusual or hitherto unidentified mineral. When making a choice between what "ought" to be present versus what the evidence seems to indicate, the evidence should be heeded.

The following lists show minerals likely to be found in a variety of common rocks. This compilation is far from complete and does not include any of the unusual associations that both frustrate and delight petrographers nor does it include any of the myriad products of alteration or weathering that may be present. The lists are organized so that the major minerals come first, followed by accessory or less common minerals.

IGNEOUS ROCKS

Felsic

quartz	epidote group
plagioclase	zircon
K-feldspar	Fe–Ti oxides
biotite	titanite
hornblende	rutile
muscovite	sillimanite
apatite	tourmaline

IGNEOUS ROCKS—*continued*

garnet	topaz
fluorite	xenotime
spinel	monazite

Intermediate

plagioclase	Fe–Ti oxides
calcic amphibole	zircon
K-feldspar	titanite
quartz	rutile
biotite	spinel
calcic clinopyroxene	tourmaline
orthopyroxene	xenotime
apatite	monazite
epidote group	allanite

Mafic

plagioclase	Fe–Ti oxides
orthopyroxene	apatite
calcic clinopyroxene	epidote group
olivine	zircon
pigeonite	titanite
calcic amphibole	rutile
biotite	spinel
cummingtonite	

Ultramafic

orthopyroxene	Fe–Ti oxides
calcic clinopyroxene	perovskite
olivine	apatite
garnet	spinel
calcic–sodic clinopyroxene	epidote group
plagioclase	

Feldspathoidal and related rocks (syenites, nepheline syenite, feldspathoidal volcanics, etc.)

plagioclase	calcite
K-feldspar	analcime
nepheline	cancrinite–vishnevite
leucite	perovskite
sodalite group	melilite
biotite	apatite
calcic amphibole	epidote group
calcic–sodic amphibole	Fe–Ti oxides
sodic amphibole	garnet
calcic–sodic clinopyroxene	vesuvianite
sodic clinopyroxene	spinel
olivine	corundum

Vesicle fillings

calcite	pectolite
aragonite	celadonite
zeolites	prehnite
chlorite	apophyllite
anhydrite	siderite

METAMORPHIC ROCKS

Pelitic

quartz	chloritoid
plagioclase	corundum
muscovite	Fe–Ti oxides
chlorite	scapolite
biotite	apatite
garnet	zircon
andalusite	epidote group
sillimanite	spinel
kyanite	tourmaline
staurolite	graphite
cordierite	dumortierite
K-feldspar	calcite

Mafic

plagioclase	gedrite
epidote group	chlorite
hornblende	stilpnomelane
anthophyllite	titanite
cummingtonite	omphacite
cordierite	apatite
garnet	zircon
calcic clinopyroxene	spinel
biotite	Fe–Ti oxides

Carbonate

calcite	monticellite
aragonite	humite group
dolomite	brucite
quartz	periclase
wollastonite	talc
tremolite–actinolite	graphite
richterite	scapolite
hornblende	prehnite
olivine	axinite
garnet	titanite
diopside–hedenbergite	corundum
biotite	perovskite
epidote group	Fe–Ti oxides
vesuvianite	

Blueschist and related

quartz	calcic–sodic pyroxene
plagioclase	aragonite
muscovite	calcite
biotite	titanite
lawsonite	serpentine
pumpellyite	epidote group
sodic amphibole	prehnite
sodic–calcic amphibole	apatite
garnet	zircon
chlorite	Fe–Ti oxides
sodic pyroxene	

SEDIMENTARY ROCKS

Clastic

quartz	muscovite
chalcedony	biotite
K-feldspar	glauconite
plagioclase	hematite
clay	zeolites
calcite	plus many others

Carbonate

calcite	glauconite
dolomite	clastic material and
clay	skeletal remains

Evaporite

calcite	sylvite
dolomite	sulfur
gypsum	chalcedony
anhydrite	clastic material
halite	(see also Table 10.2)

Tactics

With practice, optical data can be measured quite rapidly, but because of small grain size or lack of grains in appropriate orientations, it is often not possible to measure all of the optical information that might be desired. However, in routine work, identifications can usually be made without complete data, and for many common minerals only selected data needs to be obtained to confirm a tentative identification. It is important to avoid blindly accumulating optical data in hopes that an identification will somehow turn up in the end. Use each piece of information as a guide in each successive step in the identification process. If a positive identification cannot be made with the available information, then other techniques such as X-ray diffraction or chemical tests should be employed.

The procedures described here can be used for identifying a mineral, either in thin section or grain mount if allowance is made for the fact that indices cannot be measured in thin section and only detrital habits and textures can be seen in grain mounts. With grain mounts, the first mount should be used to obtain as much optical information as possible, and subsequent mounts should be used to pin down indices of refraction. Proceed as follows.

1. Examine the hand sample of the mineral to determine as many of the following characteristics as possible.
   a. Color
   b. Luster

c. Streak

d. Hardness

e. Cleavage or fracture

f. Specific gravity

g. Crystal habit

Provide a tentative identification or list of possibilities based on this information.

2. Based on the identity of associated minerals, rock type, or type of mineral deposit, make a mental list of minerals that the unknown might be.

3. Scan the slide to examine different grains of the unknown mineral. Color, relief, twinning, crystal shape, textures, and alteration usually provide the best bases for distinguishing different minerals. Cross and uncross the polars and rotate the stage as needed. Recall that different grains of pleochroic minerals may have quite different colors depending on how they are oriented. Record the following information.

a. Color and pleochroism (if any)

b. Relief relative to cement or immersion oil

c. Mineral habit, textures, and alteration

d. Whether the mineral is isotropic or anisotropic

e. Nature of twinning, if present

f. Nature of cleavage or fracture

4. If the mineral is isotropic:

a. For grain mounts, determine index of refraction with additional mounts, then go to identification tables and mineral descriptions to determine the identity.

b. For thin sections, go to identification tables and mineral descriptions.

5. If the mineral is anisotropic:

a. Scan the slide to find a grain of the unknown with the lowest interference color. Obtain an optic axis interference figure and determine if the mineral is uniaxial or biaxial.

(i) If the mineral is uniaxial (a) determine optic sign, (b) return to orthoscopic illumination and record the color and relief associated with n_ω and check the Becke line.

(ii) If the mineral is biaxial (a) determine optic sign, (b) determine $2V$, (c) determine dispersion characteristics, if any, (d) rotate the stage to place the optic plane at right angles to the lower polar vibration direction. Return to orthoscopic illumination. Record the color and relief of the mineral associated with n_β and check the Becke line.

b. Scan the slide to find a grain of the unknown with maximum interference color.

(i) In thin section, determine maximum birefringence based on thickness and interference color.

(ii) Determine the relief and check the Becke line associated with n_ϵ (uniaxial) or n_α and n_γ (biaxial) by placing the appropriate vibration direction parallel to the polarizer. Use the accessory plate to distinguish slow and fast directions. At the same time determine the color associated with ϵ (uniaxial) or X and Z (biaxial).

c. If the mineral is elongate or has cleavage:

(i) Measure extinction angles on a number of grains.

(ii) Determine sign of elongation (length fast or slow).

d. If crystallographic directions can be identified, determine optic orientation.

e. Determine indices:

(i) For grain mounts, prepare additional mounts.

(ii) For thin sections, estimate indices based on relief and Becke line, if possible.

f. Go to identification tables and mineral descriptions to determine identity of the mineral. Allow for the possibility that some of the optical data may be atypical or incorrectly measured.

Use of the Identification Tables (Appendix C)

The optical data for minerals described in Chapters 9 through 15 are presented in Appendix C as follows:

Table C.1	Color in thin section
Table C.2	Opaque minerals
Table C.3	Index of refraction of isotropic or nearly isotropic minerals
Table C.4	Indices of refraction of uniaxial minerals
Table C.5	Indices of refraction of biaxial negative minerals
Table C.6	Indices of refraction of biaxial positive minerals
Table C.7	Birefringence
Table C.8	Minerals that may display anomalous interference colors
Table C.9	Isometric minerals that may display anomalous birefringence
Table C.10	Tetragonal and hexagonal minerals that may be anomalously biaxial
Table C.11	Normally birefringent minerals that may be sensibly isotropic
Table C.12	Biaxial minerals that may be sensibly uniaxial
Table C.13	Minerals that may produce pleochroic halos in surrounding minerals
Figure C.1	$2V$ and birefringence for biaxial minerals
Figure C.2	Average index of refraction versus average birefringence

Note that Figure C.2 is printed on the reverse of the interference color chart at the back of the book.

Table C.7, Figure C.2 and the mineral descriptions in the chapters that follow all refer to the relief in thin section. It is assumed that the cement used in thin sections has an index of refraction of 1.54, which is the index of refraction of Canada balsam and commercially available epoxies such as Petropoxy (Palouse Petro Products, Route 1 Box 92, Palouse, WA 99161, USA). However, some cements have an index of refraction either higher or lower than 1.54. If the cement is significantly different than 1.54, the relief of minerals in thin section may be different than what is reported in this book. If the index of refraction of a mineral is being compared with the cement in a thin section, it is important to know what the index of refraction of the cement actually is. If in doubt, consult the manufacturer.

If grain mount or spindle stage techniques are employed, the most useful data are the tabulations of the indices of refraction (Tables C.3 through C.7). A cross-check of color, birefringence, and 2V, as appropriate, may help narrow the list of possibilities.

For isotropic minerals in thin section, the tabulations of color (Table C.1) and index of refraction (Table C.3), supplemented by Figure C.2, are typically most useful. Note, however, that only a rough estimate of index of refraction can be made from the relief in thin section.

For anisotropic minerals in thin section, Figure C.2 is a useful starting point. The tabulations of birefringence (Table C.7) and color (Table C.1) help narrow the list of possibilities. For biaxial minerals. Figure C.1, showing the optic angle, can be quite useful.

Opaque Minerals

Some of the common opaque minerals are listed in the identification tables (Appendix C, Table C.2) and described in Chapter 9. These minerals can sometimes be identified by shining a light on the top surface of the sample and observing the color of the reflected light through the microscope. This method has serious drawbacks and it is not encouraged. Small grains may easily be overlooked, the reflected color may be imperfectly perceived, and intergrowths between minerals like magnetite and ilmenite may not be recognized. If it is important to know the identity of the opaque minerals, techniques fitted to the task (reflected light microscope with polished sections, X ray, or other) should be employed, otherwise they should just be identified as opaque minerals. Note that the reflected light colors of the opaque minerals described in Chapter 9 may be different than the color seen through a ref-

lected light microscope. The interested reader is encouraged to read Cameron (1961) or Craig and Vaughn (1981) to learn more about reflected light microscopy.

Non-minerals

Imperfectly prepared thin sections may contain a variety of materials that can be mistaken for minerals or that complicate the identification process. Some of the more common are bubbles, grinding abrasive, and textile fibers (Figure 8.1).

Unless care is taken in preparing thin sections, bubbles trapped in the cement are almost inevitable.

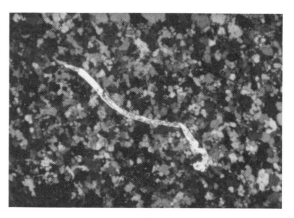

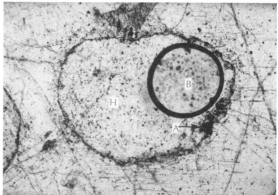

Figure 8.1 Non-minerals found in thin sections. (*Top*) Textile fiber. (*Bottom*) Hole (H) in a thin section partially occupied by a bubble (B). The dark material around the edges of the hole and scattered through the epoxy in the hole is grinding abrasive (A). Field of view of both photographs is 1.2 mm wide.

If a bubble has been trapped where a grain has been plucked out, it can be mistaken for a high-relief isotropic mineral. Small spherical bubbles may display what looks like a uniaxial cross in orthoscopic illumination. Larger bubbles may be quite irregular, but typically display curved or rounded boundaries due to the surface tension of the cement.

Silicon carbide grinding abrasive appears as fine angular opaque grains distributed throughout the slide or concentrated in cracks or void spaces in the sample. If the cement is not entirely cured before grinding, abrasive may be embedded in the cement.

Textile materials from paper towels, clothing, and other sources appear as elongate fibers that are typically kinked. These materials may display interference colors between crossed polars because many varieties are anisotropic. By carefully adjusting the focus, it can usually be determined that these fibers are in the cement either above or below the sample.

REFERENCES

Cameron, E. N., 1961, Ore microscopy: Wiley, New York, 239 p.

Craig, J. R., and Vaughn, D. J., 1981, Ore microscopy and ore petrography: Wiley, New York, 406 p.

Fleischer, M., Wilcox, R. E., and Matzko, J. J., 1984, Microscopic determination of the nonopaque minerals: U. S. Geological Survey Bulletin 1627, 453 p.

Phillips, W. R., and Griffen, D. T., 1981, Optical mineralogy, The nonopaque minerals: W. H. Freeman and Company, San Franscisco, 677 p.

Troger, W. E., 1979, Optical determination of rock-forming minerals, Part I, Determinative tables. English edition of the 4th German edition by Bambaur, H. U., Taborgzky, F., and Trochim, H. D.: Schwiezerbart'sche Verlagsbuchhandlung, Stuttgart, 188 p.

9

Native Elements, Sulfides, Halides, Oxides, and Hydroxides

NATIVE ELEMENTS

Sulfur

S
Orthorhombic
Biaxial (+)
$n_\alpha = 1.958$
$n_\beta = 2.037$
$n_\gamma = 2.245$
$\delta = 0.287$
$2V_z \cong 69°$

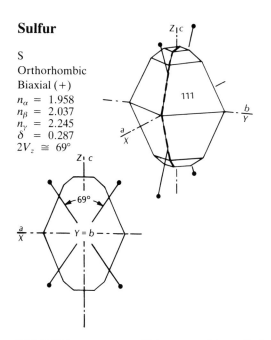

RELIEF IN THIN SECTION: Extreme positive relief.

COMPOSITION AND STRUCTURE: Structure consists of eightfold rings of Sulfur atoms that are stacked one atop the other along the c axis and that are bonded to each other by weak electrostatic bonds. Small amounts of Se or Te may substitute for S.

PHYSICAL PROPERTIES: H = $1\frac{1}{2}$–$2\frac{1}{2}$; G = 2.05–2.09; yellow in hand sample, sometimes with a reddish, brownish, or greenish cast; white streak; resinous to adamantine luster.

COLOR AND PLEOCHROISM: Pale yellow or yellowish gray in thin section and grain mount. Pleochroism is weak; from light to dark yellow.

FORM: Crystals may be skeletal or hopper types with dypyramidal or thick tabular forms. Sulfur also forms fine-grained aggregates and colloform or encrusting masses.

CLEAVAGE: Cleavages on {001} and {110} are fair to poor, but partings parallel to the four {111} dipyramid faces may be prominent.

TWINNING: Twins are rare, but if present are usually simple contact twins on {110}, {001}, or {101}.

OPTICAL ORIENTATION: $X = a$, $Y = b$, $Z = c$, optic plane = (010). Extinction is symmetrical to the partings and parallel or symmetrical to the cleavages in the principal sections. Diamond-shaped cross sections are length slow to the long diagonal.

INDICES OF REFRACTION AND BIREFRINGENCE: Indices do not vary significantly and birefringence is extreme (0.287), so interference colors are high-order white.

INTERFERENCE FIGURE: Basal sections yield acute bisectrix figures with innumerable isochromes, but the extreme birefringence makes figures difficult to interpret. Optic axis dispersion is weak, $v > r$.

ALTERATION: Sulfur may alter to sulfate minerals such as gypsum or anhydrite.

DISTINGUISHING FEATURES: Extreme relief and birefringence and limited occurrence are characteristic.

Sulfur melts at 112.8°C, so cements requiring high temperatures to cure should be avoided.

OCCURRENCE: Found around fumaroles, volcanic vents, and in hot spring deposits. Also found in evaporite deposits with sulfates, carbonates, and halides, where it is probably produced by bacterial reduction of sulfate minerals. Small amounts formed by alteration of sulfide or sulfate minerals may be found in hydrothermal ore deposits.

Graphite

C
Hexagonal
Opaque

COMPOSITION AND STRUCTURE: Graphite consists of hexagonally packed sheets of C atoms that are bonded to adjacent sheets with weak electrostatic bonds. Graphite is usually quite pure.

PHYSICAL PROPERTIES: H = 1–2; G = 2.23; black in hand sample; gray streak; metallic luster.

COLOR: Opaque. Black with metallic luster in reflected light.

FORM: Figure 9.1. Graphite forms tabular hexagonal crystals, scaly masses, fine-grained aggregates, or fine disseminated flakes or scales.

CLEAVAGE: One perfect cleavage on {0001} is generally difficult to recognize in thin section because the mineral is opaque.

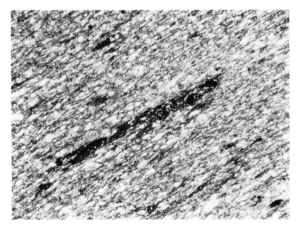

Figure 9.1 Graphite (black) in slate. Field of view is 0.6 mm wide.

DISTINGUISHING FEATURES: Opaque nature, black color, softness (H = 1–2) with soapy feel. May be mistaken for magnetite or ilmenite, but softness, lack of magnetism, and cleavage distinguish it.

OCCURRENCE: A relatively common constituent in regional and contact metamorphic rocks such as marble, skarn deposits, gneiss, schist, phyllite, and slate that originally contained carbonates or organic material. Much less common as an accessory mineral in igneous rocks.

SULFIDES

Pyrite

FeS$_2$
Isometric
Opaque

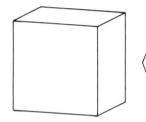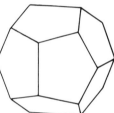

COMPOSITION AND STRUCTURE: Pyrite has a cubic structure similar to halite with S$_2$ pairs and Fe instead of Cl and Na. Pyrite displays relatively little substitution of other metals for Fe, but small amounts of Ni, Co, and Mn may be present. Marcasite is the orthorhombic dimorph of pyrite.

PHYSICAL PROPERTIES: H = 6–6$\frac{1}{2}$; G = 5.02; metallic brassy yellow in hand sample; greenish or brownish black streak; iridescent tarnish.

COLOR: Pyrite is completely opaque. It is pale brassy yellow in reflected light as seen through a petrographic microscope.

FORM: Usually forms cubes {001}, pentagonal dodecahedrons (pyritohedrons) {210}, or occasionally octahedrons {111}. Pyrite also forms granular and reniform encrusting masses that may have a radiating structure. Small, raspberry like aggregates composed of innumerable spherical grains of pyrite are found in shale, coal, and other sediments and are called framboids.

CLEAVAGE: Cleavage on {001}, {011}, {111}, and {311} has been reported, but it usually displays conchoidal fracture.

TWINNING: Penetration twins called iron crosses with twin axis [001] are relatively common, but unless the crystal outline betrays it, twinning will not be recognized.

ALTERATION: Pyrite commonly alters to Fe oxides (hematite, limonite, goethite, etc.); and these minerals may form pseudomorphs after pyrite.

DISTINGUISHING FEATURES: Sections through cubic pyrite crystals are usually triangular, rectangular, or square. Yellowish color in reflected light distinguishes it from magnetite, ilmenite, and hematite, but it is easily mistaken for marcasite, pyrrhotite, chalcopyrite, or other sulfides. Examining hand samples with a hand lens or binocular microscope is usually more productive than examining either thin sections or grain mounts. Examining a polished section with a reflected light microscope is encouraged if the identity is important.

OCCURRENCE: Pyrite is the most common sulfide mineral, is nearly ubiquitous in hydrothermal mineral deposits, and is a common accessory mineral in many igneous and metamorphic rocks. Pyrite is also formed in such sediments as shale and coal deposited in a reducing environment, where it may occur as framboids.

Sphalerite

(Zn,Fe)S
Isometric
Isotropic
$n = 2.37–2.50$

RELIEF IN THIN SECTION: Very high positive relief.

COMPOSITION AND STRUCTURE: The structure of the zinc sulfides consists of layers of S atoms stacked one atop the other with Zn and Fe placed between the layers so that they coordinate with four S atoms. In sphalerite, the layers of S are stacked in a slightly opened version of cubic closest packing, whereas in the polymorph wurtzite, the layers are stacked in an open version of hexagonal closest packing. Samples of natural sphalerite containing submicroscopic domains with wurtzite stucture are anisotropic. The amount of Fe that can be accepted in the structure increases with increasing temperature and can be in excess of 40 mole percent.

PHYSICAL PROPERTIES: H = 3–4; G = 3.9–4.1; light brown, brown, or nearly black in hand sample, darker with increasing Fe; white, yellow, light brown streak; adamantine, resinous, or submetallic luster.

COLOR: Colorless, pale yellow, or pale brown in thin section or grain mount without pleochroism.

FORM: Crystals show tetrahedral or dodecahedral forms, often with curved faces. It also forms coarse to fine granular aggregates and is rarely fibrous.

CLEAVAGE: Dodecahedral {110} cleavage in six directions is perfect.

TWINNING: Multiple contact twins on octahedral {111} faces are common but will not be seen unless the sample displays birefringence.

INDICES OF REFRACTION AND BIREFRINGENCE: The index of refraction increases from 2.37 for pure ZnS to 2.50 with 40 percent Fe replacing Zn. Sphalerite may display birefringence because of the intergrowth of the sphalerite and wurtzite structures. The birefringence increases in a linear fashion from zero for 100 percent sphalerite structure to 0.022 for 100 percent wurtzite structure. Wurtzite is uniaxial positive with $n_\omega \cong 2.356$, and $n_\epsilon \cong 2.378$, and has a similar appearance.

ALTERATION: Sphalerite may alter to oxides, hydroxides, sulfates, or carbonates or Fe and Zn (limonite, goethite, smithsonite, hydrozincite, goslarite, siderite, etc.).

DISTINGUISHING FEATURES: Extreme relief, isotropism, and six perfect cleavages are characteristic. Garnet lacks cleavage.

OCCURRENCE: Sphalerite is a common constituent of hydrothermal sulfide deposits in association with galena, pyrite, chalcopyrite, and other sulfides. It rarely occurs 'as an accessory mineral in felsic igneous rocks and is occasionally found in coal beds.

Pyrrhotite

Fe$_{1-x}$S   (x = 0–0.17)
Monoclinic (pseudohexagonal)
Opaque

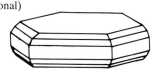

COMPOSITION AND STRUCTURE: Pyrrhotite consists of a slightly distorted hexagonal close-packed array of S with Fe atoms in sixfold coordination between the layers of S. As indicated by the formula, the Fe:S ratio is slightly less than 1 and may vary because up to 17 percent of the sites that would otherwise contain Fe atoms may be vacant. Troilite, which is found in meteorites, has a composition close to the ideal FeS and is hexagonal. Above about 300°C, pyrrhotite possesses a hexagonal structure but reverts to a monoclinic or occasionally orthorhombic structure at lower temperatures.

PHYSICAL PROPERTIES: H = $3\frac{1}{2}$–$4\frac{1}{2}$; G $\cong$ 4.6; metallic bronze-yellow, sometimes with a brownish or reddish cast; dark grayish black streak; weakly magnetic.

COLOR AND PLEOCHROISM: Opaque, even on thin edges. Has a bronze color in reflected light as viewed through the petrographic microscope.

FORM: Usually in granular aggregates or irregular grains; tabular hexagonal crystals are uncommon.

CLEAVAGE: None. There is a {0001} parting, which is generally not visible in grain mount or thin section.

TWINNING: Twins on {10$\bar{1}$2} are rare.

ALTERATION: May be replaced by pyrite, marcasite, or other sulfides or oxidized to oxides, hydroxides, sulfates, and carbonates of iron.

DISTINGUISHING FEATURES: The bronze color in reflected light is different from that of pyrite, chalcopyrite, or marcasite but the difference may be difficult to discern. Identification based on hand-sample properties, with a reflected light microscope or with X-ray diffraction techniques, is more reliable than with a petrographic microscope.

OCCURRENCE: Usually found in mafic igneous rocks or in high-temperature hydrothermal sulfide deposits.

Chalcopyrite

CuFeS$_2$
Tetragonal
Opaque

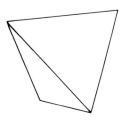

COMPOSITION AND STRUCTURE: The structure of chalcopyrite is very similar to sphalerite, consisting of essentially cubic close-packed S atoms. Cu and Fe atoms are in fourfold coordination between the layers of S atoms. Exhibits relatively little solid solution but inclusions of other minerals containing Ag, Au, Pt, Pb, Co, Ni, Mn, Sn, Zn, or other elements are common.

PHYSICAL PROPERTIES: H = $3\frac{1}{2}$–4; G = 4.1–4.3; metallic brassy yellow in hand sample; greenish black streak; often tarnished.

COLOR: Opaque even on thin edges. Yellow-gold or brassy yellow when viewed through the petrographic microscope with reflected light.

FORM: Crystals are commonly tetrahedral, although irregular grains and granular masses are more common.

CLEAVAGE: Cleavages on {011} and {111} are generally poor and not usually visible because the mineral is opaque.

TWINNING: Multiple lamellar twins on {112} are common but not usually visible.

ALTERATION: Often altered to oxides, carbonates, sulfates, and hydroxides of Cu and Fe.

DISTINGUISHING FEATURES: The color of pyrite in reflected light is typically lighter. Identification based on hand-sample characteristics, with a reflected light microscope, or X-ray diffraction techniques is more reliable than identification with a petrographic microscope.

OCCURRENCE: Chalcopyrite is a common mineral in hydrothermal mineral deposits. It is occasionally found as an accessory mineral in mafic igneous rocks and rarely as a primary precipitate or secondary mineral in sedimentary rocks.

HALIDES

Halite

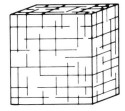

NaCl
Isometric
Isotropic
$n = 1.544$

RELIEF IN THIN SECTION: Low positive relief.

COMPOSITION AND STRUCTURE: The structure of halite consists of alternating Na^+ and Cl^- ions arranged in a face-centered cubic lattice. The composition is nearly pure NaCl. Although halite and sylvite are isostructural, there is very little substitution of K^+ for Na^+. Bromine or iodine may substitute for chlorine in small amounts.

PHYSICAL PROPERTIES: $H = 2\frac{1}{2}$; $G = 2.16$; colorless or white in hand sample, may be orange, red, gray, brown, blue, or yellow due to lattice imperfections or inclusions; white streak; vitreous luster; has salty taste and is soluble in water.

COLOR: Usually colorless in thin section; may be faintly colored if hand sample displays a strong color. Zoning parallel to crystal faces may be apparent.

FORM: Crystals are cubes. Halite also forms granular aggregates and anhedral grains.

CLEAVAGE: Perfect cubic {001} cleavage forms cubic or rectangular fragments.

TWINNING: May twin on {111} in synthetic crystals, but twinning will not be apparent in thin section or fragments due to isotropism.

INDICES OF REFRACTION: Intimate intergrowths of halite ($n = 1.544$) and sylvite ($n = 1.490$) may have intermediate indices.

ALTERATION: Due to its solubility, halite is often removed from rocks exposed to weathering, leaving casts of cubic crystals.

DISTINGUISHING FEATURES: Low relief, cubic shape, and cleavage distinguish halite from most other isotropic minerals. Index of sylvite (1.490) is lower than the cement usually used in thin sections, whereas

that of halite (1.544) is higher. Halite will usually be lost from conventionally made thin sections because it is highly soluble in water.

OCCURRENCE: Abundant in marine evaporite deposits and commonly associated with calcite, dolomite, gypsum, anhydrite, and sylvite, along with clay and other detrital material. It is also found in deposits from saline lakes (cf. Table 10.2).

Sylvite

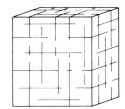

KCl
Isometric
Isotropic
$n = 1.490$

RELIEF IN THIN SECTION: Moderate negative relief.

COMPOSITION AND STRUCTURE: The structure of sylvite is the same as halite and consists of alternating K^+ and Cl^- ions in a face-centered cubic lattice. Sylvite is almost pure KCl with very little substitution of Na^+ for K^+ or Br^- for Cl^-.

PHYSICAL PROPERTIES: $H = 2\frac{1}{2}$; $G = 1.99$; colorless or white in hand sample; may be grayish, bluish, or yellowish. Hematite inclusions give it a red or orange color. White streak; somewhat sectile; has a bitter salty taste and is very soluble in water.

COLOR: Usually colorless in thin section or grain mount.

FORM: Crystals are usually cubes. Inclusions may be arranged parallel to crystal faces. Sylvite also forms anhedral grains or granular masses.

INDICES OF REFRACTION: Pure sylvite has an index $n = 1.490$. Indices intermediate between 1.490 and 1.544 are possible for submicroscopic mixtures of sylvite and halite.

ALTERATION: Easily dissolved in water, sylvite will usually be lost from conventionally prepared thin sections. If rocks containing sylvite are exposed to weathering most or all of it will be dissolved. Casts of sylvite crystals are sometimes present but those of halite are more common.

DISTINGUISHING FEATURES: Bitter taste and index lower than cement distinguish sylvite from halite.

OCCURRENCE: Sylvite is primarily restricted to evaporite deposits and is much less common than halite. It is commonly associated with calcite, dolomite, halite, gypsum, anhydrite, or other evaporite minerals.

Fluorite

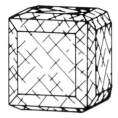

CaF$_2$
Isometric
Isotropic
$n = 1.433–1.435$

RELIEF IN THIN SECTION: Moderately high negative relief.

COMPOSITION AND STRUCTURE: Fluorite has a face-centered cubic structure with Ca^{2+} occupying the corners and center of each face and F$^-$ at the center of the eight smaller cubes formed by cutting the cube in half in all three directions. Most fluorite is relatively pure CaF$_2$ although substitution of Sr, Y, or Ce for Ca is possible.

PHYSICAL PROPERTIES: H = 4; G = 3.18; commonly colorless, blue, purple, or green, but almost any color is possible; streak white; vitreous luster; usually fluoresces in both short- and long-wave UV light. Some may be phosphorescent.

COLOR: Usually colorless in thin section or grain mount, but pale colors corresponding to the hand-sample color may be visible in some cases. Colored samples may be blotchy or zoned parallel to the crystal faces. Because the mineral is isotropic, there is no pleochroism.

FORM: Cubes or cubes modified by octahedral or dodecahedral faces are common. Fluorite also forms anhedral grains or granular masses, or occasionally columnar or fibrous aggregates.

CLEAVAGE: Perfect octahedral {111} cleavage (four directions).

TWINNING: Penetration twins by rotation on [111] are common but will not be visible in thin section or fragments unless belied by crystal shape or color zoning.

INDICES OF REFRACTION: Most fluorite has index $n \cong$ 1.434 although substituting Y for Ca increases the index to as high as 1.457.

DISTINGUISHING FEATURES: The moderately high relief, with $n <$ cement, and octahedral cleavage distinguish fluorite from most similar minerals. Cryolite (biaxial ($+$), $2V_z = 43°$, $n_\beta = 1.338$, $\delta = 0.001$) is the most likely alternative, but it is birefringent and has even lower indices of refraction. A simple staining technique to aid identification has been described by Sharp and others (1977).

OCCURRENCE: Fluorite is relatively common in some hydrothermal mineral deposits associated with sulfide minerals and also is found as an accessory mineral in granite, syenite, granite pegmatite, and related igneous rocks. Sometimes occurs in veins in carbonate sediments and may be found as detrital grains or (rarely) as a cementing agent in clastic sediments.

OXIDES

Periclase

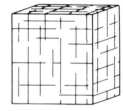

MgO
Isometric
Isotropic
$n = 1.735–1.756$

RELIEF IN THIN SECTION: High positive relief.

COMPOSITION AND STRUCTURE: The structure of periclase is the same as that of NaCl with alternating Mg and O arranged in a face-centered cubic lattice. Most periclase is relatively pure MgO, but up to 10 mole percent Fe may substitute for Mg.

PHYSICAL PROPERTIES: H = 5$\frac{1}{2}$; G = 3.56–3.68; colorless to grayish white in hand sample, sometimes yellow, brown, or black due to inclusions or high Fe content; white streak; vitreous luster.

COLOR: Colorless in thin section or grain mount. High Fe varieties may be pale yellow or brown without pleochroism.

FORM: Crystals are cubes or octahedrons but periclase is more common as anhedral grains, surrounded by fibrous brucite.

CLEAVAGE: Perfect cubic {001} cleavage.

INDICES OF REFRACTION: Pure periclase has $n = 1.736$. With increasing Fe content the index rises to around 1.756.

ALTERATION: Typically alters to fibrous brucite or other Mg-bearing hydroxides or carbonates.

DISTINGUISHING FEATURES: Isotropism with cubic cleavage and high relief in thin section are characteristic as is the alteration to fibrous brucite.

OCCURRENCE: Relatively uncommon mineral found predominantly in high-temperature contact metamorphosed siliceous dolomite or dolomitic limestone. It is usually associated with calcite, forsterite, humite group minerals, serpentine, brucite, and spinel.

Rutile

TiO$_2$
Tetragonal
Uniaxial (+)
n_ω = 2.61–2.65
n_ϵ = 2.90–2.80
δ = 0.29–0.15

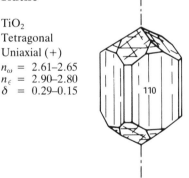

RELIEF IN THIN SECTION: Extreme positive relief.

COMPOSITION AND STRUCTURE: The rutile structure consists of Ti at the center and corners of the tetragonal unit cell with O arranged in between so that each Ti is in octahedral coordination. Although most rutile is nearly pure TiO$_2$, significant amounts of Fe, Ta, and Nb may substitute for Ti. Minor amounts of V, Cr, or Sn also may be present.

PHYSICAL PROPERTIES: H = 6–6$\frac{1}{2}$; G = 4.23–5.5; adamantine to metallic reddish brown in hand sample, sometimes black, bluish, or yellowish; streak pale brown or grayish black.

COLOR AND PLEOCHROISM: Rutile is reddish brown, pale brown, or sometimes almost opaque in thin section and grain mount. Pleochroism is usually weak, $\epsilon > \omega$. Darker colors are associated with higher Fe, Nb, or Ta content. May be color zoned.

FORM: Crystals are elongate tetragonal prisms with square or octagonal cross sections. Also as slender, acicular, or hair like crystals, and anhedral grains. Rounded and abraided grains are common in some clastic sediments.

CLEAVAGE: Good prismatic {110} cleavage and fair {100} cleavage may control the orientation of fragments.

TWINNING: Contact twinning on {101} is common, producing knee-shaped twins or, in some cases, cyclic twins with six or eight segments.

OPTICAL ORIENTATION: Elongate sections show parallel extinction. Crystals are length slow, but this usually cannot be recognized due to the extreme birefringence.

INDICES OF REFRACTION AND BIREFRINGENCE: Indices of refraction are substantially higher than indices of routinely used index oils. Birefringence is extreme, yielding upper-order white interference colors that are generally masked by the color of the mineral. Even basal sections will show high colors due to the moderately converging nature of orthoscopic illumination.

INTERFERENCE FIGURE: Basal sections yield uniaxial interference figures with innumerable isochromes that are often brownish due to the inherent color of the mineral. May be anomalously biaxial (+) with $2V_z = 10°$.

ALTERATION: Rutile generally shows no alteration although alteration to titanite or leucoxene is possible. It often forms as an alteration product of other Ti-bearing minerals.

DISTINGUISHING FEATURES: Color, extreme birefringence, extreme relief, and habit are characteristic. Anatase and hematite are uniaxial negative, and hematite is deep red and has a flaky habit. Cassiterite is quite similar but has lower birefringence and indices, and lighter color. Limonite may have a similar color but is sensibly isotropic.

OCCURRENCE: Commonly occurs as an accessory mineral in many igneous and metamorphic rocks. Is also a common detrital mineral in clastic sediments.

Anatase

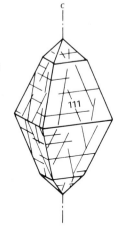

TiO_2
Tetragonal
Uniaxial $(-)$
$n_\omega = 2.561$
$n_\epsilon = 2.488$
$\delta = 0.073$

RELIEF IN THIN SECTION: Extreme positive relief.

COMPOSITION AND STRUCTURE: Anatase, like *brookite*, is a polymorph of rutile, and, like rutile, the structure consists of TiO_2 octahedra. The difference between the two structures is in which edges of the octahedra are shared with neighboring octahedra. Minor amounts of Fe may substitute for Ti, and some substitution of Nb or Ta is reported.

PHYSICAL PROPERTIES: H = $5\frac{1}{2}$–6; G = 3.90; yellowish brown, red-brown, or black in hand sample, less commonly green, blue, gray, or colorless; colorless to pale yellow streak; adamantine to submetallic luster.

COLOR AND PLEOCHROISM: Usually pale brown, red-brown, or deep brown; sometimes green or blue; rarely colorless. Pleochroism is usually weak, $\omega > \epsilon$ or $\epsilon > \omega$.

FORM: Crystals are elongate dipyramids, so sections are usually rectangular or diamond shaped. Also forms anhedral grains.

CLEAVAGE: Perfect basal {001} and pyramidal {111} cleavages. Fragments are commonly triangular in cross section.

TWINNING: Generally not found, although it is possible on {112}.

OPTICAL ORIENTATION: Diamond-shaped cross sections are fast parallel to the long diagonal. Extinction is parallel to basal cleavage {001} in all sections and parallel, symmetrical, or inclined to {111} dipyramid faces and cleavage, depending on how the section is cut.

INDICES OF REFRACTION AND BIREFRINGENCE: Indices are above the range of routinely used index oils and produce extreme relief in thin section. The published range of n_ω is from about 2.55 to 2.60. Birefringence is very high (0.07) and produces creamy upper-order colors, which may be masked by the mineral color.

INTERFERENCE FIGURE: Basal sections and fragments on (001) yield uniaxial negative interference figures with numerous isochromes that are often masked by the color of the mineral. Dark-colored varieties may be anomalously biaxial negative with a small $2V$.

ALTERATION: May alter to rutile or be replaced by leucoxene, which is opaque amorphous titanium oxide. Leucoxene is gray or white in reflected light.

DISTINGUISHING FEATURES: Resembles rutile but is distinguished by a different optic sign. Brookite is another polymorph of rutile with similar appearance, but it is biaxial positive, with indices $n_\alpha = 2.58$, $n_\beta = 2.59$, $n_\gamma = 2.70$ and has strong crossed axial plane dispersion with a small $2V$ ($\sim$0–30°).

OCCURRENCE: Minor constituent of igneous rocks such as granite, granite pegmatite, and felsic volcanic rocks and occasionally found in hydrothermal vein deposits. Is also a common alteration product after other Ti-bearing minerals such as titanite or ilmenite. Because anatase is relatively stable in the weathering environment, it may be found as detrital grains in clastic sediments.

Cassiterite

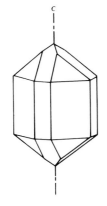

SnO_2
Tetragonal
Uniaxial (+)
n_ω = 1.990–2.010
n_ϵ = 2.091–2.100
δ = 0.09–0.10

RELIEF IN THIN SECTION: Very high to extreme.

COMPOSITION AND STRUCTURE: Has the same structure as rutile with Sn atoms at the corners and center of the tetragonal unit cell with O atoms in between, so that the Sn is in octahedral coordination. Composition is usually close to SnO_2, with minor to significant substitution of Fe, Nb, or Ta.

PHYSICAL PROPERTIES: H = 6–7; G = 6.8–7.1; brown to black in hand sample, rarely red, yellow or nearly white; white, gray, or brownish streak; dull, or adamantine to submetallic luster.

COLOR AND PLEOCHROISM: Usually colored in shades of brown, yellow, or red, occasionally green or almost colorless. Fragments often appear almost opaque due to the strong color. Pleochroism is weak to strong in shades of yellow, brown, and red, $\epsilon >$ ω. Color zoning is sometimes found.

FORM: Crystals are stubby tetragonal prisms with dipyramid terminations, or acicular needles. Also forms colloform or encrusting masses (wood tin) of radiating or parallel subhedral to anhedral grains.

CLEAVAGE: Prismatic cleavages on {110} and {010} are poor. Partings parallel to the pyramid faces {111} are sometimes found.

TWINNING: Twinning is common on {101} forming elbow twins or, in some cases, cyclic or polysynthetic twins.

OPTICAL ORIENTATION: Crystals are length slow, and extinction is parallel to prismatic crystal faces or cleavages in sections cut parallel to the c axis. Extinction is inclined to twin planes.

INDICES OF REFRACTION AND BIREFRINGENCE: Indices of refraction vary in a small range with higher indices reflecting higher Fe content but are outside of the range of routinely used index oils. Maximum birefringence is very high (0.09–0.10) and in thin sections gives up to fifth-order pale pastel colors that are usually masked by the color of the mineral.

INTERFERENCE FIGURE: Basal sections give uniaxial positive figures with numerous isochromes, and isogyres that fan out toward the edge of the field of view. Biaxial positive cassiterite with a small 2V has been reported but is probably not common.

ALTERATION: Cassiterite is resistant to weathering or other alteration and, consequently, is sometimes found as a heavy mineral in clastic sediments.

DISTINGUISHING FEATURES: Very similar to rutile, but can be distinguished because rutile has even higher relief in thin section, higher birefringence, and is commonly more strongly colored. Anatase is uniaxial negative and length fast.

OCCURRENCE: Usually occurs as an accessory mineral in silicic plutonic igneous rocks, and in high-temperature hydrothermal deposits where it is typically associated with tungsten and molybdenum minerals. It is also found in granitic pegmatites and as a heavy mineral in clastic sediments.

Corundum

Al_2O_3
Hexagonal (trigonal)
Uniaxial (−)
n_ω = 1.766–1.794
n_ϵ = 1.758–1.785
δ = 0.008–0.009

RELIEF IN THIN SECTION: High positive relief.

COMPOSITION AND STRUCTURE: Consists of layers of hexagonal close-packed O atoms with aluminum occupying two-thirds of the octahedral sites between the layers of oxygen atoms. Most corundum is nearly pure Al_2O_3 but small amounts of other metals may substitute for aluminum and are responsible for the colored varieties of corundum. Sapphire (blue) contains Fe^{2+} and Ti^{4+}, ruby (red) contains Cr^{3+}, and yellow and green corundum contain varying amounts of Fe^{2+} and Fe^{3+}.

PHYSICAL PROPERTIES: H = 9; G = 3.98–4.10; gray, blue, red, green, yellow, and many other colors in hand sample; white streak; vitreous to adamantine luster.

COLOR AND PLEOCHROISM: Usually colorless in fragments or thin section, although pale colors corresponding to hand-sample color may be displayed. Colored samples usually display weak pleochroism, $\omega > \epsilon$. ω = shades of blue and purple, ϵ = blue, blue-green, yellowish green, yellow. Color zoning and systematic distribution of inclusions are common.

FORM: Figure 9.2. Commonly occurs as well-formed hexagonal crystals that taper at both ends or as tabular hexagonal crystals, less commonly as anhedral grains.

Figure 9.2 Anhedral high relief corundum showing the rhombohedral parting, with bladed biotite. Field of view is 1.7 mm wide.

CLEAVAGE: No cleavage is present but there is a prominent basal parting on $\{0001\}$ and a rhombohedral parting on $\{10\bar{1}1\}$.

TWINNING: Multiple twinning is common on $\{0001\}$ and $\{10\bar{1}1\}$, producing a lamellar structure.

OPTICAL ORIENTATION: Elongate crystals are length fast. Extinction is parallel to basal partings and inclined to rhombohedral partings, except for certain orientations, where it is symmetrical.

INDICES OF REFRACTION AND BIREFRINGENCE: Relatively pure corundum has indices $n_\omega = 1.768$, $n_\epsilon = 1.760$. Presence of small amounts of Cr or Fe increases the indices slightly. Very Fe-rich corundum is reported with indices as high as $n_\omega = 1.794$, $n_\epsilon = 1.785$. Birefringence of 0.008 to 0.009 produces first-order gray interference colors in standard thin sections. However, most thin sections containing corundum are thicker than normal owing to its great hardness, and higher-order colors are usually encountered.

INTERFERENCE FIGURE: Basal sections yield uniaxial negative interference figures with low first-order colors. Upper first-order color isochromes may be visible with thick sections or fragments. Some corundum is biaxial, possibly because of twinning, and $2V_x$ may be as high as 50 or 60°.

ALTERATION: Corundum is commonly altered to fine-grained aggregates of other aluminous minerals such as muscovite, margarite, diaspore, gibbsite, andalusite, kyanite, and sillimanite.

DISTINGUISHING FEATURES: The high relief, low birefringence, and uniaxial negative character are characteristic. Twin lamellae are usually present.

OCCURRENCE: Although not very abundant, corundum is widespread. Most commonly found in silica-poor igneous rocks such as syenite and associated pegmatites and in high-grade Al-rich pelitic metamorphic rocks. Is also found in Si-poor hornfelses, metamorphosed bauxite deposits (emery), xenoliths in mafic igneous rocks, and metamorphosed carbonate rocks. In igneous rocks, corundum is not usually associated with quartz and is often found with spinel or other aluminous minerals. In pelitic metamorphic rocks, corundum is often associated with sillimanite, kyanite, or other aluminous minerals.

Locally, detrital corundum is an important constituent of the heavy fraction of clastic sediments.

Hematite

Fe_2O_3
Hexagonal (trigonal)
Uniaxial ($-$) often practically opaque
n_ω = 3.15–3.22
n_ϵ = 2.87–2.94
δ = 0.28

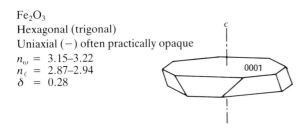

RELIEF IN THIN SECTION: Extreme positive relief.

COMPOSITION AND STRUCTURE: The structure of hematite consists of layers of hexagonal close-packed O atoms with Fe^{3+} in two-thirds of the octahedral sites between the layers of O. Some Ti may substitute for Fe as may minor amounts of Al or Mn.

PHYSICAL PROPERTIES: H = 5–6 (earthy varieties softer); G = 5.26 (sometimes lower); color in hand sample is metallic steel gray (specular hematite) or red to red-brown (earthy varieties); red-brown streak; weakly magnetic.

COLOR AND PLEOCHROISM: Deep red-brown in very small crystals or along thin edges. Usually opaque. Thin crystals are pleochroic with ω = brownish red, and ϵ = yellowish red or brown ($\omega > \epsilon$). Metallic black with red internal reflection in reflected light using a petrographic microscope.

FORM: Well-formed crystals of specular hematite are usually flat hexagonal plates or scales that look micaceous in hand sample. More commonly found as fine-grained anhedral aggregated or grains. Also forms oolites in sedimentary rocks and reniform masses with a fibrous or radiating structure.

CLEAVAGE: No cleavage but may have a pronounced basal parting {0001}, or a rhombohedral parting on {1011}, both of which are related to the twinning.

TWINNING: Lamellar twinning on {0001} and {10$\bar{1}$1}. However, lamellar twinning is not usually visible due to hematite's opacity.

OPTICAL ORIENTATION: In grains that are sufficiently thin to be transparent, extinction should be parallel to the basal parting and to the top and bottom faces of tabular crystals. Tabular crystals also should be length fast. However, the extreme dispersion commonly prevents the grains from becoming extinct with stage rotation, and the mineral color and extreme birefringence make determining the sign of elongation impossible.

INDICES OF REFRACTION AND BIREFRINGENCE: Indices of refraction are well beyond the range of routinely used index oils. Dispersion of the indices is extreme. Birefringence is extreme (0.28) but the color of the mineral usually masks the interference color in grains that are thin enough to be transparent.

INTERFERENCE FIGURE: Although hematite is uniaxial negative, useful interference figures are generally not obtainable due to the opacity, extreme birefringence, and extreme dispersion.

ALTERATION: Relatively stable in the weathering environment and is often the product of weathering or other alteration of Fe-bearing minerals. May be altered to Fe hydroxide minerals (limonite, goethite, etc.) or siderite.

DISTINGUISHING FEATURES: Earthy varieties of hematite resemble limonite and goethite, but hematite's anisotropic character distinguishes it from limonite and its uniaxial character distinguishes it from goethite, which is biaxial. These characteristics usually cannot be determined, however. It is distinguished from other opaque minerals by red color on thin edges. Identification with reflected light microscope, X-ray diffraction, or other techniques is more reliable than identification with the petrographic microscope.

OCCURRENCE: Hematite is a common alteration product of other mafic minerals and is common as finely disseminated grains in many clastic sedimentary rocks, to which it gives a red color. Locally, it may be a major constituent of sedimentary units. It is a major constituent of Precambrian banded iron formations, is a common mineral in some hydrothermal mineral deposits, and (infrequently) is a primary accessory mineral in igneous rocks.

Ilmenite

FeTiO$_3$
Hexagonal (trigonal)
Opaque

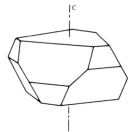

COMPOSITION AND STRUCTURE: Ilmenite has the same structure as hematite and corundum and consists of hexagonal close-packed O atoms with Fe^{2+} and Ti^{4+} in two-thirds of the octahedral sites between the layers of O. There is continuous solid solution to hematite with substitution of Fe^{3+} for Ti^{4+} and Fe^{3+} for Fe^{2+}. Mg^{2+} and Mn^{2+} also may substitute for Fe^{2+}.

PHYSICAL PROPERTIES: H = 5–6; G = 4.70–4.79; metallic iron black in hand sample; black or reddish black streak (red due to intergrown hematite); often weakly magnetic due to intergrown magnetite.

COLOR: Opaque in thin section and fragments. Very thin edges may transmit light and have a deep red or brown color. Metallic black in reflected light as viewed through the petrographic microcope unless altered to leucoxene, which is grayish white and looks somewhat like cotton wool.

FORM: Crystals are usually tabular parallel to (0001) and may be skeletal. Sections tend to be elongate rectangles. Also forms anhedral grains or granular masses and is common as lamellar intergrowths in magnetite. Lamellae of magnetite also are common in ilmenite. Detrital grains are more or less equant.

CLEAVAGE: No cleavage, but partings parallel to {0001} and {10$\bar{1}$1} associated with twinning are common. Partings are generally not observed in thin sections or fragments.

TWINNING: Simple twins on {0001} and multiple lamellar twins on {10$\bar{1}$1} are not visible in thin section or fragments because the mineral is opaque.

INDICES OF REFRACTION AND BIREFRINGENCE: Although opaque in thin section or grain mounts, ilmenite is uniaxial negative with n_ω in the vicinity of 2.7 and has very strong birefringence.

ALTERATION: Ilmenite may be altered to leucoxene, which is an aggregate of titanium oxides. In hand sample, leucoxene is usually grayish white although it may show shades of red, yellow, and brown due to the presence of Fe oxide and hydroxides.

DISTINGUISHING FEATURES: Ilmenite most closely resembles magnetite, from which it may be difficult to distinguish if grains are small or anhedral. Magnetite is strongly magnetic and euhedral crystals usually show as triangular-, square-, or diamond-shaped sections in thin section rather than elongate rectangles. Hematite also is similar to ilmenite because both may display rectangular sections. The leucoxene alteration is distinctive, but other Ti-bearing minerals also may alter to leucoxene. Identification with conventional petrographic microscope is uncertain; a reflected light microscope or other techniques are much more reliable.

OCCURRENCE: Widespread accessory mineral in many igneous and metamorphic rocks and is a common heavy mineral in clastic sediments. Less common constituent of hydrothermal mineral deposits.

Perovskite

CaTiO$_3$
Pseudoisometric (monoclinic or orthorhombic)
Essentially isotropic
n = 2.27–2.40

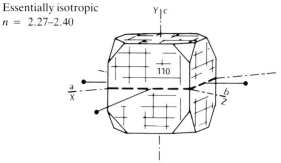

RELIEF IN THIN SECTION: Extreme.

COMPOSITION AND STRUCTURE: The idealized structure of perovskite consists of a cube with Ca at the center, Ti at each corner, and O at the midpoint of each edge. Most perovskite is close to the idealized composition, although substitution of Na, Fe^{2+}, or Ce for Ca, and Nb or Ta for Ti is possible. Other rare earth elements are commonly found in small amounts.

PHYSICAL PROPERTIES: $H = 5\frac{1}{2}$; $G = 3.98$–4.26; yellow, brown, or black in hand sample, white or gray streak; adamantine to metallic luster.

COLOR AND PLEOCHROISM: Colorless to shades of brown in thin section or as fragments, rarely gray or green. Colored varieties may display weak pleochroism $Z > X$.

FORM: Usually forms tiny cubic or octahedral crystals.

CLEAVAGE: Poor cubic cleavage {001} is usually not seen due to the small size of the grains but may be seen on large grains.

TWINNING: Usually displays complex lamellar twinning on {111} which is visible only in birefringent varieties.

OPTICAL ORIENTATION: Although it appears to be essentially isotropic, perovskite is biaxial positive with $2V$ variable up to 90°. If perovskite is considered monoclinic, then $b = Y$, $c \wedge Z \cong 45°$, and the optic plane is {010}. If it is considered orthorhombic, then the crystal axes are assigned differently and $X = a$, $Y = c$, $Z = b$, and the optic plane is (001). The axes shown in the figure above are based on the orthorhombic interpretation.

INDICES OF REFRACTION AND BIREFRINGENCE: Birefringence is very low (0.000–0.002) and yields dark first-order gray colors in thin section. The interference color usually is noticed because of variation due to twinning. Small grains appear isotropic.

INTERFERENCE FIGURE: Because of small grain size and very low birefringence, interference figures are usually not obtainable. Perovskite is reported to be biaxial positive with $2V \cong 90°$, and $r > v$ optic axis dispersion.

ALTERATION: May alter to leucoxene, a fine-grained white or gray aggregate consisting of Ti-bearing minerals such as rutile, brookite, and anatase. May be an alteration product after titanite and is commonly a constituent of leucoxene.

DISTINGUISHING FEATURES: Distinguished by extreme relief, habit, and possible weak birefringence and lamellar twinning. Brown spinel and garnet have lower relief and a different habit and rutile has extreme birefringence and a different habit.

OCCURRENCE: Perovskite is usually found as tiny crystals in silica-deficient igneous rocks such as peridotites, pyroxenites, syenites, ijolites, and leucite-bearing volcanic rocks. Is also found in carbonatites and in metamorphosed carbonate rocks.

THE SPINEL GROUP

The members of the spinel group include the spinel, magnetite, and chromite series. All have the general formula $A^{2+}B^{3+}_2O_4$ and consist of cubic close-packed oxygen atoms with the metal cations in tetrahedral and octahedral sites between the layers of oxygen. The spinel and chromite series have normal spinel structure with the divalent cations (A^{2+}) in tetrahedral (fourfold) coordination and the trivalent cations (B^{3+}) in octahedral (sixfold) coordination with oxygen, even though the trivalent cations are smaller. The magnetite series has an inverse spinel structure with all of the larger divalent cations (A^{2+}) in octahedral sites and half of the trivalent cations (B^{3+}) in octahedral sites and half in tetrahedral sites.

The three major series of the spinel group are defined based on the identity of the trivalent (B) cation. In the spinel series, it is Al, in the magnetite series Fe^{3+}, and in the chromite series Cr. The members of each series are as follows:

Spinel Series
Spinel	$MgAl_2O_4$
Hercynite	$FeAl_2O_4$
Gahnite	$ZnAl_2O_4$
Galaxite	$MnAl_2O_4$

Magnetite Series
Magnetite	$FeFe_2O_4$
Magnesioferrite	$MgFe_2O_4$
Ulvöspinel	$FeFeTiO_4$
Franklinite	$ZnFe_2O_4$
Jacobsite	$MnFe_2O_4$
Trevorite	$NiFe_2O_4$

Chromite Series
Chromite	$FeCr_2O_4$
Magnesiochromite	$MgCr_2O_4$

Of these, only magnetite, chromite and the members of the spinel series are common.

Spinel Series

Isometric
Isotropic

Spinel	$MgAl_2O_4$	$n = 1.714$
Hercynite	$FeAl_2O_4$	$n = 1.835$
Gahnite	$ZnAl_2O_4$	$n = 1.805$
Galaxite	$MnAl_2O_4$	$n = 1.920$

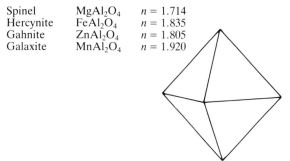

RELIEF IN THIN SECTION: Very high positive relief.

COMPOSITION AND STRUCTURE: There may be extensive solid solution among the members of the series. Spinel is the most common member of the series, and if it contains substantial Fe^{2+} it is called pleonaste. Picotite is a variety of hercynite in which substantial Cr has been substituted for Al.

PHYSICAL PROPERTIES: H = $7\frac{1}{2}$–8; G = 3.55–4.62; hand-sample color is quite variable; white, gray, green, or brown streak; vitreous to submetallic luster.

Spinel: colorless, green, blue, red
Pleonaste: green to blue-green
Hercynite: dark green
Picotite: olive brown to brown
Gahnite: blue-green, yellow, brown
Galaxite: red-brown, black

COLOR AND PLEOCHROISM: Color in thin sections or fragments generally corresponds with hand sample color. Some deeply colored samples may be practically opaque. Gahnite may show anomalous pleochroism.

FORM: Crystals are usually octahedrons yielding triangular, square, or diamond-shaped cross sections in thin section. Subequant anhedral grains also are common.

CLEAVAGE: None. A prominent octahedral parting on {111} parallel to the octahedron faces is sometimes found.

TWINNING: Single (or sometimes multiple) twins on the {111} contact plane (spinel law) are commonly found. Unless betrayed by crystal shape, twinning will not be visible because the mineral is isotropic.

INDICES OF REFRACTION: Indices of refraction vary in linear fashion between the end member compositions. There are too many variables in the solid solution to allow the index of refraction alone to be used to determine composition. Some samples of gahnite may show weak anomalous birefringence.

ALTERATION: Spinel minerals are relatively resistant to weathering or other alteration although alteration to various phyllosilicates and other minerals is reported. Zn-bearing spinels may alter to sphalerite and various phyllosilicates.

DISTINGUISHING FEATURES: Generally distinguished by high relief in thin section, strong color, and isotropic character. Different crystal shape and lighter color distinguishes garnet from spinel. In grain mounts it may be difficult to differentiate between the spinel and garnet groups, and X-ray or other techniques may be required.

OCCURRENCE: Common spinel (including pleonaste) is relatively common in highly aluminous or silica-poor metamorphic rocks associated with andalusite, kyanite, sillimanite, corundum, cordierite, or orthopyroxene, and in regionally metamorphosed carbonate rocks associated with chondrodite, phlogopite, calcite, and forsterite. Hercynite is commonly found in metamorphosed iron-rich argillaceous sediments and in some mafic and ultramafic igneous rocks as well as in some granulites. Gahnite is usually found in granitic pegmatite and in some hydrothermal deposits. Galaxite is relatively rare and is found in Mn-rich hydrothermal vein deposits associated with Mn-bearing minerals such as spessartine and rhodonite.

Magnetite

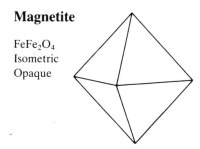

$FeFe_2O_4$
Isometric
Opaque

COMPOSITION: Substantial amounts of Ti^{4+} may substitute for Fe^{3+}, and there is complete solid solution to ulvöspinel ($FeFeTiO_4$). Although extensive substitution of Mg, Mn, Zn, and Ni is possible, most natural magnetite contains little of these elements.

PHYSICAL PROPERTIES: $H = 5\frac{1}{2}$–6; $G = 5.20$; iron black with metallic luster and black streak; strongly magnetic.

COLOR: Opaque in thin section or as fragments. Gray in reflected light as viewed through the petrographic microscope.

FORM: Crystals are usually octahedral, and in thin section they produce triangular-, square-, trapezoidal-, and diamond-shaped cross sections; dodecahedral crystals are rare. Also found as anhedral grains and granular masses.

CLEAVAGE: None, but a distinct octahedral parting {111} is common although it generally will not be visible because of magnetite's opacity.

TWINNING: Single or multiple twins on {111} (spinel law) are common, but unless betrayed by crystal shape, twins will not be visible.

ALTERATION: Magnetite commonly alters to hematite, limonite, or goethite in the weathering environment. Low-temperature oxidation of Ti-bearing magnetite may result in the exsolution of lamellae of ilmenite.

DISTINGUISHING FEATURES: Strong magnetism is characteristic, but with the small grains found in most rocks, it is hard to detect unless the grains are separated from the rock. Magnetite is easily confused with ilmenite and chromite, although chromite is transparent along very thin edges, and ilmenite's crystal shape is different. Pyrite is usually cube shaped and is yellowish in reflected light; hematite is often reddish. Identification with a reflected light microscope or hand-sample properties is more reliable than with the conventional petrographic microscope.

OCCURRENCE: Magnetite is a widespread mineral and is common as an accessory in most igneous and metamorphic rocks. It may also form massive deposits in contact metasomatic skarns and in mafic layered igneous intrusions and anorthosites. Although magnetite is susceptible to weathering, it often forms a major constituent of the heavy fraction of many sands and sandstones.

Chromite

$FeCr_2O_4$
Isometric
Isotropic, opaque except along thin edges
$n = 1.90$–2.12

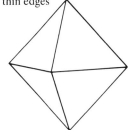

COMPOSITION: Most chromite contains substantial amounts of Mg substituting for Fe^{2+} and there is extensive solid solution to magnesiochromite ($MgCr_2O_4$). Zn, Al, Mn, and Fe^{3+} may also be present.

PHYSICAL PROPERTIES: $H = 5\frac{1}{2}$; $G = 5.09$–4.43 (decreases with increasing Mg content); metallic iron black, frequently pitchy; streak brownish black; may be weakly magnetic.

COLOR: Opaque. Magnesiochromite may be transparent along thin edges, which are dark brown and nonpleochroic. Light gray with slight brownish cast as viewed through the petrographic microscope with reflected light.

FORM: Chromite crystals usually have an octahedral shape, which produces triangular-, trapezoidal-, square-, or diamond-shaped sections. More common as anhedral grains or granular masses.

CLEAVAGE: None, but there may be a parting on {111}. Fracture is uneven.

TWINNING: May be twinned on {111} (spinel law), but twinning generally will not be visible.

INDICES OF REFRACTION: The index of refraction varies between 2.12 ($FeCr_2O_4$) and 1.90 ($MgCr_2O_4$) for the synthetic end members. However, the essentially opaque character makes measuring the index quite difficult.

ALTERATION: Chromite alters to limonite and other oxides and hydroxides of Fe, Mg, and Cr, and to chrome-bearing clays.

DISTINGUISHING FEATURES: Easily confused with magnetite and ilmenite. Chromite is much less magnetic than magnetite, and ilmenite has a different crystal shape. Some chromite is transparent along thin edges. Positive identification with the conventional petrographic microscope may be difficult and use of reflected light microscope or other techniques is recommended.

OCCURRENCE: Chromite is usually restricted to mafic and ultramafic igneous rocks, such as peridotite, pyroxenite, or dunite. It also may be found as a detrital mineral.

HYDROXIDES

Brucite

$Mg(OH)_2$
Hexagonal (trigonal)
Uniaxial (+)
n_ω = 1.559–1.590
n_ϵ = 1.580–1.600
δ = 0.010–0.021

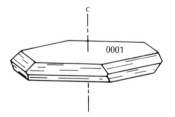

RELIEF IN THIN SECTION: Moderate positive relief.

COMPOSITION AND STRUCTURE: Brucite has a layered structure consisting of two close-packed layers of $(OH)^-$ with Mg^{2+} occupying three out of three of the octahedral sites in between. These double layers of $(OH)^-$; are bonded to adjacent layers with weak electrostatic bonds. A significant amount of Fe^{2+} and limited amount of Mn^{2+} may substitute for Mg^{2+}.

PHYSICAL PROPERTIES: H = $2\frac{1}{2}$; G = 2.39; white, gray, pale green, brown or blue in hand sample; white streak; vitreous to pearly or waxy luster; slightly greasy feel; somewhat sectile.

COLOR AND PLEOCHROISM: Colorless in thin section or fragments.

FORM: Brucite usually occurs as foliated or swirled masses and aggregates or as fibrous masses. Fibrous brucite is called *hemalite* and is relatively uncommon. The folia are tabular parallel to {0001}; and the fibers are elongate at right angles to the c axis.

CLEAVAGE: Perfect cleavage on {0001}; folia are flexible.

TWINNING: None reported.

OPTICAL ORIENTATION: Extinction measured to tabular grains and the basal cleavage is parallel. Sections through tabular crystals are length fast. Because fibers are elongate at right angles to the c axis, they are length fast.

INDICES OF REFRACTION AND BIREFRINGENCE: Indices of refraction increase with increasing amounts of Fe^{2+} and with Mn^{2+}. A manganiferous brucite with n_ω = 1.59 and n_ϵ = 1.60 has been reported. *Amakinite* [(Fe, Mg) OH_2)] has n_ω = 1.707 and n_ϵ = 1.722. Brucite displays relatively strong dispersion yielding anomalous red-brown or blue first-order interference colors in thin section.

INTERFERENCE FIGURE: Uniaxial positive figure may be difficult to obtain due to small grain size. Cleavage flakes yield a centered figure. Some samples, particularly fibrous varieties, may be biaxial positive with a small 2V. Interference colors in the figure may be anomalous due to strong dispersion.

ALTERATION: Brucite alters readily to hydromagnesite ($3MgCO_3 \cdot Mg(OH)_2 \cdot 3H_2O$), or occasionally to serpentine, periclase, or other Mg-bearing minerals.

DISTINGUISHING FEATURES: Brucite is most commonly mistaken for talc, micas, or gypsum. Both talc and the micas are biaxial negative and have higher birefringence, and gypsum is biaxial positive. Chlorite and serpentine are also similar but are usually a pale green color, are usually length slow, and are biaxial. Brucite may be stained to aid in rapid identification (Haines, 1968).

OCCURRENCE: Brucite is most commonly found in marble, resulting from the alteration of periclase. It also may be found in serpentinite and chlorite schist, usually as small veins, along with talc, magnesite, and other Mg-bearing minerals.

Gibbsite

$Al(OH)_3$
Monoclinic
$\angle \beta = 94.5°$
Biaxial (+)
n_α = 1.568–1.578
n_β = 1.568–1.579
n_γ = 1.587–1.590
$\delta \cong$ 0.019–0.012
$2V_z$ = 0–40°

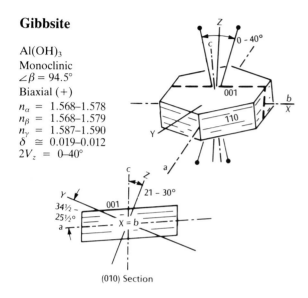

(010) Section

RELIEF IN THIN SECTION: Moderate positive relief.

COMPOSITION AND STRUCTURE: The structure of gibbsite is very similar to that of brucite and consists of Al^{3+} in octahedral coordination between two layers of $(OH)^-$. Because aluminum is trivalent, only two-thirds of the octahedral sites are occupied to maintain charge balance. The double layers of $(OH)^-$ are bonded to each other by weak electrostatic bonds which allows for the perfect cleavage between the layers. The layers of $(OH)^-$ are offset slightly to produce the monoclinic structure. Small amounts of Fe^{3+} may substitute for Al^{3+}.

PHYSICAL PROPERTIES: H = $2\frac{1}{2}$–$3\frac{1}{2}$; G = 2.38–2.42; white or gray to pale pink, green, or brown in hand sample; white streak; vitreous to pearly or earthy luster.

COLOR AND PLEOCHROISM: Usually colorless or less commonly pale brown in thin section or as fragments. Pleochroism is either absent or very weak.

FORM: Crystals are usually pseudohexagonal plates. Also found as lamellar aggregates, concretionary or encrusting masses, and as fine-grained earthy masses. Gibbsite is usually very fine grained.

CLEAVAGE: Perfect basal cleavage {001}. However, fine grain size may preclude seeing it. Cleavage and crystal shape may control fragment orientation.

TWINNING: Contact twins on {001}, {110}, or {100}. Also twins with [130] as a rotation axis. However, fine grain size generally precludes seeing the twinning.

OPTICAL ORIENTATION: $X = b$, $Y \wedge a$ = +25.5 to +34.5°, $Z \wedge c$ = −21 to −30°, the optic plane is perpendicular to (010). The trace of cleavage is length fast, and the maximum extinction angle is about 25 to 35°. However, with heating the $2V$ angle closes to 0° on the Z axis and above 56°C opens again but in an optic plane oriented parallel to (010), so that $Y = b$ and the extinction angle $X \wedge a$ is 40°.

INDICES OF REFRACTION AND BIREFRINGENCE: Most gibbsite has indices near the lower end of the range previously given. Maximum interference colors in thin section are upper first order, although fine grain size may only allow lower first-order colors.

INTERFERENCE FIGURE: Basal sections yield off-center acute bisectrix figures with a small $2V$ (0–5°) at room temperature. Optic axis dispersion is strong, $r > v$ or $v > r$. Fine grain size may make interference figures difficult to obtain. Above 56°C, $2V_z$ may be as large as 40°.

ALTERATION: Generally quite stable in the weathering environment. Gibbsite may be altered with addition of silica to form kaolinite or other clay minerals or be dehydrated to form boehmite.

DISTINGUISHING FEATURES: Gibbsite is very similar to many clay minerals and, in the usual fine-grained form, is practically impossible to distinguish from them without X ray or other information. Coarse-grained gibbsite is similar to the micas and talc, which are biaxial negative and have higher birefringence. Boehmite and diaspore with which it often is associated have parallel extinction, higher relief, and larger $2V$, while brucite displays anomalous interference colors, has parallel extinction, and is uniaxial.

OCCURRENCE: Gibbsite is most commonly formed as a weathering product of feldspars or other aluminous minerals. It is common in soils and is a major constituent of bauxite. It may also be formed as a low temperature hydrothermal mineral in veins or cavities in Al-rich igneous rocks.

Diaspore

αAlO(OH)
Orthorhombic
Biaxial (+)
n_α = 1.700–1.702
n_β = 1.715–1.722
n_γ = 1.740–1.750
δ = 0.040–0.048
$2V_z$ = 84–86°

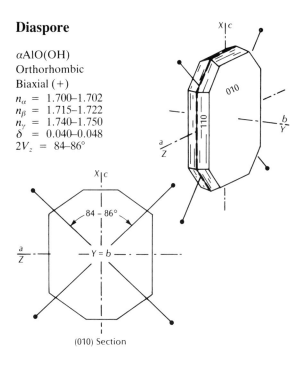

(010) Section

RELIEF IN THIN SECTION: High positive relief.

COMPOSITION AND STRUCTURE: The structure of diaspore is analogous to corundum and consists of a slightly distorted hexagonal close-packed array of O atoms with Al^{3+} in octahedral sites between the layers of oxygen. Fewer of the octahedral sites are occupied than in corundum because H^+ ions bond between two oxygens in adjacent layers to accommodate some of the negative charge. Small amounts of Fe may substitute for Al, as may lesser amounts of Mn.

PHYSICAL PROPERTIES: H = $6\frac{1}{2}$–7; G = 3.3–3.5; white, gray, colorless in hand sample, or shades of green, brown, yellow, pink, violet, or red; white streak, vitreous or pearly luster; decrepitates strongly in a flame.

COLOR AND PLEOCHROISM: Usually colorless in thin section or as fragments. Fragments of colored varie-

ties may show pale colors and pleochroism with $Z > Y > X$.

FORM: Crystals are usually tabular or occasionally fibrous. It most commonly occurs as scaly aggregates that are often very fine grained.

CLEAVAGE: Diaspore has a single perfect cleavage on {010}, imperfect prismatic cleavage on {110} and {210}, and a single poor cleavage on {100}. The {010} cleavage tends to control fragment orientation.

TWINNING: Not usually twinned although twinning on {061} or {021} is possible.

OPTICAL ORIENTATION: $X = c$, $Y = b$, $Z = a$, optic plane = (010). Extinction is parallel to the trace of cleavage in principal sections. Elongate sections may be either length fast or length slow, depending on how they are cut.

INDICES OF REFRACTION AND BIREFRINGENCE: The high birefringence generally produces vivid third-order interference colors in thin section. Cleavage flakes usually lie flat on (010), produce maximum birefringence, and allow measurement of n_α and n_γ.

INTERFERENCE FIGURE: Interference figures often are difficult to obtain due to small grain size. Cleavage flakes yield flash (optic normal) figures. Acute bisectrix and optic axis figures are obtained from grains oriented to show the cleavage distinctly. Optic axis dispersion is weak with $v > r$.

ALTERATION: Generally is not altered although it may convert to corundum on heating and dehydration, or with the addition of silica it may form kaolinite or other clay minerals.

DISTINGUISHING FEATURES: Diaspore may be mistaken for sillimanite, which has a different habit, lower birefringence, and lower indices of refraction. Gibbsite, with which diaspore is often associated, shows lower relief and inclined extinction. It is often difficult to identify due to fine grain size, particularly in bauxite and related occurrences.

OCCURRENCE: Diaspore is a major constituent of bauxite, some laterites, and aluminous clays. It is usually associated with gibbsite, boehmite, Fe oxides and hydroxides, and some quartz. It is produced by weathering of Al-bearing minerals. Diaspore is also commonly found in emery deposits and other aluminous metamorphic rocks with corundum, sillimanite, kyanite, and andalusite. Hydrothermal alteration of aluminous rocks such as syenite and felsic volcanics may also yield diaspore.

Boehmite

$\gamma AlO(OH)$
Orthorhombic
Biaxial $(+)$ or $(-)$
$n_\alpha = 1.640–1.648$
$n_\beta = 1.649–1.657$
$n_\gamma = 1.655–1.668$
$\delta = 0.006–0.020$
$2V_x > 80°$

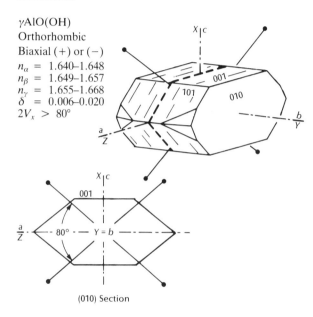

(010) Section

RELIEF IN THIN SECTION: Moderately high positive relief.

COMPOSITION AND STRUCTURE: Boehmite can be considered a slightly distorted cubic close-packed array of oxygen atoms with Al^{3+} in octahedral sites between the layers of O, similar to the spinel structure. Fewer octahedral sites are occupied than in spinels because H^+ bonds between O atoms on adjacent layers and balances part of the negative charge. Small amounts of Fe and lesser amounts of Mn and Cr may substitute for Al.

PHYSICAL PROPERTIES: H $= 3\frac{1}{2}–4$; G $= 3.01–3.06$; usually white or gray in hand sample; white streak; vitreous or earthy luster.

COLOR: Usually colorless in thin section and as fragments.

FORM: Almost invariably as fine-grained aggregates. Minute tabular crystals are usually visible only with a scanning electron microscope.

CLEAVAGE: Perfect cleavage on {010} is rarely visible.

OPTICAL ORIENTATION: The fine grain size has made gathering optical data difficult. The orientation used here is $X = c$, $Y = b$, $Z = a$, optic plane $= (010)$. Other orientations also are reported. Elongate grains show parallel extinction and are reported to be length slow or length fast.

INDICES OF REFRACTION AND BIREFRINGENCE: The indices of refraction decrease with loss of water caused by heating. Interference colors in thin section are usually low or mid first order.

INTERFERENCE FIGURE: Fine grain size usually makes it impossible to obtain an interference figure. Both optically positive and negative boehmite has been reported with a large $2V$ ($>80°$).

ALTERATION: Boehmite is not readily altered because it is stable in the weathering environment. Dehydration may result in formation of cubic alumina (γAl_2O_3), and addition of silica may result in formation of clay minerals.

DISTINGUISHING FEATURES: Due to boehmite's fine grain size, it is often difficult to positively identify by optical means, and X ray or other techniques should be employed. Identifications are usually made based on occurrence and not on optical properties. Gibbsite has lower refractive indices and inclined extinction, and diaspore has higher refractive indices and birefringence.

OCCURRENCE: Boehmite is formed by the weathering of aluminous minerals. It is characteristic of bauxite where it is associated with gibbsite, diaspore, Fe hydroxides, quartz, and some clay minerals. Boehmite may also be a constituent of laterite and of aluminous clays and shales.

Goethite

$\alpha FeO(OH)$
Orthorhombic
Biaxial $(-)$
$n_\alpha = 2.15–2.275$
$n_\beta = 2.22–2.409$
$n_\gamma = 2.23–2.415$
$\delta = 0.08–0.140$
$2V_x = 0–27°$

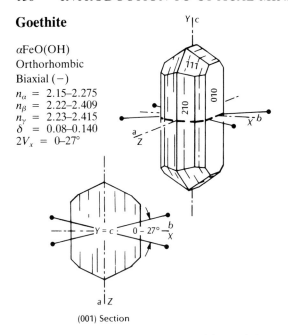

(001) Section

RELIEF IN THIN SECTION: Extreme positive relief.

COMPOSITION AND STRUCTURE: The structure of goethite is analogous to diaspore and consists of slightly distorted hexagonal close-packed O atoms with Fe^{3+} in octahedral interstices between the layers of O. The H^+ ions bond between two oxygens in adjacent layers. Small amounts of Mn^{3+} may substitute for Fe^{3+}.

PHYSICAL PROPERTIES: $H = 5–5\frac{1}{2}$; $G \cong 4.3$; dark brown with yellowish or reddish cast in hand sample; yellow-brown streak. Crystals are adamantine or metallic, aggregates are usually earthy.

COLOR AND PLEOCHROISM: Yellow, orange-red, or brownish orange in thin section; fragments may be essentially opaque except on thin edges. Distinctly pleochroic with $X > Z > Y$, or $Z > Y > X$ in shades of yellow, orange-red, or brown.

FORM: Grains are usually fibrous or acicular and may form radiating botryoidal aggregates. Also found as fine aggregates or disseminated grains.

CLEAVAGE: A single perfect cleavage {010} parallel to the length of elongate grains may control fragment orientation. Imperfect cleavage on {100}.

TWINNING: Generally not twinned, although cruciform twins are reported.

OPTICAL ORIENTATION: The optic plane is (001) with $X = b$, $Y = c$, and $Z = a$ for the yellow, green, and violet portion of the spectrum and is (100) with $X = b$, $Y = a$, and $Z = c$ for the red end of the spectrum. Because the eye is most sensitive to yellow and green light, the optical properties based on that part of the spectrum are usually observed. Elongate grains may be either length fast or slow depending on how they are cut, but the extreme birefringence and mineral color usually make determining the sign of elongation difficult. Extinction is parallel to cleavage traces or the length of elongated grains.

INDICES OF REFRACTION AND BIREFRINGENCE: Maximum birefringence is extreme (0.08–0.140) and produces a high-order white color, which is masked by the mineral color. However, sections oriented so that (010) is close to horizontal have moderate birefringence because n_β and n_γ are relatively close and may show bright first-, second-, or third-order colors.

INTERFERENCE FIGURE: Goethite displays crossed axial plane dispersion. At room temperature, the size of $2V$ varies strongly as a function of the wavelength of light. $2V$ is about 27° for green light and the optic plane lies in (001) with $b = $ Bxa. The value of $2V$ decreases to 0° at 615 nm and is uniaxial for that wavelength. For longer wavelengths, $2V$ opens about the b axis in the (100) plane to about 23° for red light. Isochromes and isogyres formed by the red end of the spectrum are difficult to see because the eye is not sensitive to those wavelengths, and the interference figure will show numerous isochromes on an orange or brownish field. Cleavage fragments yield acute bisectrix figures.

ALTERATION: Goethite is stable in the weathering environment and is generally not altered, although it may be converted to hematite by dehydration.

DISTINGUISHING FEATURES: Goethite is often difficult to distinguish from other iron oxide and hydroxide minerals due to fine grain size and common intergrowth with those minerals. Hematite is distinctly redder and is usually practically opaque, and lepidocrocite is red-brown and has a larger $2V$.

OCCURRENCE: Goethite is formed by weathering or hydrothermal alteration of other Fe-bearing minerals, particularly oxides and sulfides. Goethite is a common constituent of lateritic soils and in the supergene zone of hydrothermal sulfide deposits. It also is a primary mineral in sedimentary iron deposits. Limonite is a mixture of goethite and other Fe oxides and hydroxides.

Lepidocrocite

γ FeO(OH)
Orthorhombic
Biaxial (+)
$n_\alpha = 1.94$
$n_\beta = 2.20$
$n_\gamma = 2.51$
$\delta = 0.57$
$2V_x \cong 83°$

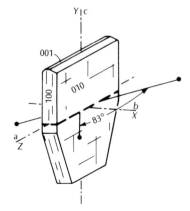

RELIEF IN THIN SECTION: Extreme positive relief.

COMPOSITION AND STRUCTURE: The structure of lepidocrocite is analogous to boehmite and consists of cubic close-packed O atoms with Fe^{3+} in octahedral interstices between the layers of O. The H^+ is bonded between two oxygens in adjacent layers. Some Mn^{3+} may replace Fe^{3+}.

PHYSICAL PROPERTIES: H = 5; G = 4.09; brown to red in hand sample; orange streak; luster adamantine to metallic for crystals, dull for aggregates.

COLOR AND PLEOCHROISM: Yellow, orange, or red in thin section; fragments may be nearly opaque. Strongly pleochroic $Z > Y > X$: X = light yellow to yellow, Y = red-orange to red, Z = orange-red to brown-red. Darkest when elongate sections are parallel to the lower polar vibration direction.

FORM: Crystals are tabular on {010}. Forms scaly aggregates or occasionally acicular crystals elongate parallel to the a axis.

CLEAVAGE: One perfect cleavage on {010}, also imperfect cleavages on {100} and {001}.

TWINNING: None reported.

OPTICAL ORIENTATION: $X = b$, $Y = c$, $Z = a$, optic plane = (001). Elongate sections are length slow although the high interference colors and strong mineral color may make determining the sign of elongation difficult. Extinction is parallel to cleavage in principal sections.

INDICES OF REFRACTION AND BIREFRINGENCE: Birefringence is extreme (~ 0.57) and yields high-order colors that are masked by the mineral color.

INTERFERENCE FIGURE: Cleavage fragments and sections cut with the b axis vertical yield acute bisectrix figures with very large $2V$ ($\sim 83°$) and numerous isochromes superimposed on an orangish field. Optic axis dispersion is weak.

ALTERATION: Lepidocrocite is stable in the weathering environment and is usually not altered.

DISTINGUISHING FEATURES: Fine-grained lepidocrocite is difficult to distinguish from other Fe oxide and hydroxide minerals with which it usually is associated. Goethite has a smaller $2V$, strong dispersion, and is more yellowish, while hematite is more red and is usually nearly opaque.

OCCURRENCE: Lepidocrocite is formed by the weathering or hydrothermal alteration of Fe-bearing minerals. It is a common constituent of lateritic soils, along with goethite and hematite, and may be found in the supergene zone of hydrothermal sulfide deposits. Limonite usually includes fine-grained lepidocrocite.

Limonite

Amorphous
Isotropic or cryptocrystalline
$n = 2.0–2.4$

RELIEF IN THIN SECTION: Extreme positive relief.

COMPOSITION AND STRUCTURE: Limonite is not a mineral but a fine-grained mixture of various Fe oxides and hydroxides. The common constituents are goethite, hematite, lepidocrocite, clay, silica, aluminum hydroxides, and manganese oxides and hydroxides. It also may be composed of an amorphous gel of hydrated Fe.

PHYSICAL PROPERTIES: Highly variable; H =1–5$\frac{1}{2}$; G = 2.7–4.3; shades of yellow-brown and brown in hand sample; yellow-brown streak; usually dull and earthy luster.

COLOR AND PLEOCHROISM: Usually dark shades of red, yellow, or brown in thin section; fragments may be nearly opaque. No pleochroism.

FORM: Usually an aggregate of extremely fine-grained material or amorphous. Often forms colloform crusts or masses.

INDICES OF REFRACTION AND BIREFRINGENCE: The index is quite variable (2.0–2.4) and may increase on standing in index liquids. Some varieties show birefringence, reflecting the fact that they contain goethite or other birefringent minerals.

DISTINGUISHING FEATURES: Limonite is the proper term to apply to aggregates of fine-grained Fe oxides and hydroxides whose mineralogy is not known.

OCCURRENCE: Limonite is produced by weathering or alteration of Fe-bearing minerals.

REFERENCES

Haines, M., 1968, Two staining tests for brucite in marble: Mineralogical Magazine, v. 36, p. 886–888.
Sharp, W. E., Carlson, E. L., and Kheoruenromne, I., 1977, A stain test for fluorite: American Mineralogist, v. 62, p. 171–172.

10

Carbonates, Borates, Sulfates, and Phosphates

CARBONATES

The carbonates are an important group of minerals found in a wide variety of environments. They all contain carbonate groups (CO_3^{2-}) which consist of a single carbon atom in the center of three oxygens arranged in a triangle. There are three important groups of carbonate minerals: the calcite group, the dolomite group, and the aragonite group (Table 10.1).

The structure of the calcite group is analogous to the structure of halite with Ca^{2+} and CO_3^{2-} in the place of Na^+ and Cl^-. The unit cell is flattened to rhombohedral symmetry by shortening the cube along a diagonal through the center, which becomes the c crystallographic axis and the optic axis (Figure 6.2). The result is a structure that consists of alternating layers of cations and carbonate groups parallel to (0001). This structure can accommodate cations up to about 1 Å ionic radius, because the cations are in sixfold coordination with oxygen in the carbonate groups. Larger cations require the orthorhombic structure of the aragonite group.

The dolomite group is also rhombohedral and has a structure that is essentially the same as the calcite group, except that it must accommodate two distinctly different sized cations. This is done by replacing one half of the Ca^{2+} layers in the calcite structure with Mg^{2+} or Fe^{2+}.

The aragonite group, which is orthorhombic, contains cations larger than about 1 Å ionic radius. The cations are arranged in layers in what amounts to an open version of hexagonal close packing, with layers of carbonate groups in between, so that the cations coordinate with nine oxygens. The symmetry is reduced to orthorhombic because the corners of the triangular carbonate groups do not all point the same way, but crystals often possess pseudohexagonal shapes.

Measuring n_ω of the rhombohedral carbonates in grain mount is relatively easy because all grains have one ray with index n_ω regardless of orientation. Determining n_ϵ is more difficult because the fragments typically lie on the $\{10\bar{1}1\}$ cleavage and allow measurement of only $n_{\epsilon}'{}_{(10\bar{1}1)}$. Note, however, if n_ω and $n_{\epsilon}'{}_{(10\bar{1}1)}$ are known, Equation 6.1 can be arranged to solve for n_ϵ if it is assumed that the angle between the extraordinary wave normal and the optic axis is 44.6°, which is close to the angle found for most carbonates lying on the $\{10\bar{1}1\}$ cleavage. Loupekine (1947) has a series of curves that allow graphical solution to the problem.

The approximate variation of n_ω with composition for the rhombohedral carbonates is shown in Figure 10.1.

Table 10.1. Carbonate minerals

Calcite Group (Rhombohedral)		Dolomite Group (Rhombohedral)		Aragonite Group (Orthorhombic)	
Calcite	$CaCO_3$	Dolomite	$CaMg(CO_3)_2$	Aragonite	$CaCO_3$
Magnesite	$MgCO_3$	Ankerite	$Ca(Mg, Fe)(CO_3)_2$	Witherite	$BaCO_3$
Siderite	$FeCO_3$	Kutnohorite	$CaMn(CO_3)_2$	Strontianite	$SrCO_3$
Rhodochrosite	$MnCO_3$				

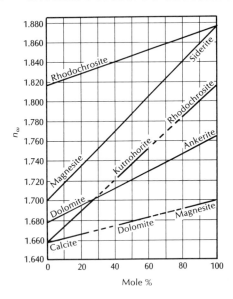

Figure 10.1 Variation of n_ω with composition in the rhombohedral carbonates. After Kennedy (1947).

Calcite

CaCO$_3$
Hexagonal (trigonal)
Uniaxial ($-$)
$n_\omega = 1.658$
$n_\epsilon = 1.486$
$\delta = 0.172$

RELIEF IN THIN SECTION: Moderate negative to high; positive relief; marked change with rotation.

COMPOSITION: Most calcite, particularly that formed in the sedimentary environment, consists of almost pure CaCO$_3$. However, elevated temperatures allow substantial amounts of Mg, Fe, Mn, or Zn to substitute for Ca. Mg-bearing calcite is the most common. Small amounts of large cations such as Sr or Ba also may substitute for Ca.

PHYSICAL PROPERTIES: H = 3; G = 2.71, higher with substitution of Fe, Mg, or Zn for Ca; usually white or colorless in hand sample, although a wide range of colors is possible; white streak; vitreous luster, effervesces vigorously in cold dilute HCl.

COLOR: Colorless in thin section and grain mount.

FORM: Figure 10.2. Crystals of calcite have many habits but usually consist of combinations of scalenohedrons and rhombohedrons. However, in most rocks, calcite forms anhedral grains or aggregates of grains. Fossil shells and thin veins may be fibrous or columnar.

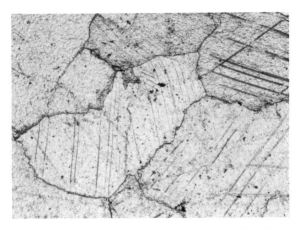

Figure 10.2 Calcite with distinct cleavage in marble. Field of view is 1.7 mm wide.

CLEAVAGE: Calcite has perfect rhombohedral cleavage $\{10\bar{1}1\}$, and fragments commonly lie flat on cleavage surfaces. The angle between the cleavages is 74°57′.

TWINNING: Calcite commonly has lamellar twins on the negative rhombohedron $\{01\bar{1}2\}$. The lamellae are usually parallel to one edge of the cleavage rhomb or along the long diagonal of the rhomb (Figure 10.3). Simple twins also may develop on $\{0001\}$ and rare twins are found on $\{10\bar{1}1\}$ and $\{0\bar{2}21\}$. The lamellar twinning is often the result of deformation and may sometimes be used to determine the orientation of the stresses that produced the deformation. Samples also may be caused to twin during cutting and grinding in the preparation of thin sections or when samples are crushed to prepare grain mounts.

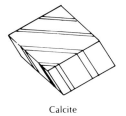 

Calcite   Dolomite

Figure 10.3 Orientation of twin lamellae in calcite and dolomite.

OPTICAL ORIENTATION: Extinction is inclined or symmetrical to cleavage traces (cf. Figure 6.11). The fast ray vibration direction is parallel to the short diagonal of the rhombohedral faces. The angle between the fast ray (ϵ') vibration direction and the trace of twin lamellae is greater than about 55°.

INDICES OF REFRACTION AND BIREFRINGENCE: Indices of refraction vary in an approximately linear manner with composition (Figure 10.1). The index $n_{\epsilon'(10\bar{1}1)}$ for fragments lying on cleavage surfaces is 1.566 for pure calcite and increases with substitution of Mg or Mn. Because the birefringence is extreme (0.172), interference colors in thin section are typically creamy high-order colors even if the optic axis is close to vertical. Twin lamellae may show as bands of pastel pink or green, and zones of overlap of inclined twin lamellae may not go entirely extinct with stage rotation. Grains in thin section usually show a marked change of relief with rotation of the stage. Calcite displays low relief when the short diagonal of the rhomb is parallel to the vibration direction of the polarizer and high relief when the long diagonal of the rhomb is in the same position.

INTERFERENCE FIGURE: Optic axis figures are uniaxial negative with numerous isochromes and thin, well-defined isogyres. Due to the extreme birefringence and the somewhat converging nature of orthoscopic illumination, grains oriented to produce usable optic axis figures will still display high-order interference colors and can often be located only by examining a number of grains. Some calcite, particularly from metamorphic rocks, is biaxial negative with $2V_x$ up to about 15° or (rarely) higher. Fragments on ($10\bar{1}1$) yield strongly off-center figures.

ALTERATION: Calcite may be altered to dolomite by diagenetic processes and may be replaced by quartz,

opal, iron or manganese oxides, or other minerals to yield pseudomorphs after calcite. It is also soluble in many natural waters and may be removed from a rock by solution.

DISTINGUISHING FEATURES: The cleavage, extreme birefringence, change of relief with rotation, and reaction with weak acid distinguish the rhombohedral carbonates from most other minerals. The orthorhombic carbonates are biaxial and do not show rhombohedral cleavage. Distinguishing among the rhombohedral carbonates may be difficult. Calcite shows lower indices of refraction than the other rhombohedral carbonates and n_ϵ is substantially lower than the index of cement ($\sim$1.537) in thin section. The twin lamellae in calcite are more common than in the other carbonates and are parallel to the edge or long diagonal of the cleavage rhomb, whereas dolomite and ankerite may show lamellae parallel to both the short and long diagonal (Figure 10.3). Dolomite more commonly forms euhedral rhombs in limestones and related sedimentary rocks and may be cloudy or stained with iron oxides. For additional criteria that may distinguish calcite from dolomite, see the description for dolomite. For rapid discrimination among calcite, dolomite, aragonite, and gypsum in thin section, a variety of staining techniques have been developed (e.g., Friedman, 1959; Wolf and others, 1967; Warne, 1962).

OCCURRENCE: Calcite is a very common and widespread mineral. It is an important mineral in many sedimentary rocks as a cementing agent, as fossil fragments, or as an essential constituent in limestone and related rocks. Calcite is also commonly found in metamorphic rocks derived from carbonate-bearing sediments and is the major mineral in marble, where it may occur with wollastonite, garnet, olivine, diopside, idocrase, tremolite, epidote, and other calc-silicate minerals. Calcite in igneous rock is relatively rare but may be found in silica-poor, alkali-rich igneous rocks containing nepheline or other feldspathoids, as vesicle fillings in volcanic rocks, and as rare calcite-rich intrusions called carbonatites. Hydrothermal deposits commonly contain calcite as a gangue mineral, where it may fill fractures and form beautiful crystals. It is also common as a joint coating or fracture filling in almost any type of rock and may be formed by the weathering or alteration of calcium-bearing minerals such as plagioclase.

Magnesite

MgCO$_3$
Hexagonal (trigonal)
Uniaxial (−)
n_ω = 1.700
n_ϵ = 1.509
δ = 0.191

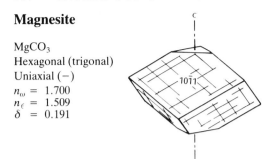

RELIEF IN THIN SECTION: Low negative to high positive relief; marked change with rotation.

COMPOSITION AND STRUCTURE: Magnesite (MgCO$_3$) forms a complete solid solution series with siderite (FeCO$_3$). There is also limited substitution of Mn and Ca for Mg.

PHYSICAL PROPERTIES: H = $3\frac{1}{2}$–$4\frac{1}{2}$; G = 3.01 (pure) to 3.48 (50 percent Fe); usually white or gray in hand sample, may be yellow or brown if iron bearing; white streak; vitreous to earthy luster. Fine granular varieties are commonly intergrown with microcrystalline quartz or opal, which makes it difficult to determine hardness and specific gravity. Reacts vigorously with cold dilute hydrochloric acid only if powdered.

COLOR: Colorless in thin section and grain mount; may be cloudy.

FORM: Magnesite usually occurs as compact granular aggregates that may appear chalky or like porcelain in hand sample. It also forms lamellar or fibrous aggregates. Crystals are rhombs {10$\bar{1}$1} sometimes modified by basal pinacoids {0001} or may be prismatic.

CLEAVAGE: Perfect rhombohedral {10$\bar{1}$1} cleavage, like the other rhombohedral carbonates. Fragments commonly lie on cleavage surfaces.

TWINNING: Typically not twinned although translation gliding on {0001} may occur.

OPTICAL ORIENTATION: Extinction is inclined or symmetrical to cleavage traces (cf. Figure 6.11). The fast ray vibration direction is parallel to the short diagonal of the rhombohedral faces.

INDICES OF REFRACTION AND BIREFRINGENCE: Indices of refraction vary in an approximately linear manner with composition (Figure 10.1), as does the birefringence. The index n_ϵ' for fragments lying on cleavage surfaces is 1.602 and increases with increasing Fe content. Interference colors seen in thin section or grain mount are typically creamy high-order colors, even if the optic axis is close to vertical. Due to the extreme birefringence, grains in thin secton typically show a marked change of relief with rotation.

INTERFERENCE FIGURE: Basal sections yield uniaxial interference figures with numerous isochromes, and isogyres which broaden substantially toward the edge of the field of view. Cleavage fragments yield strongly off-center figures.

ALTERATION: Generally not significantly altered, although Fe-bearing varieties may show red staining or be altered to iron hydroxides and oxides.

DISTINGUISHING FEATURES: Magnesite is distinguished from the other rhombohedral carbonates by lack of twin lamellae, indices of refraction, and physical properties.

OCCURRENCE: Magnesite occurs most commonly as fine- to extremely fine-grained masses produced as an alteration product of Mg-rich rocks such as serpentinite, peridotite, pyroxenite, and dunite. In metamorphic rocks, magnesite may occur as disseminated grains or stratified layers in talc, chlorite, or mica schists, or as the result of the alteration of calcite by Mg-bearing solutions. In sedimentary rocks, magnesite occasionally occurs in evaporate deposits either as the result of direct precipitation or of alteration of calcite or dolomite during diagenesis. Magnesite is infrequently reported as a primary mineral in igneous rocks where it is associated with Mg-rich silicates, and it is occasionally found as a gangue mineral in hydrothermal sulfide deposits.

Siderite

$FeCO_3$
Hexagonal (trigonal)
Uniaxial (−)
$n_\omega = 1.875$
$n_\epsilon = 1.633$
$\delta = 0.242$

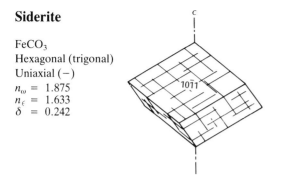

RELIEF IN THIN SECTION: Moderate to high positive relief, changes with rotation.

COMPOSITION AND STRUCTURE: Siderite ($FeCO_3$) forms a complete solid solution series with magnesite ($MgCO_3$) and rhodochrosite ($MnCO_3$), and most siderite contains significant amounts of Mg or Mn substituting for Fe. Up to about 10 mole percent Ca may substitute for Fe, as can minor amounts of Co and Zn.

PHYSICAL PROPERTIES: $H = 4-4\frac{1}{2}$; $G = 3.96$ (pure), lower with substitution of Mg, Mn, or Ca, and with 50 mole percent Mg is approximately 3.48; generally some shade of yellowish, reddish, or grayish brown, sometimes grayish green or gray in hand sample; white streak; vitreous luster. Reacts with slow effervescence in cold dilute hydrochloric acid. Reacts vigorously if powdered or if the acid is hot.

COLOR AND PLEOCHROISM: Colorless, ash gray, pale yellow, or yellowish brown in thin section or grain mount. Colored varieties may be pleochroic $\omega > \epsilon$, so samples are darker when the long diagonal of the rhomb is parallel to the lower polar vibration direction.

FORM: Usually occurs as coarse-grained anhedral aggregates, as oolites, as nodular or botryoidal forms composed of radiating coarse fibers, or as earthy aggregates. Crystals are usually rhombohedrons $\{10\bar{1}1\}$, sometimes modified by a basal pinacoid $\{0001\}$.

CLEAVAGE: Perfect rhombohedral $\{10\bar{1}1\}$ cleavage typical of the rhombohedral carbonates. Fragments typically lie on cleavage faces.

TWINNING: May show lamellar twins on $\{01\bar{1}2\}$ similar to calcite or (rarely) simple twins on $\{0001\}$. Twin lamellae are usually parallel to the edges or along the long diagonal of the rhombs.

OPTICAL ORIENTATION: Extinction is inclined or symmetrical to cleavage traces (cf. Figure 6.11). The fast ray vibration direction is parallel to the short diagonal of rhombohedral faces.

INDICES OF REFRACTION AND BIREFRINGENCE: Indices of refraction vary in an approximately linear manner with composition (Figure 10.1). Fragments lying on $\{10\bar{1}1\}$ cleavage faces show index $n_\epsilon' = 1.748$ for pure siderite. This value decreases to about 1.675 for 50 mole percent Mg and 1.725 for 50 mole percent Mn. Interference colors seen in thin section or grain mount are typically upper-order white and gray, even if the optic axis is nearly vertical.

INTERFERENCE FIGURE: Basal sections yield uniaxial negative figures with numerous isochromes, and isogyres that fan out significantly toward the edge of the field of view. Cleavage fragments yield strongly off-center figures.

ALTERATION: Siderite is often altered to goethite or less commonly to hematite or magnetite. Pseudomorphs after siderite by these minerals and others are relatively common.

DISTINGUISHING FEATURES: Siderite has higher indices than the other rhombohedral carbonates. The yellowish or brownish color, if present, may be distinctive, as is the common alteration to iron oxides or hydroxides. Siderite may be mistaken for cassiterite, which is optically positive and has higher indices and lower birefringence, or titanite, which is biaxial and usually has a darker color.

OCCURRENCE: Siderite most commonly occurs as disseminated grains or fine-grained masses in sedimentary iron formations where it is associated with clays and various iron oxides, hydroxides, and silicates. It also forms concretionary masses in many sedimentary rock types. Some hydrothermal sulfide deposits contain Mn-bearing siderite as a gangue mineral. Siderite is also found in metamorphic iron formations and in carbonate rocks altered by Fe-bearing solutions. It has rarely been reported in mica schists.

In igneous rocks, siderite may be present in carbonatites and is occasionally found in fractures and amygdules in basalt, diabase, and andesite.

Rhodochrosite

MnCO$_3$
Hexagonal (trigonal)
Uniaxial (−)
n_ω = 1.816
n_ϵ = 1.597
δ = 0.219

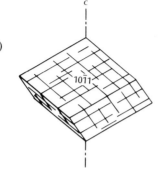

RELIEF IN THIN SECTION: Moderate to high positive relief; changes with rotation.

COMPOSITION AND STRUCTURE: Rhodochrosite (MnCO$_3$) shows complete solid solution with siderite (FeCO$_3$), and it may contain substantial amounts of Zn, Mg, Co, and Ca. Kutnohorite [CaMn(CO$_3$)$_2$] is a dolomite group mineral intermediate between rhodochrosite and calcite.

PHYSICAL PROPERTIES: H = $3\frac{1}{2}$–4; G = 3.70, decreases with Ca and Mg content, increases with Fe and Zn content; usually shades of pink in hand sample, also yellow, gray, or brown; white streak; vitreous luster. Does not readily react in cold dilute HCl, but reacts with effervescence when powdered or if the acid is hot.

COLOR AND PLEOCHROISM: Colorless to pale pink in thin section or grain mount; sometimes with color zoning. Colored varieties are pleochroic in shades of pink with ω > ε. The color is darkest when the long axis of rhomb-shaped faces is parallel to the lower polar vibration direction.

FORM: The rare crystals are rhombohedral {10$\bar{1}$1} or scalenohedral and often rounded. It usually forms coarse- to fine-grained aggregates or grains encrusting other minerals.

CLEAVAGE: Perfect rhombohedral cleavage {10$\bar{1}$1} like the other rhombohedral carbonates.

TWINNING: Lamellar twinning on {01$\bar{1}$2} (like calcite) is quite uncommon.

OPTICAL ORIENTATION: Extinction is inclined or symmetrical to cleavage traces (cf. Figure 6.11). The fast ray vibration direction is parallel to the short diagonal of rhombohedral faces.

INDICES OF REFRACTION AND BIREFRINGENCE: Indices of refraction vary in an approximately linear function of composition (Figure 10.1). Fragments of pure rhodochrosite lying on {10$\bar{1}$1} cleavage surfaces display n_ϵ' = 1.702. Indices increase with increasing amounts of Fe, Co, or Zn and decrease with Ca and Mg. Birefringence is extreme and produces creamy high-order colors in thin section and grain mount, even if the optic axis is nearly vertical. In zoned crystals, the darker-colored areas usually display higher indices of refraction.

INTERFERENCE FIGURE: Basal sections yield uniaxial negative figures with numerous isochromes, and isogyres that fan out strongly toward the edge of the field of view. Fragments lying on cleavage surfaces yield strongly off-center figures.

ALTERATION: Rhodochrosite may be altered to various dark-colored Mn oxides and hydroxides or may be replaced by quartz or other minerals to form pseudomorphs.

DISTINGUISHING FEATURES: Rhodochrosite is distinguished from the other rhombohedral carbonates by its indices of refraction, pink color (if present), association with other Mn-bearing minerals, and alteration.

OCCURRENCE: Rhodochrosite usually occurs as a gangue mineral in hydrothermal vein and replacement deposits associated with sulfide minerals, other carbonates, fluorite, barite, quartz, and other manganese-bearing minerals. It is occasionally found in Mn-rich sediments associated with siderite, and Fe-bearing silicates and clay. Rhodochrosite may also be found in the metamorphosed equivalent of the Mn-bearing sediments and in high-temperature metasomatic deposits.

Dolomite–Ankerite

Hexagonal (trigonal)
Uniaxial (−)

Dolomite
$CaMg(CO_3)_2$
n_ω = 1.679–1.690
n_ϵ = 1.500–1.510
δ = 0.179–0.182

Ankerite
$Ca(Mg,Fe)(CO_3)_2$
n_ω = 1.690–1.750
n_ϵ = 1.510–1.548
δ = 0.182–0.202

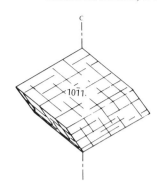

RELIEF IN THIN SECTION: Low negative to high positive relief; changes with rotation.

COMPOSITION: Dolomite and ankerite are part of a solid solution series that extends from dolomite [$CaMg(CO_3)_2$] to ferrodolomite [$CaFe(CO_3)_2$]. The boundary between dolomite and ankerite is arbitrarily set at 20 percent $CaFe(CO_3)_2$, and natural ankerite may contain as much as 75 percent ferrodolomite. More Fe-rich compositions are apparently not found in nature. A significant amount of Mn may also be substituted, and there is substantial solid solution between dolomite/ankerite and kutnohorite [$CaMn(CO_3)_2$]. Small amounts of Co and Zn also may replace Mg, and larger cations such as Pb or Ba may replace some Ca.

PHYSICAL PROPERTIES: H = $3\frac{1}{2}$–4; G = 2.86–2.93 (dolomite), 2.93–3.10 (ankerite), increasing with Fe content; white, colorless, gray, yellowish brown to brown in hand sample (darker with higher Fe content), less commonly pinkish if Mn or Co is present; white streak; vitreous to pearly luster. Will vigorously effervesce in dilute HCl only if powdered or if the acid is hot. Fe-rich varieties may become magnetic on heating.

COLOR: Usually colorless in thin section or grain mount, although weathered or altered iron-rich samples may be brownish due to the presence of iron oxides and hydroxides.

FORM: Figure 10.4. Crystals are usually rhombohedrons {10$\bar{1}$1} and may have curved faces. It is more common as coarse- to fine-grained aggregates, or occasionally as fibrous or columnar aggregates or as oolites.

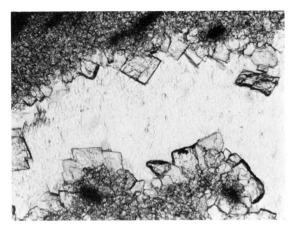

Figure 10.4 Dolomite. (*Top*) Dolomite rhombs lining a void. (*Bottom*) Subhedral dolomite in marble. Note the thin twin lamellae approximately parallel to the short diagonal of the cleavage rhomb. Field of view is 1.2 mm wide for both photographs.

CLEAVAGE: Dolomite and ankerite have perfect rhombohedral cleavage {10$\bar{1}$1} like the other rhombohedral carbonates.

TWINNING: Lamellar twinning is moderately common on {02$\bar{2}$1}, which produces lamellae that may be parallel either to the long or the short diagonal of the rhombohedral faces (Figure 10.3). Simple twins may be found on {0001}, {11$\bar{2}$0}, {10$\bar{1}$1}, or {10$\bar{1}$0}.

OPTICAL ORIENTATION: Extinction is symmetrical or inclined to cleavage traces (cf. Figure 6.11). The fast ray vibrates parallel to the long diagonal of cleavage rhombs.

INDICES OF REFRACTION AND BIREFRINGENCE: Indices of refraction and birefringence vary in an approximately linear manner with composition (Figure 10.1). Fragments lying on cleavage faces show $n_\epsilon'_{(10\bar11)} = 1.588$ for pure dolomite. The substitution of Mn for Mg in dolomite produces a somewhat smaller increase in indices than does the substitution of Fe. Because birefringence is extreme, interference colors seen in thin section or grain mount are typically creamy upper-order white or gray, even if the optic axis is nearly vertical. Pastel colors sometimes show along the trace of twin lamellae.

INTERFERENCE FIGURE: Basal sections yield uniaxial negative figures with numerous isochromes, and isogyres that fan out broadly toward the edge of the field of view. Fragments lying on cleavage surfaces yield highly off-center figures.

ALTERATION: Dolomite may be pseudomorphically replaced by other carbonates, quartz, pyrite, iron oxides, etc. Iron-rich samples may be brownish due to oxidation or weathering.

DISTINGUISHING FEATURES: Dolomite and calcite are commonly found together in limestone, dolomite, marble, and related rocks, and they may be difficult to distinguish optically. The following features may be used to help made the distinction:

1. Dolomite is more commonly euhedral.
2. Calcite is more commonly twinned.
3. Twin lamellae in calcite may be parallel or oblique to the long diagonal or parallel to the edges of cleavage rhombs but not parallel to the short diagonal. Twin lamellae in dolomite may be parallel to both the long and short diagonal of cleavage rhombs (Figure 10.3).
4. Dolomite has higher refractive indices.
5. Dolomite may be colorless, cloudy, or stained by iron oxides, whereas calcite is usually colorless.
6. Dolomite has a higher specific gravity and is less reactive with dilute cold HCl.

Iron-rich samples may be confused with magnesite because the indices overlap, but magnesite does not usually display euhedral crystals or twinning. In some cases, it may be necessary to use chemical or X-ray diffraction tests to distinguish them. It is often convenient to stain the carbonate minerals in thin sections to aid in their rapid identification (Friedman, 1959; Wolf and others, 1967; Warne, 1962).

OCCURRENCE: Dolomite is an important mineral in limestone, dolomite, and evaporate deposits and may be found in almost any carbonate-bearing sediment. In clastic and carbonate sediments, it is almost always associated with calcite. In evaporites, it may also be associated with halite, sylvite, gypsum, calcite, anhydrite and related minerals.

Dolomite also is found in marble and other metamorphosed carbonates in association with calcite, talc, forsterite, tremolite, actinolite, wollastonite, and other calc-silicate minerals. It is also found in hydrothermal mineral deposits.

Ankerite is less common than dolomite. It may be found in iron-rich sediments as disseminated grains, concretionary masses, or veins associated with siderite, iron oxides and hydroxides, and clay minerals. It may also be found in metamorphosed iron formations and in hydrothermal mineral deposits.

Carbonatites may contain dolomite or ankerite as a primary igneous mineral. Hydrothermal alteration of mafic and ultramafic igneous rocks may yield dolomite or ankerite along with serpentine, talc, magnesite, or other Mg-bearing minerals.

Aragonite

CaCO₃
Orthorhombic
Biaxial (−)
$n_\alpha = 1.530$
$n_\beta = 1.680$
$n_\gamma = 1.685$
$\delta = 0.155$
$2V_x = 18°$

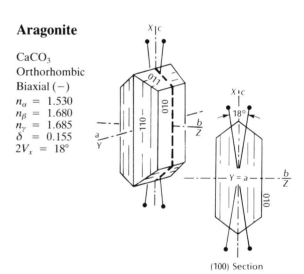

(100) Section

RELIEF IN THIN SECTION: Low negative to high positive relief, depending on orientation. May show marked change of relief with rotation.

COMPOSITION: Most aragonite is relatively pure CaCO₃, although small amounts of large cations such as Sr or Pb may substitute for Ca.

PHYSICAL PROPERTIES: H = 3½–4; G = 2.94; usually white or colorless in hand sample, also gray, yellow, blue, green, violet, or rose-red; white streak; vitreous luster. Effervesces vigorously in cold dilute HCl.

COLOR: Usually colorless in thin section or grain mount.

FORM: Common as radiating or columnar aggregates of grains elongate along c. Crystals are typically twinned, producing columnar crystals with pseudo-hexagonal cross sections. Hot springs or cave deposits are commonly stalactitic, encrusting, or colloform.

CLEAVAGE: A single cleavage on {010} is imperfect, and prismatic cleavage on {110} is poor and not usually seen.

TWINNING: Twinning is common in aragonite on {110} and produces cyclic twins with six segments which are visible in basal sections. Twinning on {110} also may be polysynthetic and produces parallel twin lamellae.

OPTICAL ORIENTATION: $X = c$, $Y = a$, $Z = b$, optic plane = (100). Extinction is parallel to crystal elongation and to the trace of the {010} cleavage in longitudinal sections. The fast ray vibrates parallel to the length of columnar crystals and parallel to cleavage traces in both longitudinal and basal sections.

INDICES OF REFRACTION AND BIREFRINGENCE: The indices of aragonite vary relatively little; however, Sr-rich varieties may have indices as low as $n_\alpha = 1.527$, $n_\beta = 1.670$, and $n_\gamma = 1.676$, and Pb-rich varieties may have indices as high as $n_\alpha = 1.540$, $n_\beta = 1.695$, and $n_\gamma = 1.703$. Birefringence is extreme and yields creamy high-order white interference colors in both thin section and grain mount for most orientations except those with an optic axis vertical.

INTERFERENCE FIGURE: Basal sections yield centered acute bisectrix figures with isogyres which are relatively thin at the melatopes and broaden substantially toward the edge of the field of view, numerous isochromes, and $2V_x = 18°$. Optic axis dispersion is weak with $v > r$. Pb-rich varieties may have $2V$ as large as $23°$. Fragments lying on cleavage surfaces yield obtuse bisectrix figures that look like flash figures because of the small $2V$.

ALTERATION: Aragonite commonly inverts to its polymorph calcite and pseudomorphs of calcite after aragonite are common. It may also be replaced by dolomite or other minerals.

DISTINGUISHING FEATURES: Aragonite is most easily confused with calcite and the two may be difficult to distinguish if fine grained. The features that distinguish it from calcite are lack of rhombohedral cleavage, biaxial character, and slightly higher indices of refraction. Aragonite cleavage fragments yield obtuse bisectrix figures, while calcite fragments yield off-center uniaxial figures. In thin section, the two minerals may be stained to tell them apart rapidly (Friedman, 1959). Aragonite also may be mistaken for zeolites in cavities and vesicles in volcanic rocks, but zeolites have low birefringence, and indices lower than the cements used in thin sections.

OCCURRENCE: Aragonite is found in carbonate-bearing blueschist facies metamorphic rocks associated with glaucophane, lawsonite, pumpellyite, and related minerals, and in cave and hot springs deposits of recent geologic origin. It also precipitates directly from sea water as fine needles and forms ooliths in calcareous muds, and is found in some evaporite deposits. The shells of some marine animals are made of aragonite. In igneous rocks, it is usually only found in cavities and vesicles in basalt and andesite, where it is associated with zeolites. Aragonite is also reported from the oxidized zone of hydrothermal sulfide mineral deposits and from iron-rich sediments associated with siderite and iron oxides.

Strontianite

SrCO$_3$
Orthorhombic
Biaxial ($-$)
$n_\alpha = 1.516–1.525$
$n_\beta = 1.664–1.686$
$n_\gamma = 1.666–1.690$
$\delta = 0.148–0.165$
$2V_x = 7–10°$

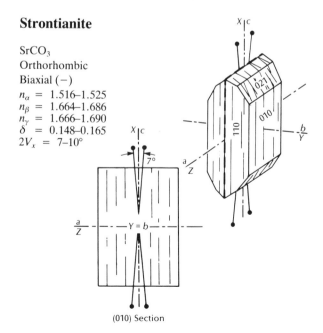

(010) Section

RELIEF IN THIN SECTION: Low negative to high positive relief, depending on orientation; may show marked change with rotation.

COMPOSITION AND STRUCTURE: Strontianite ($SrCO_3$) has the same structure as aragonite and consists of layers of triangular carbonate groups parallel to the basal plane (001) alternating with layers of Sr cations. Most strontianite contains significant amounts of Ca substituting for Sr. Substituting Ba for Sr is also possible.

PHYSICAL PROPERTIES: H = $3\frac{1}{2}$; G = 3.75; colorless, gray, or white in hand sample, also yellowish, brownish, or greenish, white streak; vitreous luster. Effervesces readily in cold dilute HCl.

COLOR: Colorless in thin section or grain mounts.

FORM: Usually as granular masses or aggregates of columnar, acicular, or fibrous crystals. The acicular and fibrous crystals are elongate parallel to the c axis, and the columnar crystals are often pseudohexagonal, like aragonite.

CLEAVAGE: Strontianite has good prismatic {110} cleavages in two directions which intersect at about 64°.

TWINNING: Twins with {110} composition planes are common and may be simple, polysynthetic, or cyclic.

OPTICAL ORIENTATION: $X = c$, $Y = b$, $Z = a$, optic plane = (010). Longitudinal sections and elongate cleavage fragments are length fast with parallel extinction to cleavage and long dimension. Extinction to cleavage in basal sections is symmetrical.

INDICES OF REFRACTION AND BIREFRINGENCE: Indices of refraction increase with substitution of Ca and Ba for Sr, up to about $n_\alpha = 1.525$, $n_\beta = 1.686$, and $n_\gamma = 1.690$, with birefringence of 0.165. High-order white interference colors are found for most orientations in both grain mount and thin section. Fragments lying on cleavage surfaces display near-maximum birefringence.

INTERFERENCE FIGURE: Basal sections yield centered acute bisectrix figures with numerous isochromes and a small optic angle. $2V_x$ is 7° for pure strontianite and only increases to about 10° for calcic samples. Fragments lying on cleavage surfaces yield off-center obtuse bisectrix figures that look like flash

figures due to the small 2V. Optic axis dispersion is weak, $v > r$.

ALTERATION: Alteration to celestite is sometimes found.

DISTINGUISHING FEATURES: Strontianite resembles the other orthorhombic carbonates, with extreme birefringence and parallel extinction. Aragonite has higher indices and only one prominent cleavage. Witherite has higher indices and a larger 2V. Specific gravity and brilliant red flame test for Sr also may be used to distinguish strontianite.

OCCURRENCE: Strontianite is usually found in veins, cavities, and irregular masses in limestone or calcareous clays where it was deposited by hydrothermal solutions. It also sometimes occurs as a gangue mineral in hydrothermal sulfide deposits.

Witherite

$BaCO_3$
Orthorhombic
Biaxial (−)
$n_\alpha = 1.529$
$n_\beta = 1.676$
$n_\gamma = 1.677$
$\delta = 0.148$
$2V_x = 16°$

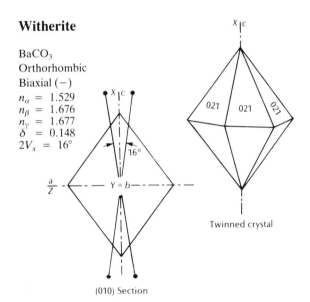

(010) Section

Twinned crystal

RELIEF IN THIN SECTION: Low negative to high positive relief, depending on orientation; may show a marked change with rotation.

COMPOSITION AND STRUCTURE: Witherite has the aragonite structure. Its composition is usually close to pure $BaCO_3$, although small amounts of Sr or Ca may substitute for Ba.

PHYSICAL PROPERTIES: H = $3\frac{1}{2}$; G = 4.30; usually colorless, white, or gray in hand sample, also pale yellow, brown, or green; white streak; vitreous luster. Effervesces in cold dilute HCl.

COLOR: Colorless in thin section and grain mount.

FORM: Crystals are stubby pseudohexagonal dipyramids. The hexagonal shape is the result of cyclic twinning on {110}, which produces six segments. It also forms granular, columnar, or fibrous masses that may have rounded or botryoidal shapes.

CLEAVAGE: There is a single distinct cleavage on {010} and poor cleavages on {110} and {012}.

TWINNING: Witherite is always twinned on {110}, forming cyclic twins with six segments.

OPTICAL ORIENTATION: $X = c$, $Y = b$, $Z = a$, optic plane = (010). Extinction is parallel to the cleavage traces in principal sections. The fast ray vibrates parallel to cleavage traces in both basal and longitudinal sections.

INDICES OF REFRACTION AND BIREFRINGENCE: Because there is relatively little compositional variation, the indices for witherite do not vary significantly. Interference colors in thin section and grain mount are usually upper-order white for most orientations except those with an optic axis nearly vertical.

INTERFERENCE FIGURE: Basal sections yield centered acute bisectrix figures with numerous isochromes and $2V_x = 16°$. Fragments lying on the dominant {010} cleavage yield centered flash figures. Optic axis dispersion is weak, $r > v$.

ALTERATION: Witherite may alter to barite or may be produced as an alteration product after barite.

DISTINGUISHING FEATURES: Witherite is similar to the other orthorhombic carbonates, which are distinguished from the rhombohedral carbonates by biaxial character and lack of rhombohedral cleavage. Strontianite has lower indices and two good cleavages. Aragonite has almost the same indices and cleavage but has lower specific gravity. Aragonite fragments float in diodomethane (1.74 index oil) and witherite fragments sink. The yellow-green flame test of Ba is diagnostic.

OCCURRENCE: Witherite is usually found in low-temperature hydrothermal veins and masses deposited in limestone or other calcareous sediment. It is commonly associated with barite and galena.

BORATES

Borax

$Na_2B_4O_7 \cdot 10H_2O$
Monoclinic
$\angle\beta = 106.6°$
Biaxial $(-)$
$n_\alpha = 1.447$
$n_\beta = 1.469$
$n_\gamma = 1.472$
$\delta = 0.025$
$2V_x = 39–40°$

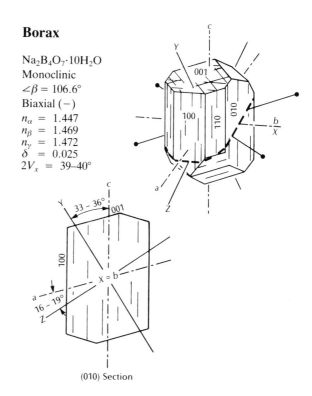

(010) Section

RELIEF IN THIN SECTION: Moderate negative relief.

COMPOSITION: The composition of borax is usually close to $Na_2B_4O_7 \cdot 10H_2O$, but it may dehydrate to a crumbly mass of *tincalconite* ($Na_2B_4O_7 \cdot 5H_2O$) on exposure to air.

PHYSICAL PROPERTIES: $H = 2–2\frac{1}{2}$; $G = 1.71$; white or gray in hand sample, less commonly greenish or bluish; white streak; vitreous luster. It is highly water soluble and has a slightly alkaline taste.

COLOR: Colorless in thin section or grain mount.

FORM: Commonly found as granular aggregates. Crystals are stubby prisms, often with eight-sided cross sections.

CLEAVAGE: Borax has three cleavages. A single cleavage on {100} is perfect and two cleavages on {110} at nearly right angles are distinct.

TWINNING: Simple twins on {100} are rare.

OPTICAL ORIENTATION: $X = b$, $Y \wedge c = +33$ to $+36°$, $Z \wedge a = -16$ to $-19°$, optic plane normal to (010). The maximum extinction angle is seen on sections cut parallel to (010) and is 33 to 36° measured to the trace of cleavage, with the fast ray vibration direction closer to the cleavage (length fast). Sections cut parallel to (100) have parallel extinction with the slow ray vibration direction parallel to the $\{110\}$ cleavage (length slow). Basal sections yield extinction parallel to the trace of the $\{100\}$ cleavage and symmetrical to the prismatic $\{110\}$ cleavage.

INDICES OF REFRACTION AND BIREFRINGENCE: Indices do not vary significantly because there is little compositional variation. Birefringence is moderate and yields up to lower second-order colors in standard

Table 10.2. Minerals found in evaporite deposits

Mineral and System	Formula	Cleavage	Indices				Sign and Dispersion	Orientation	Color in Thin Section
			n_α	n_β	n_γ	δ			
KERNITE Monoclinic $\angle\beta = 109°$	$Na_2B_4O_6(OH)_2$	$\{100\}$ perf $\{001\}$ perf	1.455	1.472	1.487	0.032	Bi $(-)$ $2V_x = 80°$ $r > v$ distinct	$X \wedge a = +38°$ $Y \wedge c = -19°$ $Z = b$	Colorless
ULEXITE Triclinic $\angle\alpha = 90.3°$ $\angle\beta = 109.2°$ $\angle\gamma = 105.1°$	$NaCaB_5O_6(OH)_6 \cdot 5H_2O$	$\{010\}$ perf $\{1\bar{1}0\}$ good $\{110\}$ poor	1.493	1.506	1.519	0.026	Bi $(+)$ $2V_z = 73–78°$	$X \wedge c = -69°$ $Y \wedge c = +21°$ $Z = b$	Colorless
TRONA Monoclinic $\angle\beta = 103°$	$Na_3H(CO_3)_2 \cdot 2H_2O$	$\{100\}$ perf $\{101\}$ good	1.416	1.494	1.542	0.126	Bi $(-)$ $2V_x = 75°$ $v > r$ distinct	$X = b$ $Y \wedge c = -7°$ $Z \wedge c = +83°$	Colorless
NAHCOLITE Monoclinic $\angle\beta = 93.3°$	$NaHCO_3$	$\{101\}$ perf $\{100\}$ good $\{111\}$ good	1.377	1.501	1.583	0.206	Bi $(-)$ $2V_x = 75°$ $v > r$ weak	$X \wedge c = +27°$ $Y = b$ $Z \wedge a = -24°$	Colorless
THENARDITE Orthorhombic	Na_2SO_4	$\{010\}$ perf $\{101\}$ good $\{100\}$ poor	1.469	1.475	1.484	0.015	Bi $(+)$ $2V_z = 83°$ $r > v$ weak	$X = c$ $Y = b$ $Z = a$	Colorless
GLAUBERITE Monoclinic $\angle\beta = 112.2°$	$Na_2Ca(SO_4)_2$	$\{001\}$ perf $\{110\}$ poor	1.515	1.535	1.536	0.021	Bi $(-)$ $2V_x = 7°$ $r > v$ strong + horizontal	$X \wedge c = +30°$ $Y \wedge a = -8°$ $Z = b$ Varies with temperature	Colorless
CARNALLITE Orthorhombic	$KMgCl_3 \cdot 6H_2O$	none	1.466	1.474	1.495	0.029	Bi $(+)$ $2V_z = 70°$ $v > r$ weak	$X = c$ $Y = b$ $Z = a$	Colorless
POLYHALITE Triclinic $\angle\alpha = 104°$ $\angle\beta = 114°$ $\angle\gamma = 101°$	$K_2MgCa_2(SO_4)_4 \cdot 2H_2O$	$\{10\bar{1}\}$ perf	1.547	1.562	1.567	0.020	Bi $(-)$ $2V_x = 62°–70°$ $v > r$	?	Colorless
KIESERITE Monoclinic $\angle\beta = 116°$	$MgSO_4 \cdot H_2O$	$\{110\}, \{111\}$ perf; $\{\bar{1}11\}$, $\{101\}, \{011\}$ fair	1.520	1.533	1.584	0.064	Bi $(+)$ $2V_z = 55°$ $r > v$ moderate	$Z \wedge c = 77°$ $Y = b$	Colorless
TINCALCO-NITE Trigonal	$Na_2B_4O_7 \cdot 5H_2O$		$n_\omega = 1.461$, $n_\epsilon = 1.473$			0.013	Uni $(+)$		Colorless
SODA NITER Trigonal	$NaNO_3$	$\{10\bar{1}1\}$ perf; $\{10\bar{1}2\}$, $\{0001\}$ fair	$n_\omega = 1.587$, $n_\epsilon = 1.336$			0.251	Uni $(-)$		Colorless

Also see halite, sylvite, colemanite, borax, calcite, dolomite, gypsum, anhydrite, and chalcedony.

thin sections. Sections cut normal to an optic axis do not go completely extinct and display anomalous brown or blue interference colors due to the strong crossed bisectrix dispersion.

INTERFERENCE FIGURE: Centered acute bisectrix figures are obtained on sections cut parallel to (010). This section produces lower first-order colors in thin section because $n_\gamma - n_\beta = 0.003$. Optic axis dispersion is strong ($r > v$), and crossed bisectrix dispersion is distinct. Isochromes typically have somewhat anomalous colors. Fragments lying on cleavage surfaces yield strongly off-center figures.

ALTERATION: Borax readily alters to white crumbly tincalconite on exposure to air.

DISTINGUISHING FEATURES: Borax is water soluble and has a very restricted occurrence, distinctive optic axis and crossed bisectrix dispersion, and anomalous interference colors in some orientations.

OCCURRENCE: Borax is found in saline lake deposits associated with other borates and halides, and in desert soils as an efflorescence. Other minerals found in evaporites are listed in Table 10.2.

Colemanite

$Ca_2B_6O_{11} \cdot 5H_2O$
Monoclinic
$\angle\beta = 110.1°$
Biaxial (+)
$n_\alpha = 1.586$
$n_\beta = 1.592$
$n_\gamma = 1.614$
$\delta = 0.028$
$2V_z \cong 56°$

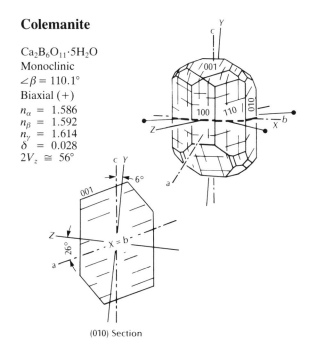

(010) Section

RELIEF IN THIN SECTION: Moderate positive relief.

COMPOSITION: The composition generally does not vary significantly, although some Mg may replace Ca.

PHYSICAL PROPERTIES: H = 4–4½; G = 2.42; colorless, white, gray, or yellowish in hand sample; white streak; vitreous luster. Decrepitates on heating.

COLOR: Colorless in thin section and grain mount.

FORM: Often as massive, cleavable, or granular masses. Crystals may be stubby prisms, equant, or rhomblike with many faces.

CLEAVAGE: A perfect cleavage on {010} is at right angles to a good cleavage on {001}.

TWINNING: None reported.

OPTICAL ORIENTATION: $X = b$, $Y \wedge c = -6°$, $Z \wedge a = +26°$, optic plane perpendicular to (010). Sections and cleavage fragments with (010) horizontal show an extinction angle of 26° to the trace of the {001} cleavage with the slow ray vibration direction closer to the cleavage (length slow). Basal sections or cleavage fragments lying on the {001} cleavage show parallel extinction with the slow ray vibration direction parallel to the trace of the {010} cleavage (length slow).

INTERFERENCE FIGURE: Sections cut parallel to (100) yield approximately centered acute bisectrix figures with $2V_z$ of about 56°. Cleavage fragments lying on (010) and (001) yield obtuse bisectrix and off-center flash figures, respectively.

INDICES OF REFRACTION AND BIREFRINGENCE: Indices of refraction do not vary significantly. Interference colors seen in standard thin sections range up to middle second order.

ALTERATION: May be altered to calcite. Colemanite also may be produced as an alteration product after other borates.

DISTINGUISHING FEATURES: Colemanite has a restricted occurrence, and higher indices of refraction than many of the other borates with which it is commonly associated.

OCCURRENCE: Colemanite is found in saline lake deposits associated with borax and other borate minerals. Some of the more common associated minerals are listed in Table 10.2.

SULFATES

Barite

$BaSO_4$
Orthorhombic
Biaxial (+)
n_α = 1.634–1.637
n_β = 1.636–1.639
n_γ = 1.646–1.649
δ = 0.012
$2V_z$ = 36–40°

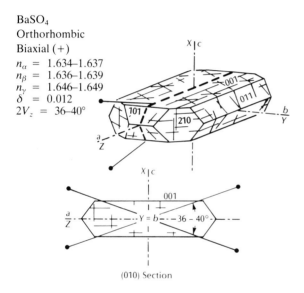

(010) Section

RELIEF IN THIN SECTION: Moderately high positive relief.

COMPOSITION AND STRUCTURE: The structure of barite consists of SO_4^{2-} tetrahedra, which are bonded laterally through Ba^{2+} cations. Each Ba^{2+} is in 12-fold coordination with the oxygens in the sulfate tetrahedra. Most barite is relatively pure $BaSO_4$, although small amounts of Sr may be present, and there is a complete solid solution series with celestite ($SrSO_4$). Up to 6 percent Ca also may substitute for Ba and minor amounts of Pb, Co, Hg, and Ra have been reported.

PHYSICAL PROPERTIES: H = 3–3$\frac{1}{2}$; G = 4.50; white, yellow, gray, light blue, light green, red, or brown in hand sample; white streak; vitreous luster.

COLOR AND PLEOCHROISM: Usually colorless in thin section and grain mount. Colored varieties may display pale colors with weak pleochroism (usually $Z > Y > X$).

FORM: Crystals are typically tabular or, less commonly, prismatic parallel to the a or b crystal axes. Crystals are often intergrown, forming rosettes or platy aggregates. Barite also forms concretionary masses with fibrous texture, and granular or cleavable masses.

CLEAVAGE: Barite has four cleavage directions; two very good cleavages on {210} which intersect at 78°, a fair to good cleavage on {010}, and a perfect basal cleavage on {001} which intersects the other three at 90°.

TWINNING: Massive varieties may have deformation-induced twinning on {110}, otherwise it is not usually twinned.

OPTICAL ORIENTATION: $X = c$, $Y = b$, $Z = a$, optic plane = (010). In sections cut parallel to the c crystal axis, extinction is parallel to cleavage and the slow ray vibrates parallel to the long dimension of the section (length slow). In basal sections, extinction is symmetrical to the prismatic cleavages. Fibrous or prismatic crystals are usually elongate along the a axis and are length slow with parallel extinction. Crystals elongate along b are relatively uncommon and show parallel extinction and may be either length fast or length slow.

INDICES OF REFRACTION AND BIREFRINGENCE: Indices of refraction decrease slightly with addition of Sr and increase with Pb. Interference colors in thin section range up to first-order yellow.

INTERFERENCE FIGURE: Sections cut parallel to (100) yield centered acute bisectrix figures with 2V of 36 to 40°, with weak optic axis dispersion ($v > r$). (100) sections show very low birefringence (~0.002) and parallel extinction. Cleavage fragments lying on the dominant {001} cleavage produce obtuse bisectrix figures.

ALTERATION: Barite may alter to witherite or may be replaced by a variety of minerals such as quartz, calcite, dolomite, or pyrite.

DISTINGUISHING FEATURES: Barite is most commonly confused with gypsum, anhydrite, or celestite. Gypsum has inclined extinction and negative relief, anhydrite has higher birefringence, and celestite has larger $2V$ and lower indices.

OCCURRENCE: Barite is commonly found as a gangue mineral in hydrothermal sulfide deposits and is also found as concretionary masses, veins, or irregular masses in limestone, dolomite, shale, or other sedimentary rocks. Barite also is found in carbonatites.

Celestite (Celestine)

$SrSO_4$
Orthorhombic
Biaxial (+)
n_α = 1.621–1.622
n_β = 1.623–1.624
n_γ = 1.630–1.632
δ = 0.009–0.010
$2V_z$ = 51°

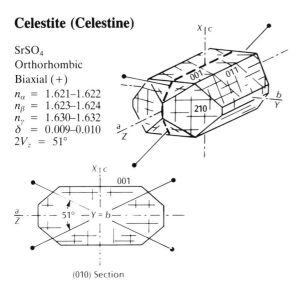

(010) Section

RELIEF IN THIN SECTION: Moderate positive relief.

COMPOSITION AND STRUCTURE: The structure of celestite is similar to barite and consists of SO_4^{2-} tetrahedra bonded laterally to Sr^{2+}, which are in 12-fold coordination with oxygen. Although there is complete solid solution to barite and limited solid solution to $CaSO_4$, natural celestite rarely contains more than 2 or 3 percent Ba or Ca substituting for Sr.

PHYSICAL PROPERTIES: H = 3; G = 3.97, pale blue, white, colorless, or less commonly reddish, greenish, or brownish in hand sample; white streak; vitreous luster.

COLOR AND PLEOCHROISM: Usually colorless in thin section or grain mount. Large grains of colored samples may display pale colors with weak pleochroism usually in shades of blue, lavender, blue-green, or violet with $Z > Y > X$.

FORM: Celestite crystals occur in a variety of habits but are commonly either tabular parallel to (001) or elongate along the a axis, less commonly elongate along the b or c axis. It also occurs as cleavable, granular, earthy, or fibrous masses.

CLEAVAGE: Celestite has four cleavage directions. The single perfect {001} cleavage is the most prominent, two good prismatic {210} cleavages intersect at 75°, and the single {010} cleavage is poor.

TWINNING: Twinning is very rare in celestite.

OPTICAL ORIENTATION: $X = c$, $Y = b$, $Z = a$, optic plane = (010). Cleavage fragments and sections with (001) horizontal show symmetrical extinction to the trace of the prismatic cleavages. Sections cut parallel to the c axis and cleavage fragments lying on one of the prismatic {210} cleavage surfaces show parallel extinction to the trace of cleavages. Sections cut through tabular crystals are length slow as are grains elongate along the a axis. Grains elongate along c are length fast. Grains that are elongate along b may be either length fast or length slow, depending on how they are cut.

INDICES OF REFRACTION AND BIREFRINGENCE: The indices of refraction display relatively little variation. Interference colors seen in thin section are usually first-order gray and white, sometimes with a slight tinge of yellow.

INTERFERENCE FIGURE: Sections cut parallel to (100) yield centered acute bisectrix figures with $2V = 51°$. Optic axis dispersion is weak with $v > r$. (100) sections show parallel extinction and low (~ 0.002) birefringence. Cleavage fragments lying on the dominant {001} cleavage yield centered obtuse bisectrix figures.

ALTERATION: Celestite may alter to strontianite or be replaced by calcite, quartz, witherite, barite, or sulfur, sometimes pseudomorphically.

DISTINGUISHING FEATURES: Celestite is most easily confused with barite or gypsum. Barite has higher indices, slightly higher birefringence, and smaller $2V$. Gypsum has inclined extinction.

OCCURRENCE: Celestite is usually found in sedimentary rocks either in evaporite deposits associated with gypsum, anhydrite and halite or as disseminated grains, irregular masses, or veins in limestone and dolomite. It also may be found in hydrothermal vein deposits.

Gypsum

$CaSO_4 \cdot 2H_2O$
Monoclinic
$\angle \beta = 127.4°$
Biaxial $(+)$
$n_\alpha = 1.519–1.521$
$n_\beta = 1.522–1.526$
$n_\gamma = 1.529–1.531$
$\delta = 0.010$
$2V_z = 58°$

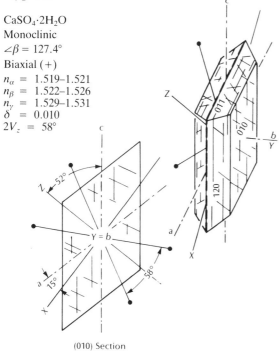

(010) Section

RELIEF IN THIN SECTION: Low negative relief.

COMPOSITION AND STRUCTURE: The structure of gypsum is layered parallel to (010) and consists of strongly bonded layers of $Ca^{2+} + SO_4^{2-}$ alternating with layers of H_2O molecules. Hydrogen bonding holds the water molecules to the $CaSO_4$ layers and, because it is quite weak, produces the excellent cleavage on {010}. The composition deviates very little from the ideal $CaSO_4 \cdot 2H_2O$, although trace amounts of Sr or Ba may replace Ca.

PHYSICAL PROPERTIES: H = 2; G = 2.31; colorless, white or sometimes gray, red, yellow, or blue in hand sample; white streak; vitreous to pearly luster.

COLOR: Colorless in thin section or grain mount.

FORM: Figure 10.5. Crystals occur in numerous habits but are usually tabular parallel to (010); less commonly prismatic or acicular parallel to the c crystal axis. Gypsum also occurs as granular aggregates and as fibrous masses called satin spar.

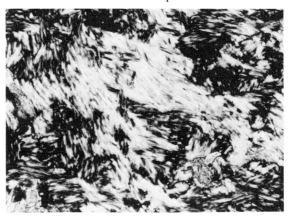

Figure 10.5 Foliated mass of gypsum (crossed polars). Field of view is 1.7 mm wide.

CLEAVAGE: There are four cleavage directions in gypsum. Cleavage on {010} is the most prominent, a second cleavage on {100} is good, and two prismatic cleavages on {$\bar{1}11$} are good and intersect at angles of 42 and 138°.

TWINNING: Simple contact twins on {100} are common; twinning on {001} is less common.

OPTICAL ORIENTATION: $X \wedge a = -15°$, $Y = b$, $Z \wedge c = +52°$, optic plane = (010). Above 91°C, $X = b$ and the optic plane is perpendicular to (010). Sections or cleavage fragments lying with (010) horizontal yield inclined extinction of 38° to the trace of the {100} cleavage, with the fast ray vibration direction closer to the cleavage. Extinction to the trace of the {$\bar{1}11$} cleavage is 14°, with the slow ray vibration direction closer to the cleavage. Sections through tabular crystals cut with (010) vertical produce parallel extinction to the trace of the {010} cleavage and may be either length fast or length slow.

INDICES OF REFRACTION AND BIREFRINGENCE: Interference colors in standard thin sections are like quartz (i.e., first-order gray and white).

INTERFERENCE FIGURE: Interference figures obtained from thin sections usually have relatively broad isogyres superimposed on a first-order white field. $2V$ is about 58° at room temperature, but decreases to 0° (uniaxial) at 91°C. Above that temperature, $2V_z$ opens into an optic plane oriented perpendicular to

(010). Optic axis dispersion is strong ($r > v$), as is inclined bisectrix dispersion. Fragments lying on (010) yield flash figures. Fragments on the other cleavages yield off-center figures.

ALTERATION: Gypsum may be replaced by quartz, opal, calcite, or celestite, forming pseudomorphs. Gypsum is often produced as the result of hydrating anhydrite.

DISTINGUISHING FEATURES: Gypsum may be confused with anhydrite or barite. Anhydrite has higher birefringence and positive relief in thin section, whereas barite has parallel extinction and moderately high positive relief in thin section.

OCCURRENCE: Gypsum is a common mineral in evaporate deposits, where it is associated with halite, sylvite, anhydrite, calcite, and dolomite, as well as clay and detrital material. In some cases, it is formed by hydrating anhydrite, which results in a volume increase that may crumple or disrupt bedding. It is found as a precipitate from saline lakes, as an efflorescence on desert soils, and may precipitate around fumaroles and volcanic vents. Occasionally, it is found in the gossan or oxidized zone of hydrothermal sulfide deposits.

Anhydrite

CaSO₄
Orthorhombic
Biaxial (+)
$n_\alpha = 1.570$
$n_\beta = 1.576$
$n_\gamma = 1.614$
$\delta = 0.044$
$2V_z = 44°$

(100) Section

RELIEF IN THIN SECTION: Moderate positive relief.

COMPOSITION AND STRUCTURE: The structure of anhydrite consists of SO_4^{2-} tetrahedra that are bonded laterally through Ca^{2+}, which are in eight-fold coordination with the oxygen. The composition typically shows little variation, although small amounts of Ba or Sr may substitute for Ca.

PHYSICAL PROPERTIES: H = 3–3½; G = 2.98; usually colorless, white, or gray, less commonly blue, red, or brown in hand sample; white streak; vitreous luster.

COLOR AND PLEOCHROISM: Usually colorless in thin section. Colored varieties may show light colors in grain mount with weak pleochroism, $Z > Y > X$.

FORM: Figure 10.6. Anhydrite usually occurs as massive aggregates and as fibrous or radiating aggregates. Crystals are usually blocky or thickly tabular and elongate parallel to either the a or c crystal axis.

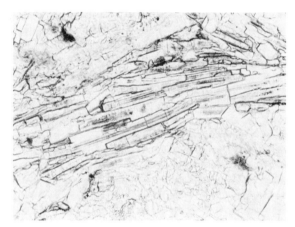

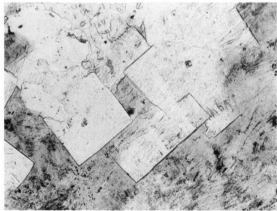

Figure 10.6 Anhydrite. (*Top*) Tabular crystals. (*Bottom*) Blocky crystals in chalcedony. Field of view is 1.7 mm wide for both photographs.

CLEAVAGE: There are three pinacoidal cleavages at right angles. The {010} cleavage is perfect, {100} is very good, and {001} is good.

TWINNING: Simple and repeated twins on {011} are relatively common.

OPTICAL ORIENTATION: $X = b$, $Y = a$, $Z = c$, optic plane = (100). Extinction is parallel to cleavages and crystal faces in all principal sections. Fibers have parallel extinction and if elongate parallel to the a axis are either length slow or length fast, and if elongate parallel to the c axis are length slow.

INDICES OF REFRACTION AND BIREFRINGENCE: Indices of refraction show little variation. Sections cut parallel to (100) show the highest interference colors, which in thin section are lower third order. Randomly cut grains usually display vivid second-order colors.

INTERFERENCE FIGURE: Sections and cleavage fragments oriented with (001) horizontal produce centered acute bisectrix figures with $2V_z \cong 44°$ and several orders of isochromes. Optic axis dispersion is distinct, $v > r$.

ALTERATION: Anhydrite is readily altered by hydration to form gypsum and may be replaced by quartz, calcite, dolomite, and other minerals.

DISTINGUISHING FEATURES: Anhydrite resembles gypsum and barite. Gypsum has lower birefringence, inclined extinction, and negative relief in thin section. Barite has lower birefringence, higher indices and cleavage is not at right angles.

OCCURRENCE: Anhydrite is a very common mineral in evaporite deposits and is often associated with halite, sylvite, calcite, dolomite, gypsum, and related minerals, as well as clay and other detrital material. It may either precipitate directly from highly saline sea water or be produced by dehydration of primary gypsum. It also is found in the oxidized zone of hydrothermal sulfide deposits associated with oxides, hydroxides, and carbonates of iron, lead, copper, zinc, and silver. Infrequently, anhydrite is found as an amygdule filling in basaltic or andesitic volcanics where it may be associated with zeolites, or in deposits around fumaroles or volcanic vents.

Alunite

$KAl_3(SO_4)_2(OH)_6$
Hexagonal (trigonal)
Uniaxial (+)
$n_\omega = 1.572–1.620$
$n_\epsilon = 1.592–1.641$
$\delta = 0.010–0.021$

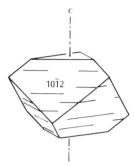

RELIEF IN THIN SECTION: Moderate positive relief.

COMPOSITION AND STRUCTURE: There is extensive solid solution between alunite and *natroalunite* [$NaAl_3(SO_4)_2(OH)_6$], and most alunite contains significant amounts of Na. Small amounts of Fe^{3+} may substitute for Al^{3+}, and solid solution with *jarosite* [$KFe_3(SO_4)_2(OH)_6$] is limited.

PHYSICAL PROPERTIES: H = $3\frac{1}{2}$–4; G = 2.6–2.9; usually white or grayish in hand sample, also yellowish, reddish, or brownish; white streak; vitreous to pearly luster.

COLOR: Colorless in thin section or grain mount.

FORM: Rare crystals are tabular parallel to (0001), or rhombohedral yielding square or diamond-shaped cross sections. The rhombohedrons are cubelike, because the angle more between faces is 90.8°. Alunite is more commonly found as granular, plumose, or flaky aggregates and as disseminated grains. The grain size is often too fine to obtain reliable optical data.

CLEAVAGE: The basal cleavage on {0001} is distinct and usually controls grain orientation in grain mounts to some degree. A very poor rhombohedral cleavage on {10$\bar{1}$2} also is present.

TWINNING: None reported.

OPTICAL ORIENTATION: Extinction is parallel to the trace of the basal cleavage {0001} in all sections and the fast ray vibrates parallel to the cleavage (length fast).

INDICES OF REFRACTION AND BIREFRINGENCE: The indices vary irregularly with composition. For relatively pure alunite, $n_\omega = 1.572$ and $n_\epsilon = 1.592$. Iron-bearing samples have higher indices. Natroalunite has indices in the same range as alunite. Interference colors seen in standard thin section range up to mid second-order for most alunite, although some has lower birefringence and produces only first-order colors. It may be difficult to measure n_ϵ in grain mount because cleavage strongly influences grain orientation.

INTERFERENCE FIGURE: Basal sections and cleavage flakes yield a uniaxial positive optic axis figure.

ALTERATION: Alunite is generally produced by the alteration of alkali feldspar.

DISTINGUISHING FEATURES: Alunite is often relatively fine grained, which makes it difficult to identify, and X-ray diffraction or other techniques may be needed. It is not unlike brucite but has a different occurrence, and brucite commonly displays anomalous interference colors.

OCCURRENCE: Alunite is a relatively common mineral in altered volcanic rocks of intermediate and felsic composition. It is usually produced by the action of sulfur-bearing fluids on feldspar and is associated with clay minerals and other products of feldspar alteration, as well as fine-grained quartz. Alunite may also be found in the oxidized and weathered zone of hydrothermal sulfide deposits.

PHOSPHATES

Apatite

$Ca_5(PO_4)_3(F,OH,Cl)$
Hexagonal
Uniaxial (−)
$n_\omega = 1.633–1.667$
$n_\epsilon = 1.629–1.665$
$\delta = 0.001–0.007$

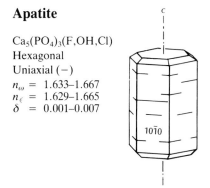

RELIEF IN THIN SECTION: Moderately high positive relief.

COMPOSITION AND STRUCTURE: The structure of apatite consists of PO_4^{3-} tetrahedra bonded laterally through Ca^{2+}, forming a hexagonal framework. The F^-, OH^-, or Cl^- are located in interstices between the phosphate tetrahedra. The most important compositional variation is in the occupancy of the hydroxyl site. There appears to be essentially complete solid solution between *fluorapatite* $[Ca_5(PO_4)_3F]$, *hydroxylapatite* $[Ca_5(PO_4)_3OH]$, and *chlorapatite* $[Ca_5(PO_4)_3Cl]$. In addition, some carbonate $[(CO_3OH)^{3-}]$ may be present, probably substituting for PO_4^{3-} groups. Carbonate-bearing apatite is called *carbonate-apatite*, or *dahllite* for OH-rich varieties and *francolite* for F-rich varieties. Most common apatite is intermediate between fluor- and hydroxylapatite, usually with more fluorine than hydroxyl. The phosphate tetrahedra is infrequently replaced by other tetrahedral groups such as SO_4^{2-}, SiO_4^{4-}, and CrO_4^{2-}. *Collophane* is a term given to fine-grained cryptocrystalline material that contains a substantial amount of apatite.

PHYSICAL PROPERTIES: H = 5; G = 2.9–3.5 (most common apatite is 3.1 to 3.2); commonly found in shades of green or gray with bluish or yellowish tints, although almost any color is possible in hand sample; white streak; vitreous luster.

COLOR AND PLEOCHROISM: Usually colorless in thin section and grain mount. Strongly colored samples may display pale colors corresponding to the hand-sample color, with weak to moderate pleochroism and absorption with $\epsilon > \omega$ so that elongate grains are darker colored when they are parallel to the vibration direction of the lower polar. Collophane is brown.

FORM: Figure 10.7. Apatite forms small euhedral to subhedral elongate prismatic crystals with hexagonal cross sections. Longitudinal sections through the crystals are usually elongate-rectangular in outline with the poor cleavage cutting across the length. Also found as anhedral grains, and granular or columnar aggregates. Collophane is usually brown and essentially isotropic; forms colloform, encrusting, spherulitic, oolitic, and related structures; and is a major constituent of some fossil bone.

CLEAVAGE: The basal cleavage {0001} is usually relatively poor, and a prismatic cleavage {10$\bar{1}$0} is very poor and is generally not observed. Cleavage does not have a strong influence on fragment orientation in grain mount.

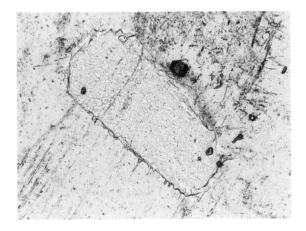

Figure 10.7 Apatite in sericitized plagioclase. Field of view is 0.4 mm wide.

TWINNING: Twinning is quite rare but has been reported on contact planes $\{11\bar{2}1\}$, $\{10\bar{1}3\}$, $\{10\bar{1}0\}$, and $\{11\bar{2}3\}$.

OPTICAL ORIENTATION: Elongate sections through crystals show parallel extinction and are length fast.

INDICES OF REFRACTION AND BIREFRINGENCE: Indices vary in a general way with composition (Figure 10.8). The indices for the end-member compositions are:

	n_ω	n_ϵ	δ
Fluorapatite	1.633	1.629	0.004
Hydroxylapatite	1.651	1.644	0.007
Chlorapatite	1.667	1.666	0.001

Most common apatite has indices $n_\omega = 1.633$–1.650, $n_\epsilon = 1.629$–1.647, and $\delta = 0.003$–0.005. Carbonate-apatite has significantly lower indices that are in the range $n_\omega = 1.603$–1.628, $n_\epsilon = 1.598$–1.619, and higher birefringence $\delta = 0.007$–0.017. Collophane may be essentially isotropic with $n = 1.58 - 1.63$. In thin section, the interference color of common apatite is first-order gray, often with a slightly anomalous blue cast to it. Small grains or grains cut so that the optic axis is within roughly 45° of being vertical may appear to be practically isotropic but can be tested with the gypsum plate to detect their weak birefringence. Carbonate-apatite has higher birefringence and may produce interference colors up to upper first or lower second order in thin section.

INTERFERENCE FIGURE: Basal sections in thin section yield uniaxial negative figures with diffuse isogyres on a first-order gray field. Some varieties, particularly carbonate-apatite, are biaxial with $2V_x$ up to 20°, although they are not common. Interference figures on small grains in thin section may be difficult to obtain.

ALTERATION: Apatite is relatively stable in most geologic environments and is not readily altered to

Figure 10.8 General variation of n_ω and birefringence (δ) with the content of F, Cl, and OH in apatite. Constructed by assuming a linear variation between end member optical properties. There is too much variability in natural apatite to allow the diagram to be used to estimate composition (cf. Taborszky, 1972; McConnell and Gruner, 1940).

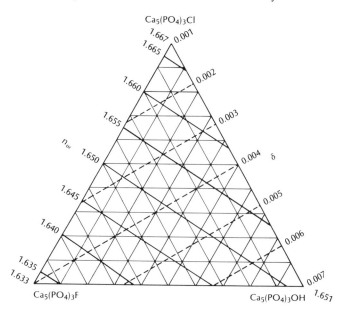

other minerals; however, pseudomorphs of clay, turquoise, serpentine, and other phosphate minerals after apatite have been reported.

DISTINGUISHING FEATURES: Apatite is distinguished by its moderate to high relief, low birefringence, and uniaxial character. In thin section it usually displays a slightly pebbly surface texture. It is distinguished from garnet by its weak birefringence. Topaz, sillimanite, and mullite have higher birefringence and are biaxial. Zoisite has significantly higher relief and often displays anomalous yellowish blue interference colors. Colored varieties of apatite may resemble tourmaline, but apatite is darker when the long axis of grains is aligned parallel to the vibration direction of the lower polar, and tourmaline is lighter. Apatite also resembles beryl, but beryl has lower indices and lower relief. Andalusite is biaxial, has somewhat higher birefringence, and shows more distinct cleavage.

OCCURRENCE: Apatite is present in a wide variety of igneous and metamorphic rocks as an accessory mineral. The grains are commonly small and are most readily spotted by carefully scanning a slide using the medium-power objective. The apatite in most igneous rocks is fluorine rich, although chlorapatite is not uncommon in mafic rocks. Carbonatites often contain substantial amounts of apatite. The apatite in skarns, marble, and calc-silicate gneiss may be coarsely crystalline.

Apatite also can be found in sedimentary rocks. It is relatively common as detrital grains in clastic sediments and often has an oval or somewhat elongate shape. Chlorapatite or carbonate-apatite is the major constituent of collophane in phosphatic limestone, shale, and ironstone, and in some nearly pure phosphate beds. Collophane commonly forms oolitic, spherulitic, and related structures.

Monazite

$(Ce,La,Th)PO_4$
Monoclinic
$\angle\beta = 104°$
Biaxial (+)
$n_\alpha = 1.777–1.800$
$n_\beta = 1.778–1.801$
$n_\gamma = 1.823–1.849$
$\delta = 0.045–0.052$
$2V_z = 6–19°$

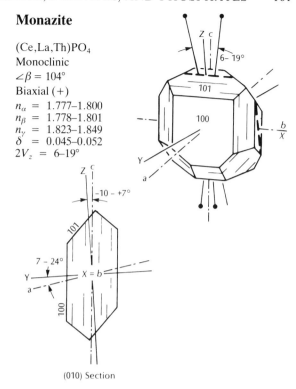

(010) Section

RELIEF IN THIN SECTION: High positive relief.

COMPOSITION AND STRUCTURE: The structure of monazite consists of slightly distorted PO_4^{3-} tetrahedra bonded laterally through cations that are in ninefold coordination with oxygen. Any of the rare earth elements may be present in monazite, but Ce usually predominates. Si or S may substitute for some of the P.

PHYSICAL PROPERTIES: H = 5; G = 4.6–5.4; yellow to reddish-yellow or reddish-brown in hand sample; white streak; resinous to waxy luster.

COLOR AND PLEOCHROISM: Pale yellow, pale green, colorless, or gray in thin section. Pleochroism in thin section is generally weak from lighter to slightly darker yellow or greenish yellow. Fragments in grain mount may be more strongly colored corresponding to the hand-sample color and show the pleochroism somewhat better: $Y > X \cong Z$, with X = light yellow, Y = dark yellow, and Z = greenish yellow. Because monazite often contains radioactive elements, it may produce pleochroic halos in enclosing biotite or other minerals.

FORM: Monazite commonly occurs as small euhedral crystals that may be equant, flattened parallel to (100), or elongate parallel to the *b* axis. Longitudinal sections often appear as four-, six-, or sometimes eight-sided parallelograms.

CLEAVAGE: A single cleavage on {100} is distinct, and a second at right angles on {010} is fair to poor. Other poor to very poor cleavages are reported and there is a basal parting on {001}. In samples without significant alteration, the cleavages are usually not very distinct, whereas relatively strongly altered grains may show both cleavages and the basal parting relatively well.

TWINNING: Simple twins on a {100} twin plane are common; lamellar twinning on {001} is rare.

OPTICAL ORIENTATION: $X = b$, $Y \wedge a = +7$ to $+24°$, $Z \wedge c = +7$ to $-10°$, optic plane is perpendicular to (010). Maximum extinction angle seen in sections parallel to (010) is 0 to 10° with the slow ray nearly parallel to the trace of the {100} cleavage. Cleavage fragments lying on the {100} cleavage surface usually do not have an elongation, although extinction is parallel to the {010} cleavage if it is displayed.

INDICES OF REFRACTION AND BIREFRINGENCE: Indices of refraction generally increase with substitution of Th and Si and decrease with alteration. Interference colors in standard thin section range up to the upper third or lower fourth order. Fragments lying on the {100} cleavage show maximum birefringence, and those resting on the basal parting show low interference colors.

INTERFERENCE FIGURE: Basal sections and fragments lying on the basal parting produce slightly eccentric acute bisectrix figures with $2V = 6-19°$. Fragments lying on the {100} cleavage produce flash figures. The high birefringence allows for numerous isochromes. Optic axis dispersion is typically weak with either $r < v$ or (rarely), $v < r$. Horizontal dispersion also is weak.

ALTERATION: Monazite is generally relatively stable in the weathering environment and occurs as detrital grains. It may alter to brownish limonitelike material along cleavages and grain boundaries and sometimes becomes clouded by minute opaque particles, which are possibly limonite.

DISTINGUISHING FEATURES: Monazite is recognized by its high birefringence, high relief, and pale yellow color. In small grains, it may be difficult to distinguish from zircon, epidote, titanite, or xenotime. Zircon is uniaxial and generally not colored. Epidote has a larger $2V$ and sometimes lower birefringence. Titanite is often more strongly colored, has higher birefringence, and has a larger $2V$ with extreme dispersion. Xenotime has higher birefringence and is uniaxial.

OCCURRENCE: Monazite is an accessory mineral in granitic rocks, syenite, granitic pegmatites, and carbonatites. It also is formed in metamorphosed dolomites and in mica schists, gneiss, and granulites. Because monazite is relatively resistant to weathering, it may be found as detrital grains in clastic sediments. It is infrequently found in hydrothermal vein deposits.

Xenotime

YPO$_4$
Tetragonal
Uniaxial (+)
$n_\omega = 1.690-1.724$
$n_\epsilon = 1.760-1.827$
$\delta = 0.070-0.107$

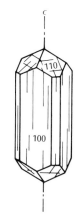

RELIEF IN THIN SECTION: High positive relief.

COMPOSITION AND STRUCTURE: The structure of xenotime consists of isolated PO$_4^{3-}$ tetrahedra bonded laterally to Y^{3+} cations, which are in strongly distorted eightfold coordination with oxygen. Rare earth elements, particularly cerium (Ce) and erbium (Er), may substitute for yttrium (Y), as may small amounts of Ca, Zr, Th, or U. Phosphorus also may be replaced by silicon or sulfur to a limited extent.

PHYSICAL PROPERTIES: H = 4–5; G = 4.3–5.1 (4.25 calculated for pure YPO$_4$); yellowish or reddish brown, or less commonly yellow, gray, salmon-pink, or greenish in hand sample; white streak; vitreous luster.

COLOR AND PLEOCHROISM: Colorless, yellow, or pale brown in thin section with weak pleochroism.

ω = pale pink, yellow-brown or yellow
ϵ = pale brownish yellow, grayish brown, or greenish

Grains are slightly darker colored when the long axis of prismatic grains is aligned parallel to the vibration direction of the lower polar.

FORM: Usually found as elongate prismatic tetragonal grains resembling zircon, and occasionally as radial aggregates or rosettes. Detrital grains are commonly elongate.

CLEAVAGE: Xenotime has good prismatic cleavage on $\{100\}$. The two cleavage directions are at right angles. The cleavages do not tend to have a strong control on fragment orientation.

TWINNING: Simple contact twins on $\{101\}$ are rare and should produce elbow-shaped twins.

OPTICAL ORIENTATION: In longitudinal section, the extinction is parallel to both the trace of cleavage and the length of crystals, and the elongation is length slow.

INDICES OF REFRACTION AND BIREFRINGENCE: The indices of refraction probably increase with the substitution of Th for Y. The birefringence is very strong and yields high-order white interference colors, which are often masked by the color of the mineral. Basal sections also usually produce high-order interference colors, due to the somewhat converging nature of the light coming from the condensor.

INTERFERENCE FIGURE: Interference figures are often difficult to obtain, due to small grain size. Basal sections on sufficiently large grains produce uniaxial positive optic axis figures with numerous isochromes, and isogyres that broaden substantially toward the edge of the field of view. Fragments lying on the $\{110\}$ cleavage yield flash figures.

ALTERATION: Not readily altered.

DISTINGUISHING FEATURES: In small grains, it may be difficult to distinguish xenotime from zircon, mona-

zite, titanite, and rutile. Zircon has higher indices and lower birefringence. Monazite is biaxial and has inclined extinction and lower birefringence. Titanite has a different habit, is biaxial, and has higher indices of refraction. Rutile has higher indices and is darker colored. Xenotime may be included in ferromagnesian minerals like biotite or hornblende and may produce dark halos in the including mineral due to the radioactive bombardment from decay of Th or other radioactive elements in the xenotime.

OCCURRENCE: Xenotime is not uncommon as an accessory mineral in granite, syenite, granodiorite, pegmatite, and related rocks. It is often incorrectly identified as zircon. In metamorphic rocks, it may be found in mica schists and gneiss or, less commonly, in marble. It also is found as detrital grains in placer deposits, beach sands, and as a heavy mineral in sandstone and related clastic sedimentary rocks.

REFERENCES

Friedman, G. M., 1959, Identification of carbonate minerals by staining methods: Journal of Sedimentary Petrology, v. 29, p. 87–97.

Kennedy, G. C., 1947, Charts for correlation of optical properties with chemical composition of some common rock forming minerals: American Mineralogist, v. 32, p. 561–573

Loupekine, I. S., 1947, Graphical derivation of refractive index ϵ for the trigonal carbonates: American Mineralogist, v. 32, p. 502–507

McConnell, D., and Gruner, J. W. 1940, The problem of the carbonate-apatites. III. Carbonate-apatite from Magnet Cove, Arkansas: American Mineralogist, v. 25, p. 157–167

Taborszky, F. K., 1972, Chemismus und Optik der apatite: Neues Jahrbuch fuer Mineralogie Monatshefte, v. 5, p. 79–91

Warne, S. St. J., 1962, A quick field or laboratory staining scheme for the differentiation of the major carbonate minerals: Journal of Sedimentary Petrology, v. 32, p. 29–38.

Wolfe, K. H., Easton, A. J., and Warne, S., 1967, Techniques of examining and analysing carbonate skeletons, minerals, and rocks. In: Chilingar, G. V., Bissell, H. J., and Fairbridge, R. W., eds. Carbonate Rocks. Developments in Sedimentology, 9B: Elsevier, Amsterdam, p. 253–342

11

Orthosilicates

Olivine

$(Fe,Mg)_2SiO_4$
Orthorhombic
Biaxial (+ or −)
$n_\alpha = 1.636–1.827$
$n_\beta = 1.651–1.869$
$n_\gamma = 1.669–1.879$
$\delta = 0.033–0.052$
$2V_x = 46–98°$

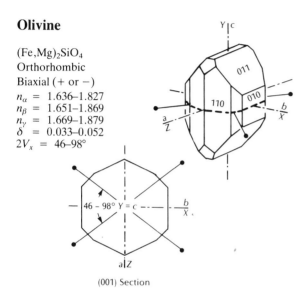

(001) Section

RELIEF IN THIN SECTION: High positive relief.

COMPOSITION AND STRUCTURE: The structure of olivine consists of isolated silicon tetrahedra bonded laterally to divalent cations. There is complete solid solution between *forsterite* (Fo, Mg_2SiO_4), and *fayalite* (Fa, Fe_2SiO_4), and most olivine samples have compositions intermediate between these two end members. There also is complete solid solution to *tephroite* (Mn_2SiO_4). Mn-bearing olivine also usually is relatively Fe rich, but most common olivine has relatively little Mn. Small amounts of Zn, Ca, Ni, Cr, or Al also may substitute for the divalent cation.

PHYSICAL PROPERTIES: H = $6\frac{1}{2}$–7; G = 3.22–4.39. Specific gravity increases and hardness decreases with increasing Fe. Usually shades of olive or yellowish green in hand sample. Oxidized samples may be reddish to nearly black. White streak; vitreous luster.

COLOR AND PLEOCHROISM: Usually colorless to pale yellow in thin section or grain mount; darker colors correspond to higher iron content. Larger fragments may be somewhat darker colored. Altered and oxidized fayalite-rich samples may be reddish to nearly black due to the presence of limonite or other impurities. Pleochroism of fayalite-rich samples is $X = Z$ = pale yellow, Y = orangish, yellowish, or reddish brown.

FORM: Figure 11.1. In most metamorphic and intrusive igneous rocks, olivine tends to form subequant, anhedral grains or aggregates of grains. Euhedral crystals are more common in volcanic rocks and sections usually are equidimensional, or slightly elongate parallel to c with six or eight sides.

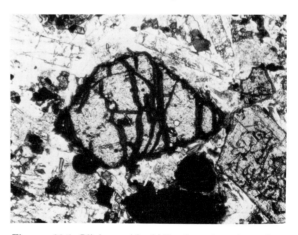

Figure 11.1 Olivine with iddingsite alteration along cracks. The tabular crystals are plagioclase and the blocky crystal to the lower right is augite. Field of view is 1.7 mm wide.

CLEAVAGE: Olivine has poor cleavage on {010} and {110}, but these are not usually seen in thin section nor do they control fragment orientation in grain mounts. Grains in thin section usually display a distinctive irregular fracture pattern.

TWINNING: Twinning is not common but may be present as simple or multiple twins with diffuse boundaries on composition planes {100}, {011}, or {012}.

OPTICAL ORIENTATION: $X = b$, $Y = c$, $Z = a$, optic plane = (001). Elongate crystals display parallel extinction and may be either length fast or length slow depending on how they are cut.

INDICES OF REFRACTION AND BIREFRINGENCE: Indices and birefringence vary continuously between forsterite and fayalite (Figure 11.2). Olivine grains from many igneous rocks are compositionally zoned with Mg-rich cores and more Fe-rich rims. The compositional zoning can sometimes be recognized by a change of birefringence. In sections cut parallel to (001) the rim will have higher birefringence than the core, and in section cut parallel to (100) the rim will have lower birefringence than the core. Interference colors in standard thin sections range up to the vivid colors of the third order.

Indices of refraction can be used to estimate composition but caution is required because olivine is often compositionally zoned and there is no assurance that all grains of olivine in a grain mount are the same composition. Other techniques employing a combination of optical data and X-ray data also have been devised, but these techniques suffer from the same problem. If the composition of olivine is needed with any accuracy, it is probably best obtained by electron microprobe techniques.

The indices of tephroite are $n_\alpha = 1.770$, $n_\beta = 1.807$, $n_\gamma = 1.817$, and $\delta = 0.047$. It is optically negative with $2V_x = 70°$. The optical properties vary continuously between fayalite and tephroite. For Fe:Mn = 1, the indices are $n_\alpha = 1.815$, $n_\beta = 1.853$, $n_\gamma = 1.867$, with $2V_x = 44°$. Substituting Mg for Mn results in a rather rapid decrease in the indices (Mossman and Pawson, 1976).

INTERFERENCE FIGURE: $2V$ varies continuously with composition (Figure 11.1). Between Fo_{100} and Fo_{85}, olivine is biaxial positive with $2V_z = 82–90°$. Between Fo_{85} and Fo_0, olivine is negative with $2V_x = 90–46°$. The $2V$ angle and optic sign can be used to provide a rough estimate of composition. Olivine from volcanic rocks may have somewhat smaller $2V$ than olivine of the same composition from plutonic rocks. Most common olivine in igneous rocks is relatively Mg rich, thus $2V$ is large, and optic axis figures provide the most reliable means of determining sign and $2V$. Optic axis dispersion is weak with $r > v$ if optically negative, and $v > r$ if positive.

ALTERATION: Olivine commonly alters to a number of distinctive products including iddingsite and chlorophaeite, which are combinations of various minerals and and are not usually subject to rigorous identification. Olivine also may alter to serpentine. The alteration usually progresses from the periphery and along cracks inward.

Iddingsite is a name given to fine-grained reddish- or yellowish-brown material that consists of goethite, clay, chlorite, quartz, talc, and other minerals. Despite the fact that iddingsite is an aggregate of

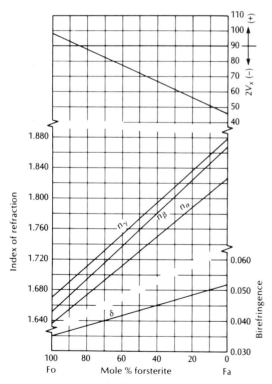

Figure 11.2 Indices, birefringence (δ), and $2V$ for olivine. After Poldervaart (1950), Deer and others (1982), and Laskowski and Scotford (1980).

several minerals, it may display birefringence of 0.04 or 0.05 with indices $n_{low} = 1.70$ and $n_{hi} \cong 1.75$ with considerable variability. The interference colors are usually well masked by the strong color of the material.

Chlorophaeite is an essentially isotropic material that is usually orangish or greenish. It also is a mixture and appears to be composed of limonite and chlorite or serpentine with other low birefringence silicates. It displays an index in the range 1.5 to 1.6.

Magnesium-rich olivine often alters to serpentine, which may either be chrysotile or antigorite. Peridotites and other rocks rich in olivine and pyroxene may be partially or completely altered to serpentinite, which consists of serpentine plus carbonates, Fe-Mg silicates, and usually some disseminated magnetite.

Olivine from igneous rocks is often mantled by pyroxene or hornblende as the result of normal magmatic reactions between the olivine and melt.

DISTINGUISHING FEATURES: Olivine is recognized by its high birefringence, distinctive fracturing, lack of cleavage, and alteration products. Clinopyroxenes show somewhat lower birefringence, have recognizable cleavage, often show lamellar twinning, and have inclined extinction. Epidote has a cleavage, inclined extinction, is optically negative, and may display a patchy, pistachio-green color.

OCCURRENCE: Relatively pure forsterite is generally restricted to metamorphosed siliceous dolomites where it is associated with calcite, dolomite, diopside, epidote group minerals, grossularite, tremolite, and related minerals.

Olivine with intermediate Fe–Mg composition is common in many mafic and ultramafic volcanic and plutonic igneous rocks. It can be found in basalt, gabbro, peridotite, pyroxenite, and is the principal constituent of dunite.

Relatively iron rich olivine is occasionally found in less mafic rocks such as syenite, nepheline syenite, phonolite, trachyte, andesite, and dacite. Fayalite is infrequently found in certain granites and rhyolites, in iron-rich metamorphosed sediments, and in vesicles and other cavities in siliceous volcanic rocks.

Mn-rich olivine in the tephroite–fayalite series is found in iron–manganese deposits, skarns, and in metamorphosed Mn-rich sediments.

Monticellite

Ca(Mg,Fe)SiO$_4$
Orthorhombic
Biaxial (−)
n_α = 1.638–1.654
n_β = 1.646–1.664
n_γ = 1.650–1.674
δ = 0.012–0.020
$2V_x$ = 69–88°

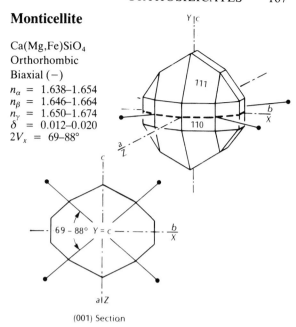

(001) Section

RELIEF IN THIN SECTION: Moderately high positive relief.

COMPOSITION AND STRUCTURE: Monticellite has the same structure as olivine with isolated silicon tetrahedra bonded laterally through octahedrally coordinated divalent cations, half of which are Ca^{2+}. Most monticellite comes close to the ideal composition CaMgSiO$_4$, but complete solid solution to synthetic *kirschsteinite* (CaFeSiO$_4$) is possible and some monticellite contains up to 20 mole percent Fe substituting for Mg. It is also possible to substitute Mn for Mg, and complete solid solution to *glaucochroite* (CaMnSiO$_4$) is probable although Mn is generally not found in abundance in monticellite.

PHYSICAL PROPERTIES: H = 5$\frac{1}{2}$; G = 3.05–3.27; colorless to gray in hand sample; white streak; vitreous luster.

COLOR: Colorless in thin section and grain mount.

FORM: Figure 11.3. Usually as anhedral, equant or irregular grains. Crystals are elongate along c and yield equant to elongate polygonal cross sections with eight or more faces.

CLEAVAGE: There is a single poor cleavage on {010} that does not significantly influence fragment orientation.

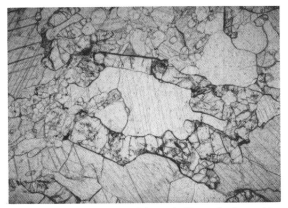

Figure 11.3 Irregular high-relief monticellite grains in marble. Field of view is 5 mm wide.

DISTINGUISHING FEATURES: Monticellite has lower birefringence than olivine. The clinopyroxenes—such as augite—display cleavage, have smaller $2V$, and are usually optically positive.

OCCURRENCE: Monticellite is found in contact metamorphosed or metasomatized siliceous dolomitic limestones, most commonly adjacent to granitic intrusions. It may be associated with calcite, dolomite, diopside, tremolite, forsterite, wollastonite, and related calc-silicate minerals. Monticellite also may be found in mafic and ultramafic igneous rocks where it may be found mantling grains or olivine, or less commonly as isolated grains.

TWINNING: Monticellite may have cyclic twins on {031} which, if euhedral, appear as six-pointed stars in cross section.

OPTICAL ORIENTATION: $X = b$, $Y = c$, $Z = a$, optic plane = (001). Elongate cross sections through euhedral crystals may be either length slow or length fast depending on how they are cut. The trace of cleavage is length slow but the cleavage is usually not visible.

INDICES OF REFRACTION AND BIREFRINGENCE: The indices for pure synthetic monticellite are $n_\alpha = 1.639$, $n_\beta = 1.646$, and $n_\gamma = 1.653$. The addition of Fe or Mn increases the indices and birefringence in an approximately linear manner. The indices of kirschsteinite are $n_\alpha = 1.674$, $n_\beta = 1.694$, $n_\gamma = 1.735$, and $2V_x = 50°$. The indices of glaucochroite are $n_\alpha = 1.686$, $n_\beta = 1.723$, $n_\gamma = 1.736$, and $2V_x = 60°$. The birefringence of most monticellite is in the vicinity of 0.012 to 0.015, so interference colors in thin section range up to first-order red.

INTERFERENCE FIGURE: Interference figures are biaxial negative with $2V_x = 69–88°$. $2V_x$ for relatively pure monticellite is generally greater than 80° and decreases with increasing Fe or Mn. Optic axis dispersion is $r > v$.

ALTERATION: Monticellite may be altered to serpentine or to other calc-silicates such as tremolite, calcic clinopyroxene, and related minerals.

Humite Group

$x(Mg_2SiO_4)\cdot Mg(OH,F)_2$

Norbergite
$Mg_2SiO_4\cdot Mg(OH,F)_2$
Orthorhombic
Biaxial (+)
n_α = 1.558–1.567
n_β = 1.563–1.579
n_γ = 1.582–1.593
δ = 0.024–0.027
$2V_z$ = 44–50°

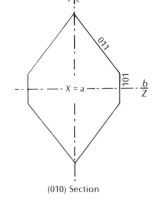

(010) Section

Chondrodite
$2(Mg_2SiO_4)\cdot Mg(OH,F)_2$
Monoclinic
$\angle\beta = 109°$
Biaxial (+)
n_α = 1.592–1.617
n_β = 1.602–1.635
n_γ = 1.621–1.646
δ = 0.028–0.034
$2V_z$ = 71–85°

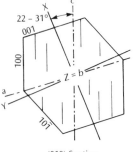

(010) Section

Humite

$3(Mg_2SiO_4)\cdot Mg(OH,F)_2$

Orthorhombic

Biaxial (+)

n_α = 1.607–1.643
n_β = 1.619–1.653
n_γ = 1.639–1.675
δ = 0.029–0.036
$2V_z$ = 65–84°

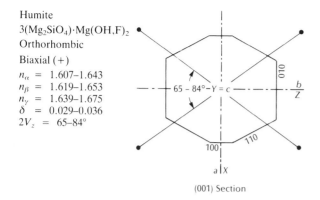

(001) Section

Clinohumite

$4(Mg_2SiO_4)\cdot Mg(OH,F)_2$

Monoclinic

$\angle\beta$ = 101°

Biaxial (+)

n_α = 1.628–1.668
n_β = 1.641–1.679
n_γ = 1.662–1.700
δ = 0.028–0.041
$2V_z$ = 73–76°

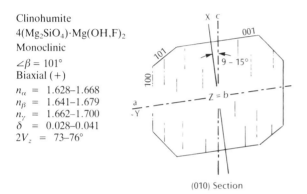

(010) Section

RELIEF IN THIN SECTION: Low to moderately high positive relief.

COMPOSITION AND STRUCTURE: The structure is similar to olivine and consists of a nearly hexagonal close-packed framework of oxygen, hydroxyl, and fluorine. The Si are in tetrahedral sites and the Mg are in octahedral sites. The distribution of the cations is arranged so that the structure consists of tabular sheets with olivine structure that are parallel to (100). The OH⁻ and F⁻ are located on the front and back surfaces of the sheets, which are bonded to adjacent sheets through octahedrally coordinated Mg^{2+}. In norbergite, the olivine sheets are one silicon tetrahedra thick, and in chondrodite, humite, and clinohumite, the olivine sheets are two, three, and four tetrahedra thick, respectively. The major substitution is between F⁻ and (OH)⁻ and substantial amounts of fluorine are always present. There is only limited substitution of Fe or Ti for Mg. Mn, Ca, Ni, and Zn may be present in minor amounts.

PHYSICAL PROPERTIES: H = 6–6½; G = 3.15–3.35; brown, yellow, orange, or red in hand sample; white streak; vitreous luster.

COLOR AND PLEOCHROISM: Colorless, pale yellow, yellow, or orangish yellow in thin section and grain mount. The darker color generally corresponds to higher Fe or Ti content. Colored varieties show weak pleochroism with $X > Y = Z$: X = pale yellow, orangish yellow, or yellow, $Y = Z$ = paler yellow, pale orangish yellow, or colorless.

FORM: Figure 11.4. Usually found as rounded or irregular shaped grains, or anhedral masses. Crystals have diverse habits but are commonly platy parallel to {100}, {010}, or {001}.

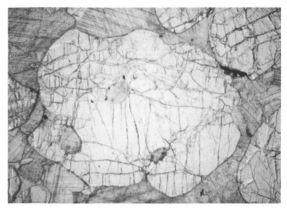

Figure 11.4 Rounded high-relief chondrodite in marble. Field of view is 5 mm wide.

CLEAVAGE: Norbergite has no cleavage, humite has fair cleavage on {001}, and clinohumite and chondrodite have a poor cleavage on {100}. The cleavage does not tend to control fragment orientation and is not usually evident in thin section.

TWINNING: Simple or multiple twins are common in clinohumite and chondrodite and are sometimes found in humite.

OPTICAL ORIENTATION: Norbergite and humite are orthorhombic with $X = a$, $Y = c$, and $Z = b$, optic plane = (001). Extinction is parallel to the trace of the cleavage, which is generally not visible, however. Elongate sections through euhedral crystals may be either length fast or length slow.

Chondrodite is monoclinic with $X \wedge c = +22$ to $+31°$, $Y \wedge a = -3$ to $-12°$, $Z = b$, and the optic plane is perpendicular to (010). Maximum extinction angle to the trace of the cleavage is 22 to 31°, but the cleavage is generally not visible. Elongate sections may be either length fast or length slow.

Clinohumite is monoclinic with $X \wedge c = +9$ to $+15°$, $Y \wedge a = +2$ to $-4°$, $Z = b$, and the optic plane is perpendicular to (010). Maximum extinction angle to the trace of the cleavage is 9 to 15° but the cleavage is generally not visible. Elongate sections may be either length fast or length slow.

INDICES OF REFRACTION AND BIREFRINGENCE: There is a general increase in the indices of refraction from norbergite to clinohumite but the indices overlap and do not provide a reliable means of distinguishing between the members of the group except for norbergite, which is quite rare. Indices for each member of the group increase with substitution of Fe^{2+}, Mn^{2+}, or Ti^{4+} for Mg^{2+}, and $(OH)^-$ for F^-.

INTERFERENCE FIGURE: All four members of the group are biaxial positive. The values of $2V_z$ overlap and, with the exception of norbergite, do not provide a reliable means of distinguishing among the members of the group. The optic axis dispersion in all members is weak, $r > v$.

ALTERATION: All members of the group may alter to serpentine or chlorite.

DISTINGUISHING FEATURES: Clear or lightly colored varieties resemble olivine, but most olivine is optically negative and the humite group has lower indices and birefringence. Forsterite, with which the humite group may be associated, is optically positive but has larger $2V$ and higher birefringence. Staurolite has similar color and pleochroism but has higher indices and lower birefringence and is usually found in mica schists. The different members of the group, with the exception of norbergite, cannot be reliably distinguished based on their optical properties. If the cleavage is visible, the inclined extinction may distinguish clinohumite and chondrodite from humite and norbergite.

OCCURRENCE: The humite group has rather restricted occurrences. It is found in contact metamorphosed or metasomatized limestones and dolomites, usually adjacent to granitic intrusions. It is commonly associated with Ca–Mg silicates such as tremolite, wollastonite, grossular, monticellite, and forsterite as well as calcite, dolomite, and related minerals commonly found in skarns and metamorphosed carbonates. It is not uncommon to find members of the humite group intergrown in a lamellar fashion, or to find lamellar intergrowths with olivine or monticellite.

Members of the humite group also are found in ultramafic igneous rocks such as peridotite and kimberlite, and in carbonatites, serpentinites, and talc schists.

Garnet Group

$X_3Y_2(SiO_4)_3$
Isometric
Isotropic

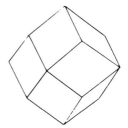

RELIEF IN THIN SECTION: High positive relief.

COMPOSITION AND STRUCTURE: The structure of garnet consists of isolated silicon tetrahedra bonded laterally to octahedrally coordinated trivalent Y cations with divalent X cations in eightfold coordination in the interstices between the tetrahedra and octahedra. The common garnets are conventionally placed into two groups. Those with Al^{3+} as the Y cation are the *pyralspite* group and include *py*rope, *al*mandine and *spes*sartine. The *ugrandite* group has Ca^{2+} as the X cation and includes *u*varovite, *gr*ossular, and *and*radite. Within the pyralspite group there can be extensive solid solution among all members, although relatively few natural garnets are found with compositions intermediate between pyrope and spessartine. Extensive solid solution is also possible among members of the ugrandite group. Compositions intermediate between grossular or andradite and members of the pyralspite group are known but are not common.

Hydrogrossular has essentially the same structure as the other garnets except that a Si^{4+} has been replaced by $4H^+$ with the hydrogens bonded to each of the four oxygens surrounding the vacant tetrahedral site. *Hibschite* [$Ca_3Al_2(H_4O_4)_3$], in which all of the silicon has been replaced by hydrogen, has been synthesized. Most natural hydrogrossular has a composition intermediate between grossular and hibschite.

Many garnets are compositionally zoned.

PHYSICAL PROPERTIES: H = 6–7$\frac{1}{2}$; G = see below. The color of the pyralspite group in hand sample is usually deep wine red or reddish brown but also may range from black to orange, dark pink, or brown. Grossular may range from nearly colorless to green, yellow, pink, or brown. Andradite ranges from yellow to green, brown, or black, and uvarovite is usually a rich emerald green. Hydrogrossular is usually white, gray, pink, pale brown, or pale green. All have a white streak and vitreous luster.

COLOR: The color in thin section or grain mount is commonly colorless or a pale version of the hand-sample color. Individual crystals may display color variation associated with compositional zonation but do not display pleochroism because they are isotropic or nearly so.

FORM: Figure 11.5. Garnets commonly occur as euhedral to subhedral dodecahedral {110} or trapezohedral {112} crystals, which in thin section yield six- or eight-sided cross sections. Garnet also occurs as granular or irregular masses.

CLEAVAGE: Garnets do not have cleavage and usually show irregular fractures. They may occasionally show a parting parallel to the dodecahedral {110} crystal faces.

TWINNING: Isotropic garnet does not display twinning, but anomalously birefringent varieties often

Figure 11.5 Garnet. Field of view is 1.7 mm wide.

display a twinning pattern consisting of six or more wedge-shaped sectors radiating from the center of the crystal.

OPTICAL ORIENTATION: Birefringent varieties show quite variable optical orientations.

INDICES OF REFRACTION AND BIREFRINGENCE: The indices of refraction vary in a systematic manner with composition, but the ranges of indices overlap, so index of refraction by itself cannot be used to estimate composition. A number of moderately successful schemes have been devised to use additional data such as specific gravity and unit cell dimension to obtain an estimate of composition (e.g., Winchell, 1958). Figure 11.6a shows one such chart that can be used for garnets in the pyralspite group, provided they do not deviate significantly from the pyrope–almandine–spessartine composition plane (e.g., by containing calcium). The garnet from most skarns and metamorphosed carbonate rocks is intermediate between grossular and andradite, and Figure 11.6b

Name	Composition	Index (Pure synthetic)	Index (Normal range)	G (calc.)
Pyrope	$Mg_3Al_2(SiO_4)_3$	1.714	1.720–1.770	3.582
Almandine	$Fe_3Al_2(SiO_4)_3$	1.830	1.770–1.820	4.315
Spessartine	$Mn_3Al_2(SiO_4)_3$	1.800	1.790–1.810	4.197
Grossular	$Ca_3Al_2(SiO_4)_3$	1.734	1.735–1.770	3.594
Andradite	$Ca_3Fe_2(SiO_4)_3$	1.887	1.850–1.890	3.859
Uvarovite	$Ca_3Cr_2(SiO_4)_3$	1.865	1.838–1.870	3.850
Hydrogrossular	$Ca_3Al_2(SiO_4)_{3-x}(OH)_{4x}$	—	1.675–1.734	3.1–3.6

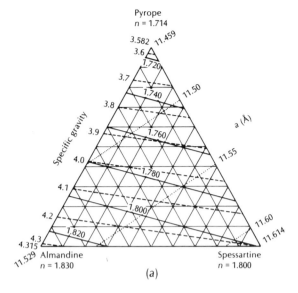

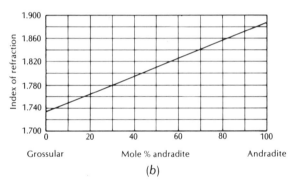

Figure 11.6 Optical properties of common garnets. (*a*) The pyralspite group. The unit cell dimension (*a*) and specific gravity also are included. (*b*) The grossular-andradite series. Adapted from Winchell (1958).

INTERFERENCE FIGURE: Anomalously birefringent varieties may be either uniaxial or biaxial with $2V$ up to 90°.

ALTERATION: Garnet may alter to chlorite or occasionally other minerals such as hornblende, epidote, or iron oxides. It is commonly involved in a variety of mineral reactions in metamorphic rocks and may display reaction or replacement textures with other minerals.

DISTINGUISHING FEATURES: Garnet is usually identified by its isotropic character, high relief, and crystal shape. The pyralspite group rarely displays birefringence, but members of the ugrandite group from metamorphosed carbonate rocks fairly frequently show weak birefringence. Spinel is usually distinctly green or brown and occurs as octahedrons. Basal sections of apatite grains may resemble garnet but they have lower relief and yield uniaxial figures. In grain mount, garnet and spinel may be difficult to distinguish, but garnet generally has higher indices and lacks the more intense color of spinel. Hydrogrossular has lower indices and specific gravity than grossular.

OCCURRENCE: The occurrence and associated minerals are usually a fairly reliable guide to which variety of garnet is present.

Pyrope is uncommon and is usually found in ultramafic igneous rocks such as peridotite and in serpentinites derived from them, and occasionally occurs in high-grade, Mg-rich metamorphic rocks.

Almandine is the common garnet in most mica schists and gneisses. It also is found in pegmatites, granite, and felsic volcanic rocks.

Spessartine or compositions intermediate between almandine and spessartine is most commonly found in granite, pegmatite, and felsic igneous rock. Spessartine also is found in Mn-rich metamorphic rocks.

Grossular is generally restricted to skarns and other metamorphosed carbonate-bearing rocks. It is usually associated with calcite, dolomite, wollastonite, tremolite, and epidote.

Andradite is found in skarns and other metamorphosed carbonate-bearing rocks. Whether grossular or andradite is present depends on the availability of Al^{3+} and Fe^{3+} during metamorphism. Andradite also is found in hydrothermal vein deposits in carbonate rocks and in such alkaline igneous rocks as nepheline syenite and their volcanic equivalents.

can be used to obtain a rough estimate of the relative amounts of the grossular and andradite components. The reader is cautioned, however, that if knowledge of the garnet composition is important, other techniques such as electron microprobe analysis should be used to obtain it, particularly because many garnets display significant compositional zoning.

Members of the ugrandite group may be anomalously birefringent and show lower first-order colors in thin section.

Uvarovite is quite rare and is usually found in chromite-bearing igneous rocks such as peridotite, or their serpentinized equivalent.

Andalusite

Al_2SiO_5
Orthorhombic
Biaxial (−)
$n_\alpha = 1.629–1.640$
$n_\beta = 1.633–1.644$
$n_\gamma = 1.638–1.650$
$\delta = 0.009–0.013$
$2V_x = 71–88°$

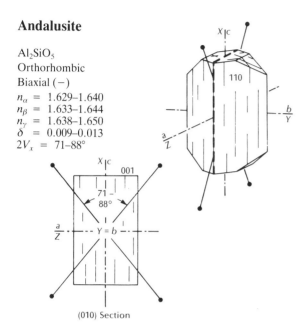

RELIEF IN THIN SECTION: Moderately high positive relief.

COMPOSITION AND STRUCTURE: The structure of andalusite consists of edge-sharing chains of aluminum octahedra that extend parallel to the c crystal axis. The chains are joined laterally through silicon tetrahedra and aluminum, which is coordinated with five oxygens. Most andalusite is relatively pure Al_2SiO_5, although a significant amount of Mn^{3+} and Fe^{3+} may substitute for the octahedral aluminum, and extensive solid solution to *kanonaite* ($MnAlSiO_5$) is possible. The term *viridine* has been used for andalusite containing substantial amounts of both Fe^{3+} and Mn^{3+}.

PHYSICAL PROPERTIES: H = $6\frac{1}{2}$–$7\frac{1}{2}$; G = 3.13–3.16; commonly white, gray, pink, or rose red in hand sample, also violet, yellow, or green; white streak; vitreous luster.

COLOR AND PLEOCHROISM: Usually colorless in thin section or grain mount; less commonly pink, reddish pink, or slightly greenish. Colored varieties show weak pleochroism, usually X = pink, $Y = Z$ = colorless. Other colored varieties are pleochroic in shades of green and yellow. Color is darker for increasing amounts of Mn^{3+} and Fe^{3+}.

FORM: Figure 11.7. Crystals are usually elongate prisms with a nearly square cross section. A variety called chiastolite contains dark inclusions that form a cross along the diagonals of the prism. Andalusite also forms anhedral grains or highly irregular masses. Numerous inclusions of quartz, fine opaques, or other minerals are common.

CLEAVAGE: Andalusite has two good prismatic {110} cleavages that are nearly at right angles and that are

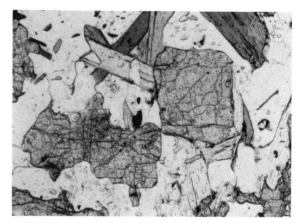

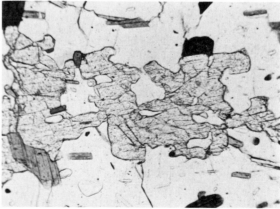

Figure 11.7 Andalusite. (*Top*) Blocky basal sections. (*Bottom*) Longitudinal section of highly irregular porphyroblast. Field of view is 1.7 mm wide for both photographs.

parallel to the prism faces in cross sections. The cleavage tends to control fragment orientation in grain mounts. Cleavage on {100} is poor.

TWINNING: Twinning on {101} is rare.

OPTICAL ORIENTATION: $X = c$, $Y = b$, $Z = a$, optic plane = (010). Extinction is parallel to the trace of cleavage and the long dimension of elongate crystals. In basal sections, extinction is symmetrical. The fast ray vibrates parallel to the trace of cleavage and long dimension of grains in elongate sections for most andalusite. However, samples that contain more than about $6\frac{1}{2}$ mole percent $Mn^{3+} + Fe^{3+}$ substituting for aluminum are length slow and have optic orientation $X = a$, $Y = b$, $Z = c$, optic plane (010), and are biaxial positive.

INDICES OF REFRACTION AND BIREFRINGENCE: The indices of refraction increase with substitution of Mn^{3+} or Fe^{3+} (Figure 11.8). Birefringence decreases to nearly zero at about $6\frac{1}{2}$ mole percent $(Mn^{3+} + Fe^{3+})$ and increases above that point. Most common andalusite has birefringence in the range 0.009 to 0.013, which gives it first-order gray and white interference colors in standard thin sections.

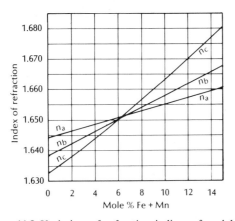

Figure 11.8 Variation of refractive indices of andalusite with mole percent (Fe + Mn). The indices n_a, n_b, and n_c are for light vibrating along the a, b, and c crystal axes, respectively. From Gunter and Bloss (1982). Used with permission of the Mineralogical Society of America.

INTERFERENCE FIGURE: Most andalusite is biaxial negative with $2V_x$ greater than 80°. Basal sections yield acute bisectrix figures with the melatopes outside the field of view. Cleavage fragments produce strongly off-center obtuse bisectrix figures. Optic axis dispersion is weak with $v > r$, occasionally $r > v$. Andalusite with more than about $6\frac{1}{2}$ mole percent $(Fe^{3+} + Mn^{3+})$ is biaxial positive with $2V_z$ between 65 and 85°. In the composition range where the transition from optically negative to positive is made, $2V$ could be quite variable.

ALTERATION: Andalusite may alter to sericite, which is mostly fine-grained white mica. Andalusite also may be altered to chlorite or other phyllosilicates. It is commonly involved in metamorphic mineral reactions and may show reaction relations with minerals such as cordierite, staurolite, garnet, sillimanite and kyanite.

DISTINGUISHING FEATURES: Moderately high relief, large 2V, parallel extinction, and length fast character distinguish andalusite. Sillimanite is length slow, biaxial positive, and has higher birefringence and a slender prismatic to fibrous habit. Kyanite has inclined extinction and higher birefringence. Staurolite has a distinctly honey-yellow color and higher indices. Orthopyroxene may have higher birefringence and is length slow. Apatite is uniaxial, has lower birefringence, and may have slightly anomalous interference colors.

OCCURRENCE: Andalusite is a common mineral in contact and regional metamorphic rocks such as hornfels and mica schists. It may be associated with cordierite, staurolite, garnet, sillimanite, kyanite, chlorite, muscovite, biotite, and plagioclase. It is less commonly found in granitic pegmatites and granite.

Sillimanite

Al_2SiO_5
Orthorhombic
Biaxial (+)
n_α = 1.653–1.661
n_β = 1.657–1.662
n_γ = 1.672–1.683
δ = 0.018–0.022
$2V_z$ = 20–30°

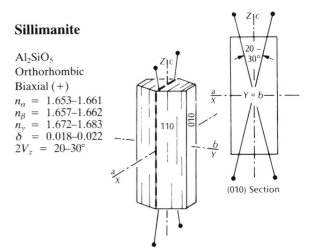

RELIEF IN THIN SECTION: High positive relief.

COMPOSITION AND STRUCTURE: The structure of sillimanite consists of octahedra containing aluminum that are joined along edges to form chains parallel to the *c* axis. The chains are bonded laterally through silicon and aluminum tetrahedra. Most sillimanite is relatively pure Al_2SiO_5, although minor amounts of Fe^{3+}, Cr^{3+}, or Ti^{4+} may be present.

PHYSICAL PROPERTIES: H = $6\frac{1}{2}$–$7\frac{1}{2}$; G = 3.23–3.27; usually colorless or white but also yellow, brown, or blue in hand sample; white streak; vitreous luster. The colored varieties usually contain iron or chromium.

COLOR AND PLEOCHROISM: Usually colorless in thin section and grain mount, although mats of fibrolite may be pale brown. Thick fragments or the rare, more darkly colored varieties may be pleochroic as follows:

Hand-sample color	X	Y	Z
Yellow	Yellow	Green-yellow	Colorless
Brown	Colorless or pale yellow	Colorless or pale yellow	Violet brown
Blue	Colorless or pale yellow	Colorless or pale yellow	Blue

The color may be patchy.

FORM: Figure 11.9. Sillimanite commonly occurs as slender prismatic crystals or as fine fibrous crystals called fibrolite. Fibrolite commonly forms radiating, swirled, or matted aggregates. Cross sections through crystals are usually more or less diamond shaped.

CLEAVAGE: Sillimanite has a single good cleavage {010} parallel to the length of crystals. Elongate crystals are commonly fractured across their length. The cleavage is usually visible in basal sections parallel to one diagonal of the diamond.

TWINNING: None reported.

OPTICAL ORIENTATION: $X = a$, $Y = b$, $Z = c$, optic plane = (010). Elongate grains show parallel extinction and are length slow.

INDICES OF REFRACTION AND BIREFRINGENCE: The indices and birefringence vary within a relatively narrow range. In thin section, interference colors

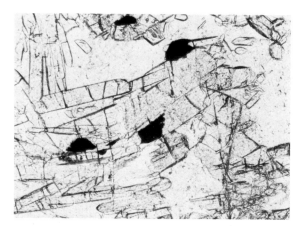

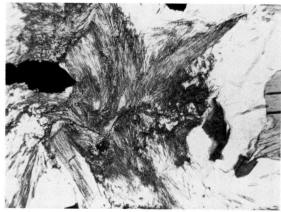

Figure 11.9 Sillimanite. (*Top*) Elongate prismatic crystals. (*Bottom*) Fibrolite. Field of view is 0.9 mm (top) and 1.7 mm (bottom) wide.

range up to lower second order, although most grains will show first-order colors. Fine fibers of fibrolite rarely show higher than first-order yellow colors.

INTERFERENCE FIGURE: If they are large enough, basal sections yield biaxial positive figures with $2V_z$ between 20 and 30°. Fragments lying on the {010} cleavage yield flash figures. Optic axis dispersion is strong with $r > v$.

ALTERATION: Sillimanite may alter to sericite, which consists predominantly of fine-grained white mica. Sillimanite is involved in a variety of metamorphic mineral reactions and may show reaction textures with associated minerals.

DISTINGUISHING FEATURES: Sillimanite is distinguished by high relief, moderate birefringence, parallel extinction, and habit. Kyanite shows inclined extinction and is optically negative with a large 2V, and andalusite is length fast and optically negative with a large 2V. Mullite and sillimanite cannot be distinguished optically, but mullite is quite rare.

OCCURRENCE: Sillimanite is a common mineral in medium- and high-grade mica schist, gneiss, hornfels, and related rocks. It is commonly associated with kyanite, andalusite, staurolite, muscovite, biotite, K-feldspar, cordierite, corundum, and garnet. Sillimanite is not uncommon as an accessory mineral in granitic rocks where the fine needles are sometimes mistakenly identified as rutile. It also is rarely found in pegmatites and quartz veins and can locally be an important detrital mineral.

Mullite $[Al(Al_{1+2x}Si_{1-2x})O_{5-x}]$ is closely related to sillimanite and can be distinguished only with X-ray diffraction examination. Mullite is very rare in nature so there is not a significant likelihood of confusing the two in common rocks. Mullite is restricted to very high temperature porcellainite hornfels in the contact zone adjacent to mafic intrusions or in pelitic inclusions (buchites) in those rocks. It also is commonly found in furnace slags and in a variety of ceramic and refractory materials.

RELIEF IN THIN SECTION: High positive relief.

COMPOSITION AND STRUCTURE: The structure of kyanite consists of chains of edge-sharing aluminum octahedra that are bonded laterally through silicon tetrahedra and aluminum in octahedral coordination. Kyanite is a relatively dense structure and the oxygens are nearly in cubic close packing. The aluminum is distributed so that the octahedral chains run parallel to the diagonal across one of the cubic faces. Most kyanite is relatively pure Al_2SiO_5, although minor substitution of Fe^{3+}, Ti^{4+}, or Cr^{3+} may be present.

PHYSICAL PROPERTIES: H = 4–5 parallel to c, $\sim 7\frac{1}{2}$ at right angles to c; G = 3.53–3.67; patchy blue, gray, or white in hand sample, less commonly light green; white streak; vitreous to pearly luster.

COLOR AND PLEOCHROISM: Usually colorless in thin section although some may show light blue colors that are often patchy. Color is usually more distinctive in thicker fragments used for grain mounts. Pleochroism is usually weak with X = colorless, Y = light violet blue, Z = light cobalt blue ($X < Y < Z$).

FORM: Figure 11.10. Kyanite commonly forms elongate bladed or columnar crystals that may appear to be bent. It is rarely fibrous.

Kyanite

Al_2SiO_5
Triclinic
$\angle\alpha = 90°05'$
$\angle\beta = 101°02'$
$\angle\gamma = 105°44'$

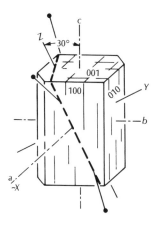

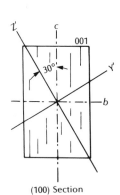

(100) Section

Biaxial (−)
$n_\alpha = 1.710–1.718$
$n_\beta = 1.719–1.725$
$n_\gamma = 1.724–1.734$
$\delta = 0.012–0.016$
$2V_x = 78–84°$

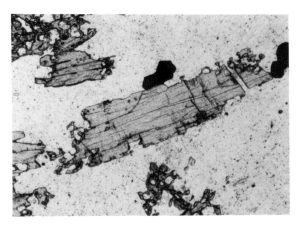

Figure 11.10 Kyanite. Note the basal parting and the trace of cleavage parallel to the length. Field of view is 1.7 mm wide.

CLEAVAGE: Kyanite has one perfect cleavage on {100} which tends to control fragment orientation, and a good cleavage on {010}. Cleavage fragments tend to be splintery. A basal parting {001} may be conspicious, cutting at about 85° to the length of crystals.

TWINNING: Single or multiple twins with {100} composition planes parallel to the length of crystals are common. Multiple twins on {001} are less common.

OPTICAL ORIENTATION: The optic plane is within a few degrees of being at right angles to the (100) face, and Z is inclined 27 to 32° to the c axis, Y is inclined about the same amount to the b axis, and X is inclined only a few degrees from a. The trace of the optic plane on (100) faces is inclined 27 to 32° to the c axis, and on (010) faces the projection of the Z axis is inclined 5 to 8° from the c axis. The maximum extinction angle of about 30° is produced in grains and cleavage fragments oriented so that the (100) face is horizontal. Basal sections yield extinction angles of less than about 3° and sections parallel to (010) yield extinction angles of around 7°. Both crystals and fragments of kyanite are length slow because the Z axis is closer to the crystal length.

INDICES OF REFRACTION AND BIREFRINGENCE: The indices vary in a rather small range. Birefringence of between 0.012 and 0.016 yields first order interference colors in thin sections. Grains in grain mount are usually not oriented to yield maximum birefringence because of the cleavages.

INTERFERENCE FIGURE: Grains oriented with (100) horizontal yield acute bisectrix figures with the melatopes well outside the field of view. These grains show the 30° extinction angle. Optic axis dispersion is weak with $r > v$.

ALTERATION: Kyanite may alter to sericite, which is fine-grained white mica. Alteration to chlorite or aluminous phyllosilicates, such as margarite or pyrophyllite, also may be found. Kyanite is involved in a variety of metamorphic mineral reactions and may display reaction textures with associated minerals such as andalusite, sillimanite, staurolite, cordierite, and garnet.

DISTINGUISHING FEATURES: Higher relief, inclined extinction, optical character, and elongation distinguish kyanite from andalusite and sillimanite. Detrital grains tend to be elongate and may have step-like features related to the cleavages.

OCCURRENCE: Kyanite is a common mineral in pelitic schist, gneiss, and related metamorphic rocks. It is commonly associated with staurolite, andalusite, sillimanite, chloritoid and garnet. Kyanite is rarely found in granitic pegmatite, eclogite, and kimberlite. Locally, it may be a common detrital mineral.

Staurolite

$Fe_2Al_9O_6(SiO_4)_4(OH)_2$
Monoclinic
$\angle\beta \cong 90°$
Biaxial (+)
$n_\alpha = 1.736–1.747$
$n_\beta = 1.740–1.754$
$n_\gamma = 1.745–1.762$
$\delta = 0.009–0.015$
$2V_z = 80–90°$

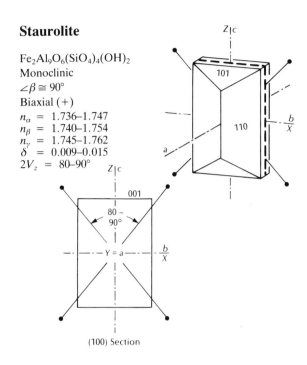

(100) Section

RELIEF IN THIN SECTION: High positive relief.

COMPOSITION AND STRUCTURE: The structure of staurolite consists of nearly cubic close-packed oxygen with the metal cations in the octahedral and tetrahedral sites between the oxygen layers. The cations are distributed so that the structure is layered parallel to (010) with slabs of kyanite structure alternating with layers of $Al_{0.7}Fe_2O_2(OH)_2$. The similarity of the structure accounts for the relatively common layered intergrowth of staurolite and kyanite. The number of Fe per 24 oxygens commonly ranges from 1.1 to 1.7 because of poorly understood substitutions involving Mg, Al, Ti, Zn, Si, and other metals. The amount of hydroxyl is also variable.

PHYSICAL PROPERTIES: H = 7; G = 3.74–3.83; dark brown, reddish brown, or yellowish brown in hand sample; gray streak; vitreous luster.

COLOR AND PLEOCHROISM: Pale honey yellow or brown in thin section with somewhat darker colors in the thicker fragments found in grain mounts. Distinctly pleochroic with X = colorless or pale yellow, Y = pale yellow to yellowish brown, and Z = golden yellow to reddish brown. Staurolite may show color zoning.

FORM: Figure 11.11. Crystals are usually prismatic and elongate parallel to the c axis. Basal sections are usually six sided with the {110} prism faces dominant. Penetration twins forming a cross at either 90° or about 60° are common but may not be obvious in thin sections. It also forms anhedral masses. Porphyroblasts in metamorphic rocks commonly are riddled with numerous rounded inclusions of quartz or other minerals.

Figure 11.11 Strongly sieved staurolite porphyroblast with quartz and opaque inclusions. The dark grains are biotite, and the fine foliated grains in the groundmass are muscovite and chlorite. Field of view is 5 mm wide.

CLEAVAGE: Staurolite has one poor cleavage on {010}, which is not usually observed in thin section nor does it control grain orientation in grain mount.

TWINNING: Penetration twins on {031} produce a right-angle cross and on {231} produce a cross at about 60°. Neither is usually obvious in thin section. Contact twins on {031} are possible.

OPTICAL ORIENTATION: $X = b$, $Y \cong a$, $Z \cong c$, optic plane = (100). Extinction is parallel to longitudinal sections and symmetrical in basal sections. Longitudinal sections are length slow, and in basal sections the long diagonal is parallel to the fast ray vibration direction.

INDICES OF REFRACTION AND BIREFRINGENCE: Indices of refraction and birefringence increase with increasing iron content (Figure 11.12). In thin section, the interference colors range up to first-order white or yellow. The white interference colors often look yellow due to the color of the mineral. Occasionally, the interference colors may be anomalously bluish.

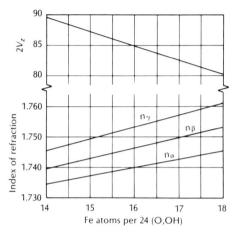

Figure 11.12 Variation of indices of refraction and $2V_z$ with iron content of staurolite. From Ribbe (1982). Used with permission of the Mineralogical Society of America.

INTERFERENCE FIGURE: $2V_z$ increases with decreasing iron content (Figure 11.12). Some samples may be optically negative since $2V_z$ may exceed 90°. Basal sections yield acute bisectrix figures with the melatopes well outside the field of view. Optic axis dispersion is weak to moderate, $r > v$.

ALTERATION: Staurolite may be altered to sericite, which is fine-grained white mica, or chlorite. Staurolite is also involved in a variety of metamorphic mineral reactions and may display reaction relations with such associated minerals as andalusite, kyanite, garnet and cordierite.

DISTINGUISHING FEATURES: The color, pleochroism, relief, and habit make staurolite relatively distinctive. It may resemble brown tourmaline, but the latter is uniaxial and is darker colored when its long axis is oriented at right angles to the vibration direction of the lower polar, whereas staurolite is the opposite.

OCCURRENCE: Staurolite is a common mineral in medium-grade pelitic metamorphic rocks. It may be associated with garnet, andalusite, sillimanite, kyanite, cordierite, chloritoid, chlorite, muscovite, and biotite. It is a relatively common heavy mineral in clastic sediments.

Chloritoid

$(Fe^{2+},Mg,Mn)_2(Al,Fe^{3+})Al_3O_2(SiO_4)_2(OH)_4$
Monoclinic (also triclinic)
$\angle\beta = 101.6°$
Biaxial (+) (monoclinic)
Biaxial (−) (triclinic)
n_α = 1.705–1.730
n_β = 1.708–1.734
n_γ = 1.712–1.740
δ = 0.005–0.022, usually 0.010–0.012
$2V_z$ = 36–72° (monoclinic)
$2V_x$ = 55–88° (triclinic)

(010) Section

RELIEF IN THIN SECTION: High positive relief.

COMPOSITION AND STRUCTURE: Chloritoid has a layered structure consisting of two different octahedral layers. One of the octahedral layers consists of two layers of O + $(OH)_2$ with Fe^{2+} in two-thirds of the octahedral sites and Al^{3+} in the remaining octahedral site, and is similar to the brucite layer in trioctahedral micas. The second octahedral layer consists of two layers of oxygen with three-fourths of the octahedral sites occupied by Al^{3+} of Fe^{3+}. These

layers are bonded together through Si^{4+}, which occupies isolated tetrahedra between the layers. Most chloritoid is monoclinic, but triclinic varieties are known.

Mg^{2+} and Mn^{2+} may replace up to about 68 percent and 50 percent of the Fe^{2+} respectively, although most is relatively iron rich. The manganese-rich variety is called *ottrelite*. Up to 25 percent of the OH^- may be replaced by F^-.

PHYSICAL PROPERTIES: H = $6\frac{1}{2}$ on (001); G = 3.48–3.61; dark green in hand sample; gray streak; pearly to vitrous luster.

COLOR AND PLEOCHROISM: Chloritoid is generally green and pleochroic with X = green or gray-green, Y = blue-gray, indigo, or blue-green, and Z = colorless, yellow, or pale green. Color zoning with an hourglass pattern is common. Color and pleochroism can be quite variable, and brown and colorless varieties are known.

FORM: Figuree 11.13. Crystals are usually platy parallel to (001) with a roughly hexagonal basal section. Random cuts through porphyroblasts are usually rectangular. Due to its platy nature, chloritoid may be foliated like the micas. It commonly contains inclusions of associated minerals, particularly quartz.

Figure 11.13 Chloritoid porphyroblast. Field of view is 2.3 mm wide.

CLEAVAGE: There is a single perfect cleavage parallel to {001} and two imperfect to poor prismatic cleavages on {110}. The basal cleavage strongly controls orientation of fragments in grain mounts. A parting is possible parallel to {010}.

TWINNING: Simple and lamellar twins on {001} are relatively common. The twin lamellae are parallel to the trace of cleavage and the tabular habit.

OPTICAL ORIENTATION: The optic orientation is somewhat variable. In the monoclinic variety, the optic plane may either be perpendicular or parallel to (010) with either X or Y parallel to b. Z is typically within about 20° of the c axis and the angle between Y or X, and the a axis is less than −25°. The maximum extinction angle is therefore less than 25° and is commonly between 10° and 20°. The trace of cleavage is length fast, because Z is roughly normal to the (001) plane. The triclinic variety has similar orientations and shows the same range of extinction angles.

INDICES OF REFRACTION AND BIREFRINGENCE: The manganese- and magnesium-bearing varieties have lower indices than iron-rich varieties, and substitution of Fe^{3+} for Al^{3+} tends to increase the indices. However, the variation in indices is not sufficiently systematic to allow them to be used to estimate composition. Birefringence usually ranges from 0.010 to 0.012, so maximum interference colors are up to pale first-order yellow in thin section. Some varieties have higher birefringence and may show lower second-order colors. Interference colors are usually masked by the mineral color and are commonly anomalous.

INTERFERENCE FIGURE: The Z axis is nearly normal to the basal cleavage, so cleavage fragments and sections cut parallel to (001) yield nearly centered acute bisectrix figures (if positive) or obtuse bisectrix figures (if negative). Different samples from the same locality may yield different values of $2V$ because of submicroscopic intergrowth of monoclinic and triclinic varieties and faults associated with stacking the octahedral sheets together. Monoclinic varieties are usually optically positive and triclinic varieties are usually optically negative. Optic axis dispersion is strong, $r > v$, but may be quite variable in different grains in the same sample.

ALTERATION: Chloritoid may alter to chlorite, fine-grained white mica (sericite) or clay. It is often involved in a variety of metamorphic mineral reactions and may display reaction relations with associated minerals.

DISTINGUISHING FEATURES: Chloritoid most closely resemble chlorite, stilpnomelane, or green biotite. Biotite and stilpnomelane have smaller $2V$, higher birefringence, a single good cleavage, and essentially parallel extinction. Chlorite has lower indices and relief, and most varieties have parallel or near-parallel extinction. If present, the hourglass structure is characteristic.

OCCURRENCE: Chloritoid is common in low- to medium-grade regional metamorphic rocks such as phyllite, quartzite, and mica, chlorite, or glaucophane schists. Chloritoid is commonly associated with garnet, staurolite, chlorite, muscovite, or the aluminosilicates. It is sometimes found associated with corundum in emery deposits. Hydrothermal processes may form triclinic chloritoid in quartz-carbonate veins and in hydrothermally altered rocks.

Titanite (Sphene)

$CaTiOSiO_4$
Monoclinic
$\angle\beta = 119.7°$
Biaxial (+)
$n_\alpha = 1.843–1.950$
$n_\beta = 1.870–2.034$
$n_\gamma = 1.943–2.110$
$\delta = 0.100–0.192$
$2V_z = 17–40°$

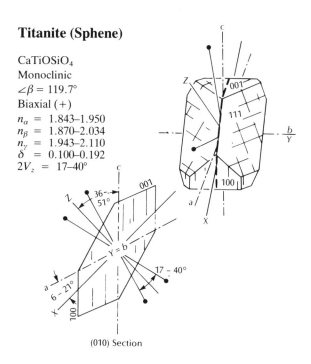

(010) Section

RELIEF IN THIN SECTION: Very high positive relief.

COMPOSITION AND STRUCTURE: The structure of titanite consists of kinked chains of TiO_6 octahedra that are joined at corners (they share one oxygen). The chains are joined laterally through silicon tetrahedra. Calcium is coordinated with seven oxygens in interstices in the structure. Both Al^{3+} and Fe^{3+} may substitute for Ti^{4+} with the charge imbalance made up by substituting OH^- or F^- for O^{2-}. A wide variety of other metal cations including rare earth elements, uranium and thorium also may substitute for calcium and titanium. As in zircon, radioactive bombardment from the decay of uranium, thorium, and other radioactive elements may produce metamict structures in which the crystal lattice is seriously disrupted.

PHYSICAL PROPERTIES: $H = 5-5\frac{1}{2}$; $G = 3.45-3.55$; hand-sample color is usually dark brown, although yellow, green, gray, or black varieties are known; white streak; vitreous luster.

COLOR AND PLEOCHROISM: Typically shades of brown in thin section, less commonly colorless or yellow. Non- to weakly pleochroic with $Z > Y > X$, for example, X = colorless or pale yellow, Y = yellow-brown, pink, or greenish yellow, and Z = orange-brown, greenish brown, green, or red. Color and pleochroism are more distinct in grain mount.

FORM: Figure 11.14. Euhedral to subhedral grains with a wedge- or diamond-shaped cross section are common, as are rounded or irregular anhedral grains. Radioactive varieties may form pleochroic halos in enclosing biotite, chlorite, or hornblende.

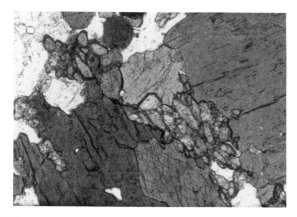

Figure 11.14 Aggregate of anhedral titanite grains between larger grains of hornblende. Field of view is 1.2 mm wide.

CLEAVAGE: There is good prismatic {110} cleavage, but it is not usually obvious in thin section. The cleavage tends to influence fragment orientation.

TWINNING: Simple twins on {100} are not uncommon and are expressed as a composition plane parallel to the long diagonal of diamond-shaped cross sections. There is occasionally lamellar twinning on {221} produced by deformation.

OPTICAL ORIENTATION: $X \wedge a$, $-6°$ to $-21°$; $Y = b$; $Z \wedge c$, $+36$ to $+51°$; optic plane = (010). The maximum extinction angle to the trace of cleavage is 36 to 51° and is seen in (010) sections. Diamond-shaped sections have symmetrical extinction, with the long diagonal usually parallel to the fast ray vibration direction.

INDICES OF REFRACTION AND BIREFRINGENCE: Neither birefringence nor indices of refraction appear to vary in a systematic way with composition. The extreme birefringence produces upper-order white interference colors in thin section but it is often masked by the color of the mineral. The strong dispersion may prevent complete extinction in white light for some sample orientations.

INTERFERENCE FIGURE: Figures are biaxial positive with numerous isochromes and small to moderate $2V$. Optic axis dispersion is very strong with $r > v$, and there is weak inclined dispersion. Fragments lying on the {110} cleavage produce off-center figures.

ALTERATION: The usual alteration product is earthy, white, or yellow leucoxene, which is an aggregate of titanium oxides, quartz, and other minerals. Titanite is sometimes formed as an alteration product after biotite or clinopyroxene.

DISTINGUISHING FEATURES: The very high relief, extreme birefringence, and form are characteristic of titanite. Monazite has lower birefringence and indices. Rutile, cassiterite, and xenotime are uniaxial.

OCCURRENCE: Titanite is a very common accessory mineral in a wide variety of igneous and metamorphic rocks. In igneous rocks, it is most common in syenite, monzonite, granodiorite, and diorite and is often closely associated with other mafic minerals like biotite and hornblende. The only volcanic rock in which it is common is phonolite. In metamorphic rocks, titanite is common in amphibolite, glaucophane and mica schists, granitic gneiss, skarns, and marble. It also may be part of the heavy mineral fraction of clastic sediments.

Topaz

$Al_2(SiO_4)(F,OH)_2$
Orthorhombic
Biaxial (+)
$n_\alpha = 1.606–1.635$
$n_\beta = 1.609–1.637$
$n_\gamma = 1.616–1.644$
$\delta = 0.008–0.011$
$2V_z = 44–68°$

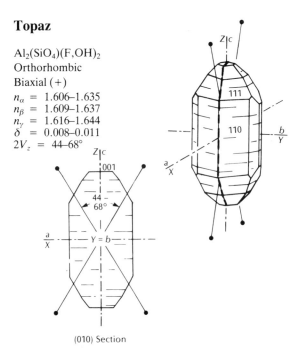

(010) Section

RELIEF IN THIN SECTION: Moderate positive relief.

COMPOSITION AND STRUCTURE: The structure of topaz consists of close packed layers of oxygen alternating with layers of O + $(F,OH)_2$. One-third of the octahedral sites between the layers are filled with aluminum and one-twelfth of the tetrahedral sites are occupied by silicon. The aluminum and silicon are distributed so that the aluminum octahedra share edges and form crankshaft-shaped chains parallel to c, which are tied laterally through the isolated tetrahedrally coordinated silicon. The only significant compositional variation is in the amount of hydroxyl that substitutes for fluorine.

PHYSICAL PROPERTIES: H = 8; G = 3.49–3.57; handsample color is colorless, white, gray, or pale shades of green, red, blue, or yellow; white streak; vitreous luster.

COLOR AND PLEOCHROISM: Usually colorless in thin section or grain mount. Thick fragments may show

pale colors with pleochroism such as $X = Y =$ shades of yellow, and $Z =$ shades of pink.

FORM: Figure 11.15. Crystals are usually stubby to elongate prisms parallel to the c axis. Basal sections range from nearly square to a rounded diamond-shape with eight faces. Topaz also forms anhedral grains and irregular masses. Detrital grains are often platy parallel to the basal cleavage. Some grains display a radial sector structure visible in basal sections.

Figure 11.15 Basal section of euhedral topaz. Field of view is 1.7 mm wide.

CLEAVAGE: The single basal cleavage {001} is perfect and tends to control the orientation of crushed fragments.

TWINNING: Twinning is rare but has been reported on {010}.

OPTICAL ORIENTATION: $X = a$, $Y = b$, $Z = c$, optic plane = (010). Longitudinal sections have parallel extinction to cleavage and elongation and are length slow relative to the elongation and length fast relative to the cleavage. The diamond-shaped basal sections show symmetrical extinction with the slow ray parallel to the long diagonal. Some topaz with sector structure is anomalously monoclinic or triclinic with variable optic orientation.

INDICES OF REFRACTION AND BIREFRINGENCE: The indices increase with increasing amounts of $(OH)^-$ substituting for F^- (Figure 11.16), but birefringence does not change significantly. Interference colors in thin section range up to first-order white or pale first-order yellow. Fragments lying on the cleavage have the Z axis vertical and show birefringence of only about 0.003.

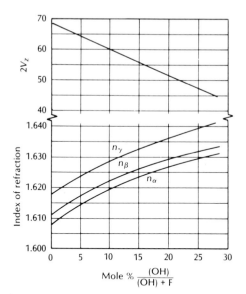

Figure 11.16 Variation of indices of refraction and $2V$ with OH content in topaz (Ribbe and Rosenberg, 1971). Used by permission of the Mineralogical Society of America.

INTERFERENCE FIGURE: $2V_z$ decreases with increasing substitution of $(OH)^-$ for F^- (Figure 11.16). Fragments lying on the {001} cleavage yield centered acute bisectrix figures. Optic axis dispersion is distinct with $r > v$.

ALTERATION: Topaz may be hydrothermally altered to sericite, clay minerals, or occasionally fluorite but is relatively stable in the weathering environment.

DISTINGUISHING FEATURES: Topaz may resemble quartz but has cleavage, higher relief, and is biaxial. Apatite has lower birefringence and is length fast. The feldspars have lower relief. Andalusite is biaxial negative with a large $2V$.

OCCURRENCE: Topaz usually occurs in both volcanic and intrusive felsic igneous rocks. Well-formed small crystals form in vesicles or other cavities in rhyolitic volcanic rocks, and large crystals or masses are found in pegmatites. It also may be found in high-temperature hydrothermal deposits associated with tungsten, tin, molybdenum, or gold mineralization, or in hydrothermally altered rocks adjacent to granitic intrusions. Topaz is occasionally found in metamorphic quartzites and schists, presumably as the result of fluorine metasomatism and may be locally abundant as detrital grains.

Zircon

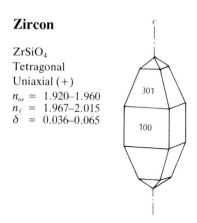

$ZrSiO_4$
Tetragonal
Uniaxial (+)
$n_\omega = 1.920–1.960$
$n_\epsilon = 1.967–2.015$
$\delta = 0.036–0.065$

RELIEF IN THIN SECTION: Very high positive relief.

COMPOSITION AND STRUCTURE: Zircon consists of isolated silicon tetrahedra bonded laterally through Zr, which is in a distorted eightfold coordination with oxygen. Zircon usually contains significant amounts of Hf substituting for Zr, and a variety of rare earth or other elements may be present in minor amounts. The radioactive decay of U, Th, and other radioactive elements causes a disruption of the crystal lattice and results in metamict zircon.

PHYSICAL PROPERTIES: H = $7\frac{1}{2}$; G = 4.67 (metamict: H $\cong$ 6, G $\cong$ 4.0); grayish, yellowish, or reddish brown in hand sample, less commonly colorless, yellow, gray, or pink, bluish green; white streak; subadamantine luster.

COLOR AND PLEOCHROISM: Usually colorless to pale brown in thin section, although high relief and small grain size may tend to make it look darker. In grain mount the color may be more apparent with weak pleochroism, $\omega < \epsilon$, so that grains are darker when the long axis is aligned parallel to the vibration direction of the lower polar. Some samples may be cloudy due to numerous inclusions or may show concentric color zoning or patchy color.

FORM: Figure 11.17. Zircon commonly occurs a euhedral to subhedral tetragonal crystals with pyramidal terminations. It is not uncommon for

Figure 11.17 High relief zircon grain (center) in quartz. The dark mineral is biotite. Field of view is 0.3 mm wide.

euhedral overgrowths to be developed on rounded or subhedral cores. As detrital particles, zircon ranges from euhedral to rounded, depending on the amount of transport. Complex forms are known but are relatively rare.

CLEAVAGE: Prismatic cleavage on {110} and dipyramidal cleavages on {111} are poor and are not usually seen in thin section and do not significantly influence grain orientation in grain mount.

TWINNING: Zircon is usually not twinned, but twinning has been reported on {111}.

OPTICAL ORIENTATION: Elongate grains display parallel extinction and are length slow.

INDICES OF REFRACTION AND BIREFRINGENCE: Pure synthetic zircon has indices of $n_\omega = 1.924$, $n_\epsilon = 1.984$, and $\delta = 0.060$. Natural zircon which is not metamict usually has indices in the range $n_\omega = 1.924–1.934$, $n_\epsilon = 1.970–1.977$ and $\delta = 0.036–0.053$. High Hf zircons have somewhat lower indices. Metamict zircons have lower indices and birefringence and, if strongly metamict, are essentially isotropic with an index around 1.80. Interference colors of relatively fresh zircon in thin section range up to third and fourth order.

INTERFERENCE FIGURE: Because the grains are usually small, interference figures may be difficult to obtain. Basal sections yield uniaxial positive optic axis figures with many isochromes. Metamict varieties are commonly essentially isotropic but may yield anomalous biaxial figures with small $2V$.

ALTERATION: Zircon is quite stable in the weathering environment and does not readily alter, although it may become metamict. Zircon grains in some igneous rocks may be partially resorbed, producing rounded shapes.

DISTINGUISHING FEATURES: Zircon is usually recognized as small, high-relief grains with bright interference colors. It can be mistaken for xenotime, monazite, titanite, or rutile. Xenotime has higher birefringence. Monazite is more commonly colored and is biaxial with a different habit. Rutile is darker colored and has higher indices and higher birefringence. Titanite has a different habit, is biaxial, and displays higher birefringence. Zircon is often included in ferromagnesian minerals such as biotite and hornblende, and may produce dark halos in the surrounding mineral due to radioactive bombardment from Th or other radioactive elements in the zircon.

OCCURRENCE: Zircon is a very common accessory mineral in granite, syenite, granodiorite, pegmatite, and related igneous rocks and is somewhat less common in more mafic rocks. It also is quite common in mica schists, gneiss, quartzite, and other metamorphic rocks derived from clastic sediments or zircon-bearing igneous rocks. It is less common in metamorphosed carbonate rocks. Zircon is a common detrital mineral and can be found in a variety of clastic sediments as part of the heavy mineral fraction.

Dumortierite

$Al_{27}O_6(BO_3)_4(SiO_4)_{12}(OH)_3$
Orthorhombic
Biaxial $(-)$
$n_\alpha = 1.655–1.686$
$n_\beta = 1.675–1.722$
$n_\gamma = 1.685–1.723$
$\delta = 0.011–0.037$
$2V_x = 13–55°$

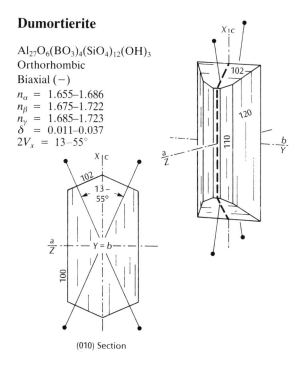

(010) Section

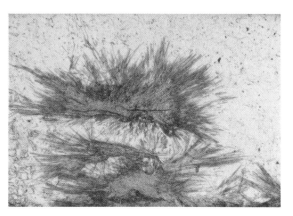

Figure 11.18 Fibrous mat of dumortierite in quartz. Field of view is 1.2 mm wide.

Detrital grains and crushed fragments tend to be elongate parallel to the c axis.

CLEAVAGE: There is a single good cleavage on {100} and two poor prismatic cleavages on {110}.

TWINNING: Cyclic twinning on {110} produces roughly hexagonal columnar crystals made up of six segments.

OPTICAL ORIENTATION: $X = c$, $Y = b$, $Z = a$, optic plane = (010). In elongate sections, extinction is parallel to the length and to the trace of cleavage, and the elongation is length fast.

INDICES OF REFRACTION AND BIREFRINGENCE: The indices of refraction and birefringence increase with substitution of Fe^{3+} for Al^{3+}. Birefringence is low to moderate and interference colors may range up to second order for the more highly birefringent varieties but are commonly no higher than first-order yellow in thin section.

INTERFERENCE FIGURE: Fragments lying on the {100} cleavage yield centered obtuse bisectrix figures. Basal sections yield negative acute bisectrix figures with $2V$ of 13 to 52°; small grain size and fibrous habit make figures difficult to obtain, however.

ALTERATION: Dumortierite may alter to fine-grained white mica.

DISTINGUISHING FEATURES: High relief, distinct bluish color, and strong pleochroism distinguish dumortierite. The sodium amphiboles glaucophane and riebeckite show inclined extinction and have amphibole cleavage. Tourmaline is uniaxial and is

RELIEF IN THIN SECTION: Moderately high positive relief.

COMPOSITION AND STRUCTURE: The composition of dumortierite is usually given as $(Al,Fe)_7O_3(BO_3)$ $(SiO_4)_3$, but it appears that hydroxyl is an essential constituent of the structure and there is a deficiency of aluminum, hence the formula given above. The structure is relatively complex and is unusual in that it includes borate groups. The main compositional variation is the substitution of Fe^{3+} and Ti for Al.

PHYSICAL PROPERTIES: H = $7–8\frac{1}{2}$; G = 3.26–3.41; bright blue, violet or greenish blue in hand sample; vitreous luster.

COLOR AND PLEOCHROISM: Distinctly colored and strongly pleochroic, with X = cobalt blue, violet, red, green, or brown; Y = colorless, yellow, lilac, pink, light blue, or light green; Z = colorless, yellow, light green, light blue, with $X \gg Y \geqslant Z$. Crystals are darkest when the long dimension is parallel to the vibration direction of the lower polar.

FORM: Figure 11.18. Usually bladed, acicular, or fibrous, and elongate parallel to the c axis. Twinned crystals may have pseudohexagonal cross sections.

darkest when the long dimension is at right angles to the vibration of the lower polar. Piemontite is biaxial positive and is usually pleochroic in shades of red and yellow. Fine fibers may resemble sillimanite, but the latter is length slow.

OCCURRENCE: Dumortierite is found in granitic pegmatites, aplites, quartz veins, and in quartz- and feldspar-bearing rocks that have been hydrothermally altered. It also may be found in medium- and high-grade gneiss, quartzite, granitic gneiss, and pelitic schist associated with other aluminous minerals such as kyanite, sillimanite, cordierite, and andalusite.

REFERENCES

Deer, W. A., Howie, R. A., and Zussman, J., 1982, Rock forming minerals, Volume 1A, Orthosilicates: Longman, London, 919 p.

Gunter, M., and Bloss, F. D., 1982, Andalusite-kanonaite series: lattice and optical parameters: American Mineralogist, v. 67, p. 1218–1228.

Laskowski, T. E., and Scotford, D. M., 1980, Rapid determination of olivine compositions in thin section using dispersion staining methodology: American Mineralogist, v. 65, p. 401–403.

Mossman, J. D., and Pawson, D. J., 1976, X-ray and optical characterization of the forsterite-fayalite-tephroite series with comments on knebelite from Bluebell Mines, British Columbia: Canadian Mineralogist, v. 14, p. 479–486.

Ribbe, P. H., 1982, Staurolite. In: Ribbe, P. H., ed., Reviews in mineralogy, Volume 5, Orthosilicates: Mineralogical Society of America, p. 171–188.

Ribbe, P. H., and Rosenberg, P., 1971, Optical and X-ray determinative methods for fluorine in topaz: American Mineralogist, v. 56, p. 1812–1821.

Poldervaart, A., 1950, Correlation of physical properties and chemical composition in the plagioclase, olivine and orthopyroxene series: American Mineralogist, v. 35, p. 1067–1079.

Winchell, H., 1958, The composition and physical properties of garnet: American Mineralogist, v. 42, p. 595–600.

12

Sorosilicates and Cyclosilicates

SOROSILICATES

Lawsonite

$CaAl_2Si_2O_7(OH)_2 \cdot H_2O$
Orthorhombic
Biaxial (+)
$n_\alpha \cong 1.665$
$n_\beta \cong 1.674$
$n_\gamma \cong 1.686$
$\delta = 0.020$
$2V_z = 76–87°$

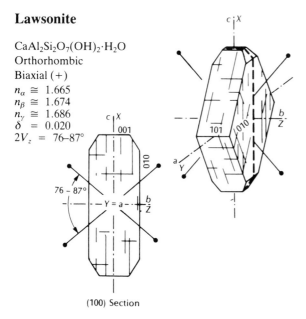

(100) Section

RELIEF IN THIN SECTION: Moderately high positive relief.

COMPOSITION AND STRUCTURE: The structure of lawsonite consists of edge-sharing chains of $AlO_4(OH)_2$ octahedra that run parallel to the c axis and that are tied laterally through Si_2O_7 double tetrahedra groups. The Ca and H_2O occupy interstices in the structure. The composition shows relatively little variation, although minor amounts of Fe, Ti, and Mg may replace Al, and minor amounts of Na may replace Ca.

PHYSICAL PROPERTIES: H = 8; G = 3.05–3.12; colorless, white, bluish green, or bluish gray in hand sample; white streak; vitreous to greasy luster.

COLOR AND PLEOCHROISM: Usually colorless in thin section, sometimes weakly colored and pleochroic. Fragments in grain mount and thick sections are colorless, or pleochroic with X = light blue or brownish yellow; Y = yellow, yellow-green, blue-green; Z = colorless or light yellow

FORM: Figure 12.1. Crystals are usually tabular on (010) with a prismatic cross section. Sections are usually rhomb shaped or rectangular. Also found as anhedral grains and granular masses. Occasionally acicular parallel to the b axis.

CLEAVAGE: There are two perfect pinacoidal cleavages on {010} and {100} that intersect at right angles, and two imperfect cleavages parallel to the {101} prism faces that intersect at about 67°. The cleavages tend to control the orientation of fragments in grain mounts.

TWINNING: Lamellar twinning with twin planes parallel to {101} prism faces is common.

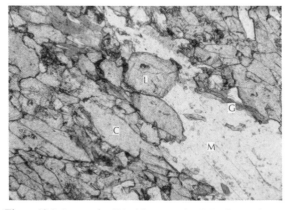

Figure 12.1 Lawsonite (L) with muscovite (M), glaucophane (G) and chlorite (C). Field of view is 1.2 mm wide.

OPTICAL ORIENTATION: $X = c$, $Y = a$, $Z = b$, optic plane = (100). Extinction in rhombic sections is symmetrical with the slow ray vibrating parallel to the long diagonal. Rectangular sections are length fast with parallel extinction. Cleavage fragments are usually elongate parallel to the c axis and are length fast. Acicular crystals elongate on b are length slow.

INDICES OF REFRACTION AND BIREFRINGENCE: Lawsonite shows little variation in indices of refraction and birefringence. Interference colors in standard thin section range up to first-order red.

INTERFERENCE FIGURE: Cleavage fragments lying on the {010} cleavage yield acute bisectrix figures with large 2V. Optic axis dispersion is strong with $r > v$.

ALTERATION: Lawsonite may be replaced by pumpellyite as the result of metamorphic mineral reactions.

DISTINGUISHING FEATURES: Lawsonite resembles zoisite and clinozoisite. Zoisite has lower birefringence, usually displays anomalous interference colors, and has smaller 2V. Clinozoisite also has anomalous interference colors, higher indices and relief, only one cleavage, and inclined extinction. Andalusite has lower birefringence and is usually optically negative. Scapolite is uniaxial. Prehnite has lower indices and relief and higher birefringence.

OCCURRENCE: Lawsonite is a common mineral in glaucophane schists and related low-temperature–high-pressure metamorphic rocks. It is commonly associated with glaucophane, pumpellyite, chlorite, and albite-rich plagioclase. Lawsonite is found in metamorphosed gabbro, diabase, and related mafic rocks and is occasionally present in marble and chlorite schist.

Pumpellyite

$Ca_2MgAl_2(SiO_4)(Si_2O_7)(OH)_2 \cdot H_2O$
Monoclinic
$\angle \beta = 97.6°$
Biaxial (+)
$n_\alpha = 1.665–1.711$
$n_\beta = 1.670–1.717$
$n_\gamma = 1.683–1.727$
$\delta = 0.012–0.018$
$2V_z = 10–85°$

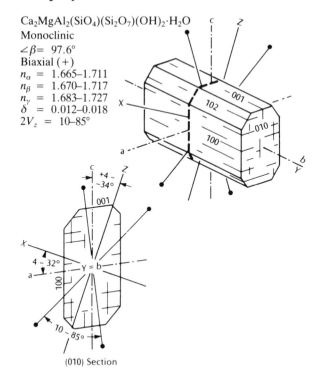

(010) Section

RELIEF IN THIN SECTION: High positive relief.

COMPOSITION AND STRUCTURE: Pumpellyite is related to the epidote group and consists of edge-sharing chains of octahedrally coordinated Mg and Al, which are joined laterally through single and double Si tetrahedra. Fe^{2+} may substitute for Mg, and Fe^{3+} may substitute for Al. Minor substitution of Na for Ca, and Mn or Ti for the octahedrally coordinated cations is also possible.

PHYSICAL PROPERTIES: H = 6; G = 3.18−3.23; usually green, blue-green, or brown in hand sample, with darker colors corresponding to higher iron content; white or gray streak; vitreous luster.

COLOR AND PLEOCHROISM: Usually pleochroic in shades of green or yellow in thin section or grain mount, with darker colors corresponding to higher iron content. Mg-rich samples may be nearly colorless. Pleochroism is strong with $Y > Z \geqslant X$: $X =$ colorless, yellow, brownish or greenish yellow; $Y =$ light green, green, blue-green, brownish yellow; $Z =$ colorless, yellow, brownish yellow, or reddish brown. Zoning is relatively common, with the outer, more iron-rich part of crystals having a darker color than the core.

FORM: Crystals are columnar, bladed, or acicular and elongate parallel to b. Pumpellyite more commonly forms subhedral or anhedral grains or radiating, sub-parallel, or randomly oriented aggregates.

CLEAVAGE: There is a single good cleavage on $\{001\}$ and a fair to poor cleavage on $\{100\}$. Fragments tend to be elongate parallel to the b axis.

TWINNING: Twins with irregular composition planes parallel to $\{001\}$ and $\{100\}$ are possible and may yield fourfold sectors in appropriately cut sections.

OPTICAL ORIENTATION: For most compositions, $X \wedge a = +4$ to $+32°$, $Y = b$, $Z \wedge c = +4$ to $-34°$, optic plane = (010). Certain iron-rich samples have $X \wedge a \cong 40°$, $Y \wedge c \cong -50°$, $Z = b$, optic plane = (001). Grains and fragments elongate parallel to b show parallel extinction and may be either length fast or length slow, depending on orientation. In the iron-rich samples, the elongate grains are length slow. In sections cut at right angles to the b axis, the maximum extinction angle to the $\{001\}$ cleavage is 0 to 34°.

INDICES OF REFRACTION AND BIREFRINGENCE: Both indices of refraction and birefringence increase with higher content of iron (Figure 12.2). The darker, more iron-rich rim of zoned crystals usually has a higher birefringence than the core. Interference colors in standard thin section may range up to first-order red or second-order blue for the more highly birefringent varieties, although middle first-order colors are more common. Interference colors may be anomalously blue or yellowish brown.

INTERFERENCE FIGURE: Most pumpellyite is biaxial positive although some iron-rich varieties may be biaxial negative. The $2V_z$ angle generally increases as the iron content increases. Fragments lying on the basal cleavage yield off-center acute bisectrix figures. Optic axis dispersion is very strong with $v > r$, also weak inclined or crossed bisectrix dispersion for $Y = b$ and $Z = b$ respectively.

ALTERATION: There is no consistent alteration, although pumpellyite may show metamorphic reaction relations with associated minerals.

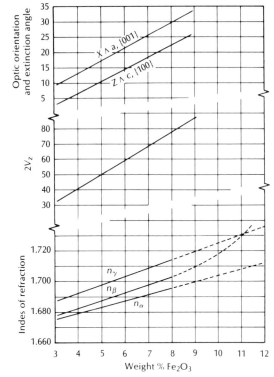

Figure 12.2 Indices of refraction, $2V_z$, and approximate maximum extinction angles to the trace of the $\{100\}$ and $\{001\}$ cleavages of pumpellyite seen in sections cut parallel to (010). After Coombs (1953) and Deer and others (1962).

DISTINGUISHING FEATURES: Light-colored pumpellyite is most easily confused with members of the epidote group, and, in some cases, X-ray diffraction techniques may be needed to distinguish them. Zoisite shows parallel extinction to cleavage. Epidote is optically negative. Clinozoisite has higher indices and lower birefringence. Iron-rich pumpellyite is more distinctly colored than the epidote group. Lawsonite has lower indices, parallel extinction, and better cleavage and is generally colorless or bluish. Chlorite has lower indices and lower birefringence.

OCCURRENCE: Pumpellyite is a common mineral in glaucophane schist and related metamorphic rocks. It is commonly associated with glaucophane, lawsonite, clinozoisite, epidote, chlorite, actinolite, calcite, and prehnite. It is also found in metamorphosed or hydrothermally altered mafic igneous rocks such as basalt and gabbro, and as a vesicle filling in basalt. Skarns and related metamorphosed carbonate rocks may infrequently contain pumpellyite. In some areas, it may be a detrital mineral but is commonly misidentified as epidote.

Melilite

$(Ca,Na)_2(Mg,Al)(Si,Al)_2O_7$
Tetragonal
Uniaxial $(-)$ or $(+)$
$n_\omega = 1.629–1.672$
$n_\epsilon = 1.624–1.661$
$\delta = 0.000–0.011$

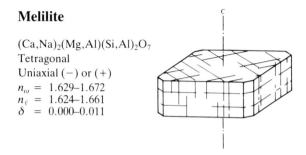

RELIEF IN THIN SECTION: Moderate to high positive relief.

COMPOSITION AND STRUCTURE: The structure of the melilite series consists of single and double tetrahedra containing Si, Al, and Mg arranged in sheets parallel to (001). The sheets are bonded together through Ca in distorted eightfold coordination. There is complete solid solution between *gehlenite* $[Ca_2Al(AlSiO_7)]$ and *åkermanite* $(Ca_2MgSi_2O_7)$, and in addition, up to about 15 percent of the Ca^{2+} may be replaced by Na^+, with the charge balanced by substitution of Al^{3+} for Mg^{2+}. Some Fe^{2+} and Fe^{3+} also may substitute for Mg^{2+} and Al^{3+}, respectively.

PHYSICAL PROPERTIES: H = 5–6; G = 3.038 (gehlenite), 2.944 (åkermanite); colorless, honey yellow, gray-green, brown, or green-brown in hand sample; white streak; vitreous to resinous luster.

COLOR AND PLEOCHROISM: Colorless or occasionally pale yellow in thin section. Unusually thick sections, or fragments in grain mount may be weakly colored and pleochroic with $\omega > \epsilon$: ω = golden brown and ϵ = faint yellow.

FORM: Crystals are usually tabular parallel to (001), with square, octagonal, or rectangular cross sections. In rectangular sections, the c axis is parallel to the short dimension. Crystals often contain numerous isotropic rodlike inclusions parallel to the c axis, forming what is called peg structure. Also found as anhedral grains, sometimes with numerous inclusions of leucite.

CLEAVAGE: A single basal cleavage {001} is fair to poor, and prismatic {110} cleavage is very poor.

TWINNING: Twinning on {100} and {001} is reported but not generally seen.

OPTICAL ORIENTATION: Extinction in rectangular sections is parallel. The long dimension of rectangular sections is length slow for gehlenite-rich compositions and length fast for åkermanite-rich compositions.

INDICES OF REFRACTION AND BIREFRINGENCE: Synthetic gehlenite is uniaxial $(-)$ with $n_\omega = 1.669$, $n_\epsilon = 1.658$, $\delta = 0.011$, and synthetic åkermanite has $n_\omega = 1.632$, $n_\epsilon = 1.640$, $\delta = 0.008$, and is uniaxial $(+)$. The indices vary systematically from gehlenite to åkermanite (Figure 12.3). In natural melilite, however, substituting iron for aluminum and magnesium causes an increase in indices and substitution of sodium for calcium decreases the indices. Neither indices of refraction nor optic sign can be reliably used to determine composition. Interference colors in thin section are usually lower to middle first order, and some samples may display anomalous Berlin blue colors.

INTERFERENCE FIGURE: Melilite is uniaxial and may be either positive or negative. Fragments in grain mount lying on the basal cleavage yield centered optic axis figures. Rectangular sections in thin section usually yield strongly off-center figures.

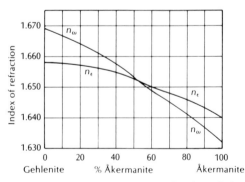

Figure 12.3 Variation of indices of refraction of the synthetic melilite series (Nurse and Midgley, 1953).

ALTERATION: The common alteration is to a brown fibrous material called cebollite [$Ca_5Al_2Si_3O_{14}(OH)_2$], other calc-silicate minerals, zeolites, or calcite.

DISTINGUISHING FEATURES: Melilite is distinguished by its habit, relief and anomalous interference colors. It resembles vesuvianite, zoisite, clinozoisite, and apatite. Vesuvianite has higher indices and relief. Zoisite and clinozoisite are biaxial. Apatite forms elongate hexagonal prisms. Melilite also has a relatively limited occurrence.

OCCURRENCE: Melilite is generally restricted to silica deficient mafic volcanic rocks. It is commonly associated with calcic plagioclase, olivine, clinopyroxene, nepheline, and leucite. It may be found in unusual mafic intrusive rocks such as alnöite, with biotite, olivine, augite, and monticellite. Åkermanite-rich compositions are found in high-grade metamorphosed carbonate rocks.

Vesuvianite (Idocrase)

$Ca_{19}(Al,Fe^{3+})_{10}(Mg,Fe^{2+})_3(Si_2O_7)_4(SiO_4)_{10}(O,OH,F)_{10}$
Tetragonal
Uniaxial ($-$)
n_ω = 1.702–1.795
n_ϵ = 1.700–1.775
δ = 0.001–0.020

RELIEF IN THIN SECTION: High positive relief.

COMPOSITION AND STRUCTURE: The structure is quite similar to garnet but differs from it in that vesuvianite has some silicons in adjacent tetrahedra, forming Si_2O_7 double tetrahedra groups and a somewhat different distribution of cations in the remaining sites. The composition can be quite variable with Na, K, Ce, or Sb substituting for Ca, Ti for (Al, Fe^{3+}), and Mn for (Mg, Fe^{2+}).

PHYSICAL PROPERTIES: H = 6–7; G = 3.32–3.43; usually yellow, green, or brown in hand sample, less commonly red or blue; light streak; vitreous luster.

COLOR AND PLEOCHROISM: Usually colorless in thin section, sometimes lightly colored in shades of yellow, brown, or green. The thicker fragments used in grain mounts are more likely to display colors. Pleochroism is weak with $\omega > \epsilon$. Crystals are frequently zoned.

FORM: Crystals are usually stubby tetragonal prisms. It also forms anhedral grains or clusters of grains in radial or columnar patterns, and may be fibrous.

CLEAVAGE: There are poor to very poor cleavages on {110}, {100}, and {001} which are seldom seen and which do not influence fragment orientation.

TWINNING: Typically not twinned, but anomalously biaxial varieties may display four sectors in basal sections.

OPTICAL ORIENTATION: Most vesuvianite is uniaxial negative, so elongate crystals are length fast with parallel extinction. It also may be biaxial with the optic plane perpendicular to (110) (i.e., normal to the edges of twin sectors).

INDICES OF REFRACTION AND BIREFRINGENCE: Indices increase with substitution of iron or titanium for magnesium or aluminum. Birefringence is usually less than 0.009 and decreases with increasing hydroxyl content. Interference colors in thin section are typically lower to middle first order. Antimony-rich samples have indices at the upper end of the range given above and birefringence may be as high as 0.020. Dispersion of the refractive indices is relatively strong and birefringence may vary as a function of wavelength of light, producing anomalous Berlin blue, olive-yellow, or brown colors.

INTERFERENCE FIGURE: Most vesuvianite is uniaxial negative, although hydroxyl-rich samples may be positive, and some may be sensibly isotropic, at least for certain wavelengths. Biaxial positive vesuvianite with $2V_z$ up to 65° also is known. Zoned grains may be uniaxial in some areas and biaxial in other.

ALTERATION: Vesuvianite is not readily altered.

DISTINGUISHING FEATURES: Crystal habit, high relief, and low birefringence often with anomalous colors are characteristic. Clinozoisite and zoisite are biaxial and have cleavage. The other members of the epidote group have higher birefringence. Andalusite is biaxial, and has lower indices and higher birefringence. Hydrogrossular is very similar and, if crystal habit is not visible, X-ray diffraction techniques may be needed for positive identification. Melilite has lower indices, better cleavage, and a different occurrence. Apatite has lower indices.

OCCURRENCE: The usual occurrence of vesuvianite is in contact metamorphosed limestone. It is often associated with garnet, diopside, wollastonite, tremolite, calcite, epidote, and related minerals. It may also be found in nepheline syenite, and in veins and cavities in hydrothermally altered mafic and ultramafic rocks (e.g., gabbro, peridotite, serpentinite).

Epidote Group

The minerals of the epidote group are the most common of the sorosilicates and have a general formula $X_2Y_3O(Si_2O_7)(SiO_4)(OH)$. The principal members of the group are:

Zoisite $Ca_2Al_3O(Si_2O_7)(SiO_4)(OH)$
Clinozoisite $Ca_2Al_3O(Si_2O_7)(SiO_4)(OH)$
Epidote $Ca_2Fe^{3+}Al_2O(Si_2O_7)(SiO_4)(OH)$
Piemontite $Ca_2(Al,Fe^{3+},Mn^{3+})_3O$ $(Si_2O_7)(SiO_4)(OH)$
Allanite $(Ca,La,Ce)_2(Fe^{2+},Fe^{3+},Al)_3O$ $(Si_2O_7)(SiO_4)(OH)$

Zoisite is orthorhombic and the others are monoclinic.

The structure of the epidote group consists of edge-sharing chains of AlO_6 and $AlO_4(OH)_2$ octahedra, which are tied laterally through single and double silicon tetrahedra. The larger Ca cations fit in the interstices between the chains and tetrahedra. The main chemical variation is the substitution of Fe^{3+} for Al^{3+} and is represented by the solid solution series between clinozoisite and epidote. Significant amounts of Mn^{3+} also may substitute in the octahedral sites yielding piemontite. Substituting trivalent Ce, La, or other rare earth elements for Ca^{2+} is balanced by substituting Fe^{2+} for Al^{3+} in the octahedral sites and yields allanite.

Zoisite

$Ca_2Al_3O(Si_2O_7)(SiO_4)(OH)$
Orthorhombic
Biaxial (+)
n_α = 1.685–1.707
n_β = 1.688–1.711
n_γ = 1.698–1.725
δ = 0.005–0.020
$2V_z$ = 0–60°

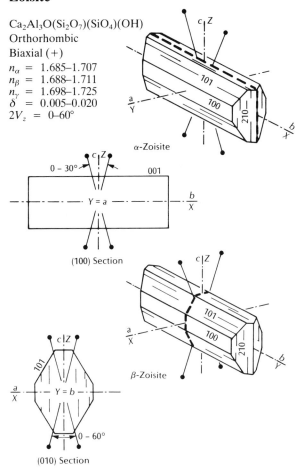

RELIEF IN THIN SECTION: High positive relief.

COMPOSITION: Most zoisite is relatively close to the ideal composition and only up to about 10 percent of the octahedral sites may be occupied by Fe^{3+}. Higher iron content apparently favors the formation of monoclinic clinozoisite. *Thulite* is a variety that has up to 2 percent of the octahedral sites occupied by Mn^{3+} and is analogous to piemontite in the monoclinic series of the epidote group. Most thulite also contains significant amounts of Fe^{3+}. Zoisite often is compositionally zoned.

PHYSICAL PROPERTIES: H = 6; G = 3.15–3.36; gray, green, brown, or pink (thulite) in hand sample; white streak; vitreous luster.

COLOR AND PLEOCHROISM: Most zoisite is colorless in thin section. Grain mounts are usually colorless but occasionally gray or gray-green. Thulite is distinctly pink and pleochroic with X = pink, Y = pale pink or colorless, and Z = pale yellow and is often zoned with a darker pink core and lighter pink rim, reflecting an outward decrease in Mn concentration.

FORM: Often found as anhedral grains or granular aggregates. Crystals are columnar, prismatic, bladed, or acicular with the long dimension parallel to the b axis. Zoisite may form oriented intergrowths with epidote.

CLEAVAGE: There is a single perfect cleavage parallel to {100} and a poor cleavage parallel to {001}. The {100} cleavage tends to orient fragments in grain mount.

TWINNING: None.

OPTICAL ORIENTATION: Zoisite has two optical orientations. In low iron or α-zoisite, X = b, Y = a, Z = c, optic plane = (100). More iron-rich β-zoisite has X = a, Y = b, Z = c, optic plane = (010). With increasing iron content, the $2V$ angle closes on the c axis in the (100) plane, then opens again in the (010) plane. Extinction is parallel to the length of elongate crystals and to the trace of cleavage, which is length fast in α-zoisite and length fast or slow in β-zoisite. Zoned crystals may show adjacent zones with different sign of elongation.

INDICES OF REFRACTION AND BIREFRINGENCE: Both indices of refraction and birefringence increase with increasing iron content. Interference colors for all but the iron-rich varieties are lower than middle first order and are often anomalously blue or greenish yellow. Zoning in the iron content is usually reflected by variation in birefringence and interference colors.

INTERFERENCE FIGURE: Increasing iron content causes $2V$ to close from about 30 to 0° on the c axis in the (100) plane for α-zoisite, and then open from 0 to about 60° in the (010) plane for β-zoisite. Cleavage fragments yield optic normal figures (α-zoisite) and obtuse bisectrix figures (β-zoisite). Optic axis dispersion is strong with $v > r$ (α-zoisite) or $r > v$ (β-zoisite). The $2V$ angle may vary from place to place on a single grain as a consequence of chemical zoning.

ALTERATION: Not commonly altered.

DISTINGUISHING FEATURES: Zoisite is distinguished by its high relief, low birefringence, and parallel extinction. It commonly has anomalous interference colors. It is similar to clinozoisite, which has inclined extinction. Epidote, piemontite, and allanite have inclined extinction and higher birefringence. Melilite resembles zoisite but is uniaxial negative. Vesuvianite is uniaxial negative and has two cleavages. Pink andalusite is very much like thulite, but is biaxial negative and has lower indices and relief. Apatite usually forms elongate crystals with hexagonal cross sections and is uniaxial negative.

OCCURRENCE: Zoisite is a common constituent of medium-grade metamorphic rocks (schist, gneiss, marble) that have been derived from magnesium-poor, carbonate-bearing sediments. It is usually associated with sodic plagioclase, garnet, biotite, hornblende, muscovite, or calcite. It is less commonly found in glaucophane schist and eclogite. Zoisite may be found as an accessory mineral in mafic or ultramafic igneous rocks, and thulite is usually restricted to pegmatites and hydrothermal veins.

Hydrothermal alteration of plagioclase sometimes yields a fine-grained material called saussurite, which is composed of albite, zoisite or clinozoisite, sericite, and other silicates.

Clinozoisite–Epidote

Monoclinic
$\angle\beta = 115.4°$

Clinozoisite
$Ca_2Al_3O(Si_2O_7)(SiO_4)(OH)$

Biaxial (+)
n_α = 1.703–1.715
n_β = 1.707–1.725
n_γ = 1.709–1.734
δ = 0.004–0.012
$2V_z$ = 14–90°

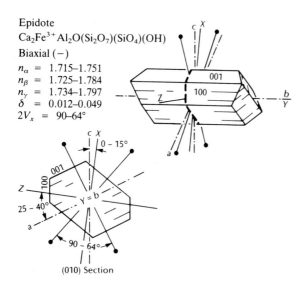

(010) Section

Epidote
$Ca_2Fe^{3+}Al_2O(Si_2O_7)(SiO_4)(OH)$

Biaxial (−)
n_α = 1.715–1.751
n_β = 1.725–1.784
n_γ = 1.734–1.797
δ = 0.012–0.049
$2V_x$ = 90–64°

(010) Section

RELIEF IN THIN SECTION: High positive relief.

COMPOSITION: Clinozoisite and epidote form a continuous solid solution series with up to about 35 percent of the octahedral Al replaced by Fe^{3+} in epidote. Most clinozoisite contains more than about 7 percent octahedral iron. Lower iron content apparently favors the formation of orthorhombic zoisite. Substitution of Mn^{3+} yields piemontite,

and there is probably continuous solid solution between clinozoisite–epidote and piemontite.

The division between clinozoisite and epidote is made at the change from optically positive to optically negative, which typically occurs at between 10 and 15 percent octahedral iron, although samples with up to 18 percent octahedral iron may be optically positive.

Zoning is common and is expressed by variation in color or birefringence.

PHYSICAL PROPERTIES: H = $6-6\frac{1}{2}$; G = 3.12–3.49 (increasing with iron); usually some shade of green in hand sample, less commonly colorless, yellow, or gray, darker with increasing iron content; white or gray streak; vitreous luster.

COLOR AND PLEOCHROISM: Clinozoisite is usually colorless in thin section and grain mount. Higher iron content in epidote produces light yellow-green colors that are pleochroic with $Y > Z > X$: X = colorless, pale yellow, or pale green; Y = yellow-green; Z = colorless, or pale yellow-green. The distribution of the color may be patchy or concentric. Even small amounts of Mn produce the pink colors characteristic of piemontite.

FORM: Figure 12.4. Commonly found as anhedral grains or granular aggregates. The crystals are columnar or bladed or, in some cases fibrous, and elongate parallel to the b axis. Columnar or radial aggregates are relatively common. Sections through

Figure 12.4 High relief epidote in biotite (dark) which is partially altered to chlorite (lighter). Hornblende is present at the lower right. Field of view is 1.7 mm wide.

crystals are usually six-sided or rectangular. Detrital grains are usually rounded or platy parallel to the basal cleavage. Oriented intergrowths with zoisite may be found, as can grains with brown allanite cores and epidote rims.

CLEAVAGE: There is a single perfect basal cleavage on {001} that tends to control fragment orientation. A fair to poor cleavage on {100} is not usually seen.

TWINNING: Lamellar twinning on {100} may be found.

OPTICAL ORIENTATION: The optical orientation varies strongly with composition (Figure 12.5). In all cases $Y = b$ and the optic plane is (010). The angle between X and c is up to about $+85°$ (usually less than $+20°$) for low-iron clinozoisite, and decreases with increasing iron content so that X and c are nearly parallel at the transition to epidote. In epidote, the angle between X and c increases from about $0°$ up to about $-10°$ or infrequently $-15°$. Elongate sections through crystals may be either length fast or length slow, since Y is parallel to the length. The maximum extinction angle to the basal {001} cleavage is seen in sections cut parallel to (010), which also show maximum birefringence and yield flash figures. In clinozoisite, the extinction angle ($Z \wedge a$) to the trace of cleavage is usually between 0 and 25° but may be as large as 60°. In epidote, the extinction angle ($Z \wedge a$) is between about 25 and 40°. In most cases, the slow ray vibration direction is closer to the trace of cleavage. There is considerable variability in orientation, particularly for low-iron clinozoisite, so the extinction angle cannot be used as a reliable guide to composition.

INDICES OF REFRACTION AND BIREFRINGENCE: Both indices of refraction and birefringence increase with increasing iron content (Figure 12.5). Clinozoisite usually shows lower to middle first-order interference colors in thin section and may be anomalously blue or greenish-yellow. Highest interference colors for epidote range from upper first order to third order, depending on composition. Grains oriented to give first-order colors may be anomalously blue or greenish yellow. Interference colors may be patchy or concentrically zoned in a single grain because of the chemical zoning.

INTERFERENCE FIGURE: The size of $2V_z$ generally increases with increasing iron content, although

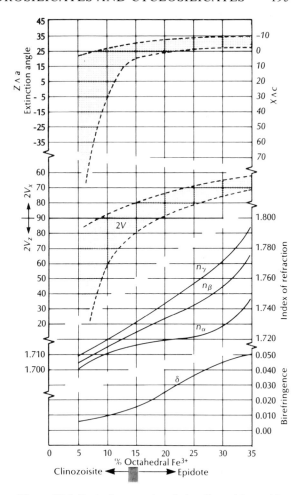

Figure 12.5 Optical properties of the clinozoisite–epidote series. Compiled from Deer and others (1962), Hormann and Raith (1971), Strens (1966), Johnson (1949), and Myer (1965).

there is considerable variability. Clinozoisite is optically positive with $2V_z$ between 14 and 90° and usually greater than about 65°. Epidote is optically negative with $2V_z$ between 90 and 116° ($2V_x = 90–64°$). The size of $2V$ may vary across a single grain because of chemical zoning. Basal sections and fragments lying on the {001} cleavage yield optic axis figures, acute bisectrix figures, obtuse bisectrix figures, or anything in between, depending on the optic orientation. Optic axis dispersion is usually strong with $v > r$ (clinozoisite) or $r > v$ (epidote).

Inclined bisectrix dispersion may be strong enough to reverse the position of the color fringes expected from the optic axis dispersion.

ALTERATION: No consistent alteration. Clinozoisite and epidote are relatively resistant in the weathering environment and are common as detrital grains in clastic sediments.

DISTINGUISHING FEATURES: Clinozoisite and epidote are distinguished from each other by optic sign, birefringence, and color. Allanite is usually brown and piemontite is usually pink, Zoisite, vesuvianite, and colorless pumpellyite all resemble clinozoisite, but zoisite has parallel extinction, vesuvianite is uniaxial negative and lacks good cleavage, and colorless pumpellyite has lower indices and higher birefringence. Small grains in thin section may be difficult to identify and may require other techniques for positive identification.

OCCURRENCE: Epidote and clinozoisite are common accessory minerals in a wide variety of regional and contact metamorphic rocks. Clinozoisite is usually favored in relatively aluminous rocks, and epidote in more iron-rich rocks. Epidote and clinozoisite may be present in quartzite, slate, phyllite, chlorite schist, mica schist, gneiss, amphibole, calc-silicate gneiss, marble, skarn deposits and hornfels.

A wide variety of igneous rocks contain epidote as a primary accessory mineral. Epidote also may be introduced as the result of deuteric or hydrothermal alteration and may occur in veins or pods. Plagioclase may be altered to saussurite, which may contain clinozoisite, and such mafic minerals as biotite, hornblende, and clinopyroxene may be altered to epidote.

Clinozoisite and epidote may be present in the heavy mineral fraction in sediments.

Piemontite

$Ca_2(Al,Fe^{3+},Mn^{3+})_3O(Si_2O_7)(SiO_4)(OH)$
Monoclinic
$\angle\beta = 115.7°$
Biaxial (+)
$n_\alpha = 1.725–1.794$
$n_\beta = 1.730–1.813$
$n_\gamma = 1.750–1.860$
$\delta = 0.025–0.088$
$2V_z = 50–86°$

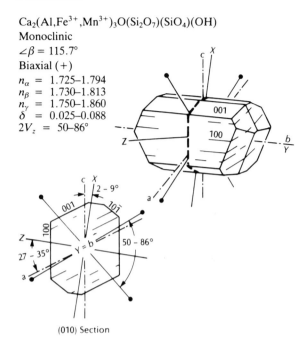

(010) Section

RELIEF IN THIN SECTION: High positive relief.

COMPOSITION: Piemontite probably forms a complete solid solution series with clinozoisite–epidote. The characteristic that distinguishes piemontite is the distinctive pink color and pleochroism, which may be produced if as little as 2 percent of the octahedral sites are occupied by Mn^{3+}. The primary compositional variation is in the relative amounts of Al, Fe^{3+}, and Mn^{3+} in the octahedral sites, although some Mn^{2+} may replace Ca. Compositional zoning is common and is expressed by variation in color, birefringence, and 2V.

PHYSICAL PROPERTIES: H = 6–$6\frac{1}{2}$; G = 3.40–3.52; reddish brown to black in hand sample; light streak; vitreous luster.

COLOR AND PLEOCHROISM: Usually colored in thin section or grain mount. Pleochroism is distinct, usually in shades of red or yellow (e.g. X = pale yellow, yellow, orange, pink; Y = pale violet, redviolet, red; Z = deep red, brownish red, pink). Some samples show distinct color zoning.

FORM: Euhedral, columnar, bladed, or acicular crystals elongate parallel to the b axis are relatively common. Sections are usually six-sided or rectangular. Also found as anhedral to subhedral grains or granular aggregates or as columnar or radiating aggregates made of elongate grains.

CLEAVAGE: There is a single perfect cleavage on {001} that tends to control fragment orientation.

TWINNING: Lamellar twinning on {100} is uncommon.

OPTICAL ORIENTATION: $X \wedge c = -2$ to $-9°$, $Y = b$, $Z \wedge a = +27$ to $+35°$, optic plane = (010). Elongate sections through crystals may be either length fast or length slow and show parallel extinction. The $Z \wedge a$ extinction angle from the trace of cleavage to the slow ray is between 27 and 35° and is seen in sections cut parallel to (010). This section also shows the highest birefringence and yields a flash figure.

INDICES OF REFRACTION AND BIREFRINGENCE: The indices of refraction and birefringence vary over a rather wide range. In general, substituting Fe^{3+} for Al causes a rapid increase in indices and birefringence, and substituting Mn^{3+} causes a somewhat less rapid increase.

INTERFERENCE FIGURE: Piemontite is optically positive with moderate to large $2V_z$. Iron-rich varieties may be optically negative and are called *manganepidote*. Fragments lying on the basal cleavage yield off-center optic axis figures. Optic axis dispersion is usually strong with $r > v$, less commonly with $v > r$.

ALTERATION: Not readily altered.

DISTINGUISHING FEATURES: Piemontite is distinguished from the other members of the epidote group by its distinctive color and pleochroism. Thulite, which has similar colors, shows parallel extinction.

OCCURRENCE: Piemontite is not particularly common, but it may be found in low- to medium-grade regional metamorphic rocks such as phyllite, chlorite schist, or glaucophane schist. It also is found in hydrothermally altered felsic or intermediate volcanic rocks as spherulites or clusters of needles, and in hydrothermal manganese deposits associated with quartz, calcite, and a variety of Mn-bearing minerals.

Allanite

$(Ca,Ce,La)_2(Fe^{2+},Fe^{3+},Al)_3O(Si_2O_7)(SiO_4)(OH)$
Monoclinic
$\angle \beta = 115°$
Biaxial ($-$) or ($+$)
$n_\alpha = 1.690–1.791$
$n_\beta = 1.700–1.815$
$n_\gamma = 1.706–1.828$
$\delta = 0.013–0.036$
$2V_x = 40–90°$ (negative)
$2V_z = 90–57°$ (positive)

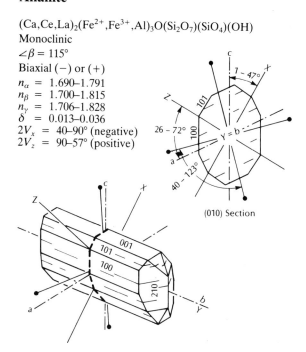

(010) Section

RELIEF IN THIN SECTION: High positive relief. Strongly metamict varieties may be low.

COMPOSITION: The primary compositional variation in allanite is the substitution of trivalent Ce, La, Y, or other rare earth elements for Ca^{2+}. The charge is balanced by substituting Fe^{2+} for Al^{3+}, so allanite is the only member of the epidote group with significant amounts of ferrous iron. A variety of other substitutions are possible, and allanite may contain significant amounts of U, Th, Mg, Ti, Na, Sn, and V. The (OH) may be replaced by F or O, and P may substitute for Si. Allanite may become metamict due to disruption of the structure caused by decay of radioactive elements in the mineral. Metamict allanite often contains substantial amounts of adsorbed water.

PHYSICAL PROPERTIES: H = 5–6$\frac{1}{2}$; G = 3.4–4.2 (lower if metamict); light brown, dark brown, or black in hand sample; gray-brown streak; vitreous to pitchy luster.

COLOR AND PLEOCHROISM: Usually some shade of brown in thin section or grain mount, less commonly greenish. Pleochroism in all but metamict varieties is distinct in various shades of brown, red-brown, yellow-brown, or less commonly, greenish brown or green. The absorption is usually $Y > Z > X$ but can be variable. Color zoning is common, often with a darker core and lighter rim.

FORM: Figure 12.6. Euhedral to subhedral crystals are columnar, bladed, or acicular and elongate parallel to the b axis. Also found as subhedral or anhedral grains or granular aggregates. Allanite may form pleochroic halos in enclosing biotite, chlorite, or hornblende because of its radioactivity. Overgrowths of epidote on allanite are common.

CLEAVAGE: There is a single fair cleavage on {001} and poor cleavages on {100} and {110}. The cleavages are generally not well developed.

TWINNING: Twinning on {100} is not common.

OPTICAL ORIENTATION: Usually $X \wedge c = -1$ to $-47°$, $Y = b$, $Z \wedge a = +26$ to $+72°$, optic plane = (010). An orientation with the optic plane normal to (010) also has been reported. Elongate sections show parallel extinction and may be either length fast or length slow, although mineral color makes determination of elongation difficult. The $Z \wedge a$ extinction angle to the trace of the basal {001} cleavage is 26 to 72°, and is seen in sections cut parallel to (010). This section also shows maximum birefringence and yields flash figures.

INDICES OF REFRACTION AND BIREFRINGENCE: In general, both indices of refraction and birefringence increase with increasing Fe, Ce, and other rare earth elements. The highest interference color in thin section is usually upper first- or second-order colors, although the mineral color often masks the interference color. Strongly metamict varieties may be sensibly isotropic and have indices as low as 1.54. Chemical zoning may be expressed by variation in birefringence.

INTERFERENCE FIGURE: Allanite is usually biaxial negative with $2V_x$ between 40 and 90°, although some varieties are biaxial positive with $2V_z$ between 57 and 90°. The strong mineral color may make the interference figure difficult to interpret. Optic axis dispersion is usually strong with $r > v$, or sometimes $v > r$.

ALTERATION: Metamict allanite is relatively common. Allanite is more readily weathered than are the other epidote group minerals.

DISTINGUISHING FEATURES: High relief, color, and pleochroism are distinctive. Brown hornblende has good cleavage and a different habit. The pleochroic halo in enclosing biotite, chlorite, and hornblende is distinctive, but other minerals such as zircon and titanite also may form pleochroic halos. Metamict varieties may be nearly isotropic with lower indices.

OCCURRENCE: Allanite is found as an accessory mineral in granite, granodiorite, diorite, syenite, nepheline syenite, and pegmatite, and in some equivalent volcanic rocks. It is usually closely associated with biotite, hornblende, or other mafic silicates. It also is found as an accessory mineral in gneiss, schist, granitic gneiss, amphibolite, or occasionally in skarns or other metamorphosed carbonate rocks.

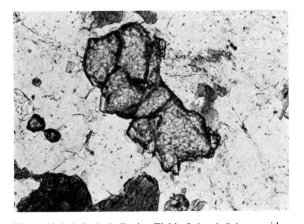

Figure 12.6 Anhedral allanite. Field of view is 0.4 mm wide.

CYCLOSILICATES

Tourmaline

Na(Mg,Fe,Li,Al)$_3$Al$_6$(Si$_6$O$_{18}$)(BO$_3$)$_3$(OH,F)$_4$
Hexagonal (trigonal)
Uniaxial (−)
n_ω = 1.631–1.698
n_ϵ = 1.610–1.675
δ = 0.015–0.035

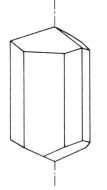

RELIEF IN THIN SECTION: Moderate to high positive relief.

COMPOSITION AND STRUCTURE: The structure of tourmaline consists of sixfold rings of silicon tetrahedra that are stacked with BO$_3$ groups between the rings to form columns parallel to the c axis. The Na resides in the center of the rings and the Mg and related elements are located along the inside edge of the rings. The stacks of rings are tied together laterally through Al in distorted octahedral coordination.

As implied by the formula given above, there is a great deal of compositional variation possible in tourmaline. The normal compositional range lies between the following end members.

Schorl	NaFe$_3$Al$_6$Si$_6$O$_{18}$(BO$_3$)$_3$(OH)$_4$
Dravite	NaMg$_3$Al$_6$Si$_6$O$_{18}$(BO$_3$)$_3$(OH)$_4$
Elbaite	Na(Li,Al)$_3$Al$_6$Si$_6$O$_{18}$(BO$_3$)$_3$(OH)$_4$

In addition, substantial amounts of Mn may substitute for Fe, Mg, and so on. F may substitute for OH, Ca or K may substitute for Na, and Fe^{3+} may substitute for Al. Concentric chemical zoning is quite common and is usually expressed by variation in color, indices, and birefringence.

PHYSICAL PROPERTIES: H = 7; G = 3.03–3.25. Schorl, which is the most common tourmaline, is black. Dravite tends to be brown and elbaite-rich tourmaline is brightly colored red, green, blue, or yellow,

occasionally colorless. Almost any color is possible. White streak; vitreous luster.

COLOR AND PLEOCHROISM: Color in thin section or grain mount can be highly variable, but colored tourmaline is consistently strongly pleochroic with ω > ϵ, so elongate crystals are darkest when the long dimension is aligned perpendicular to the vibration direction of the lower polar, and basal sections are uniformly dark. Common schorl displays shades of gray, blue, and green; less commonly pink. Elbaite-rich varieties are light colored or colorless. Dravite-rich samples are light brown, yellow, or colorless. In general, the intensity of the color increases with increasing iron content. Concentric color zoning is common.

FORM: Figure 12.7. Tourmaline commonly forms euhedral, stubby columnar to acicular crystals that show a rounded triangular or crudely hexagonal cross section. Longitudinal sections are usually roughly rectangular. Acicular crystals may form radiating masses. Also found as anhedral grains or irregular masses.

CLEAVAGE: Cleavages on {11$\bar{2}$0} and {10$\bar{1}$1} are very poor; fracture is conchoidal. There are often fractures perpendicular to the length of crystals that may be filled with quartz, feldspar, or micas.

TWINNING: Rarely twinned, though twinning on {10$\bar{1}$1} and {40$\bar{4}$1}(?) is reported.

OPTICAL ORIENTATION: Longitudinal sections show parallel extinction and are length fast.

INDICES OF REFRACTION AND BIREFRINGENCE: Both indices of refraction and birefringence increase with increasing amounts of iron (Figure 12.8), but neither can be used as a reliable guide to composition, because there is significant variability in the data. The range of values found in most common schorl is n_ω = 1.660 ± 0.010, n_ϵ = 1.635 ± 0.010, and δ = 0.024 ± 0.006. Interference colors in thin section range up to high second order but are often masked by the mineral's color. Indices as high as n_ω = 1.800 and n_ϵ = 1.743 for Fe^{3+}-rich samples and birefringence as high as 0.110 have been reported.

INTERFERENCE FIGURE: Basal sections do not change color with rotation and yield uniaxial negative figures. The dark color of the basal sections may make figures difficult to interpret. Tourmaline is

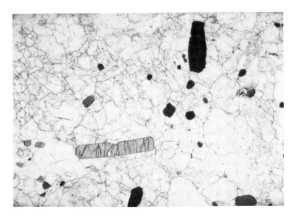

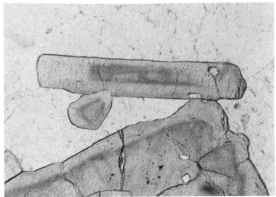

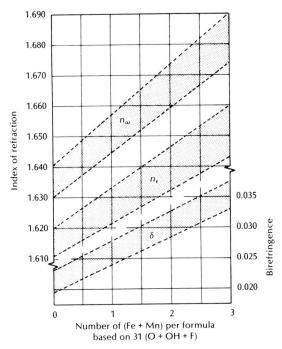

Figure 12.8 Indices of refraction and birefringence (δ) of common tourmaline. Compiled from Deer and others (1962), and Ward (1931).

Figure 12.7 Tourmaline. (*Top*) Schorl grains with strong pleochroism. Grains oriented to pass only ordinary rays are quite dark and include basal sections and longitudinal sections aligned N–S. Longitudinal sections aligned E–W pass only extraordinary rays and are light colored. (*Bottom*) Light-colored tourmaline with concentric color zoning. Width of fields of view: (*top*), 2.3 mm; (*bottom*) 1.2 mm.

occasionally biaxial with 2V up to about 10°. The biaxial character is probably caused by strain in the crystal structure and the orientation in a single crystal may be variable.

ALTERATION: Tourmaline may alter to a various phyllosilicates, including sericite, chlorite, and lepidolite. It is fairly stable in the weathering environment.

DISTINGUISHING FEATURES: Crystal habit, distinct pleochroism, and moderate birefringence distinguish tourmaline. Biotite and hornblende may display similar colors, but they have different crystal habit, good cleavage, and are darkest when their

long dimension is aligned with the vibration direction of the lower polar. Hornblende has inclined extinction. Light-colored tourmaline may resemble topaz, apatite, or corundum, but topaz is biaxial, apatite has lower birefringence, and corundum has higher indices.

OCCURRENCE: Tourmaline is a characteristic mineral in granitic pegmatites and is a common accessory mineral in granite, granodiorite, and related rocks, and in veins and alteration zones associated with these rocks. Tourmaline also is relatively common as an accessory mineral in schist, gneiss, and phyllite. Dravite-rich tourmaline may be found in metasomatically altered limestone and dolomite in contact metamorphic zones. Tourmaline may be an important detrital mineral in some areas.

Axinite

$[(MnFe^{2+},Mg,Zn)(Ca,Mn)_2(Al,Fe^{3+})_2]_2{}^{VI}$
    $[B_2(Si,Al)_8]^{IV}O_{30}(OH)_2$

Triclinic
$\angle\alpha$ = 91.85°
$\angle\beta$ = 98.1°
$\angle\gamma$ = 77.3°
Biaxial (−)
n_α = 1.654–1.694
n_β = 1.660–1.701
n_γ = 1.668–1.705
δ = 0.009–0.014
$2V_x$ = 61–88°

RELIEF IN THIN SECTION: Moderate to high positive relief.

COMPOSITION AND STRUCTURE: The structure is rather unusual and does not fall neatly into the conventional silicate classification scheme. It consists of layers of tetrahedra alternating with layers of octahedra. The tetrahedra are joined into unique $B_2Si_8O_{30}$ groups, which consists of a rectangular sixfold ring of tetrahedra with double tetrahedra groups extending from opposite ends of the rectangle. The octahedral layer consists of discontinuous chains of edge-sharing octahedra tied together laterally through Ca in octahedral coordination. The main compositional variation is substitution in the octahedral sites of Ca, Mn, Fe, Mg, Al, and Zn.

PHYSICAL PROPERTIES: H = $6\frac{1}{2}$–7; G = 3.18–3.40 (increases with Fe, Mn, and Zn); light to dark brown with purple cast in hand sample; white streak; vitreous luster.

COLOR AND PLEOCHROISM: Usually colorless or very pale yellow, brown, or violet in thin section. Grain mounts are more commonly colored due to greater thickness. Colored varieties may be weakly pleochroic with X = light brown or light green, Y = violet or yellow, and Z = colorless, yellow, pale violet, or pale green.

FORM: Crystals are commonly euhedral to subhedral with wedge-shaped, bladed, and tabular habits common; also found in clusters of radiating grains or as anhedral grains or granular aggregates.

CLEAVAGE: There is a single perfect cleavage parallel to {100}. Poor cleavages on {110}, {011}, and {001} are rarely seen.

TWINNING: Usually untwinned, although multiple twins have been reported.

OPTICAL ORIENTATION: Somewhat variable with composition, but a common orientation is X nearly perpendicular to ($\bar{1}11$), Y inclined about 35° to the a axis, and Z inclined about 40° to the c axis. Z may come close to lying in the {100} cleavage plane. The optic plane lies on a diagonal and is inclined at an acute angle to the c axis. Extinction is generally inclined for all orientations, although if Z lies in the cleavage plane, some orientations could show parallel extinction.

INDICES OF REFRACTION AND BIREFRINGENCE: Indices of refraction increase rapidly with increasing amounts of Fe, Mn, and Zn (Figure 12.9). Birefringence is low, and in thin section the highest-order colors are first-order white or yellowish white.

INTERFERENCE FIGURE: $2V_x$ ranges between 61 and 88° and shows a general increase with increasing Mg content. Optic axis dispersion is strong with $v > r$.

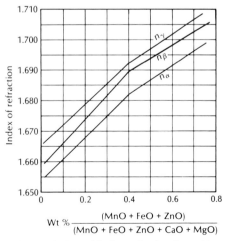

Figure 12.9 Variation of indices of refraction with composition in axinite (Lumpkin and Ribbe, 1979). Used by permission of the Mineralogical Society of America.

ALTERATION: Alteration to calcite or chlorite may occur.

DISTINGUISHING FEATURES: High relief, low birefringence, inclined extinction, and wedge-shaped crystals are characteristic. Zoisite and clinozoisite are optically positive and often have anomalous interference colors. Sapphirine has lower birefringence and does not have good cleavage.

OCCURRENCE: Axinite is usually found in contact metamorphosed limestone and dolomite where it may be associated with calc-silicate minerals (diopside, epidote, vesuvianite, grossular, etc.), tourmaline, and other boron-bearing minerals. In igneous rocks, it is found in vesicles and miarolitic cavities in granite, basalt, gabbro, and diorite. Axinite is rarely found as an accessory mineral in granitic pegmatite, hydrothermal vein deposits. gneiss, schist, or amphibolite.

REFERENCES

Coombs, D. S., 1953, The pumpellyite mineral series: Mineralogical Magazine, v. 30, p. 113–135.

Deer, W. A., Howie, R. A., and Zussman, J., 1962, Rock forming minerals, Volume 1, Ortho- and ring silicates: Longman, London, 333 p.

Hormann, P. K., and Raith, M., 1971, Optische daten, gitterkonstanten, dichte und magnetische suszeptibilitat von Al-Fe (III) epidoten: Neues Jahrbuch fuer Mineralogie Abhandlungen, v. 116, p. 41–60.

Johnston, R. W., 1949, Clinozoisite from Camaderry Mountain, Co. Wicklow: Mineralogical Magazine, v. 28, p. 505–515.

Lumpkin, G. R., and Ribbe, P. H., 1979, Chemistry and physical properties of axinite: American Mineralogist, v. 64, p. 635–645.

Myer, G. H., 1965, X-ray determination curve for epidote: American Journal of Science, v. 263, p. 78–86.

Nurse, R. W., and Midgley, H. G., 1953, Studies in the melilite solid solutions: Journal of the Iron and Steel Institute, v. 174, p. 121–131.

Strens, R. G. J., 1966, Properties of the Al–Fe–Mn epidotes: Mineralogical Magazine, v. 35, p. 928–944.

Ward, G. W., 1931, Chemical and optical study of the black tourmalines: American Mineralogist, v. 16, p. 145–190.

13

Inosilicates

PYROXENES

The pyroxenes are inosilicates with the general formula $XYSi_2O_6$. They are constructed of single chains of silicon tetrahedra that extend along the c axis (Figure 13.1). The chains are stacked in an alternating fashion, so that the bases of adjacent chains face each other. The X cations occupy the M2 structural site between the bases of adjacent chains, and are in six- or eightfold coordination, depending on exactly how the chains are arranged. The Y

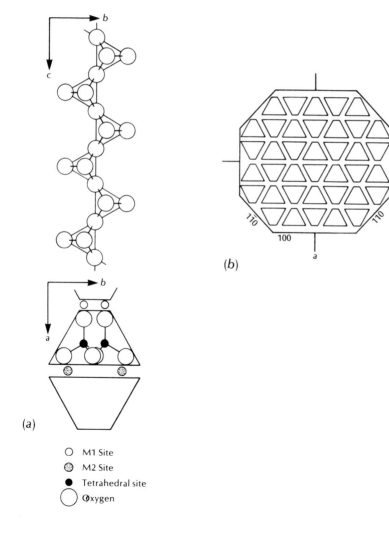

(b)

(a)

○ M1 Site
◉ M2 Site
● Tetrahedral site
○ Oxygen

Figure 13.1 Pyroxene structure. (a) Single chain of silicon tetrahedra. The chains, which are kinked somewhat in many pyroxenes, are stacked so the M1 sites are between the apices of the tetrahedra, and the M2 sites are between the bases. (b) View down the c axis schematically showing the stacking to the chains and the typical pyroxene cross section. Cleavage between the chains at about 87° and 93° is parallel to the (110) and (1$\bar{1}$0) faces.

203

cations occupy the octahedral M1 sites between the "points" of adjacent chains. This structural arrangement produces fairly blocky crystals with four- or eight-sided cross sections and cleavage between the chains at nearly right angles.

The pyroxenes are classified based on the occupancy of the M2 site (Table 13.1 and Figure 13.2). In orthopyroxenes, the only orthorhombic member of the group, the M2 site is octahedral and contains Fe and Mg. In clinopyroxene, larger cations, principally Ca or Na, are in eightfold coordination in the M2 site. The geometry required to accommodate the larger cations reduces the symmetry to monoclinic.

There appears to be a miscibility gap between the calcic clinopyroxenes and pigeonite, although intermediate compositions may be possible at high temperatures. There is extensive solid solution possible among members of the calcic, calcic-sodic, and sodic pyroxenes, whereas spodumene has a rather restricted compositional range.

Table 13.1. Classification of the pyroxenes

Orthopyroxenes
  Enstatite–
  orthoferrosilite $(Mg,Fe)_2Si_2O_6$

Clinopyroxenes
  Magnesium–iron
    Pigeonite $(Mg,Fe,Ca)_2Si_2O_6$
  Calcic
    Diopside $CaMgSi_2O_6$
    Hedenbergite $CaFeSi_2O_6$
    Augite $(Ca,Mg,Fe,Al)_2(Si,Al)_2O_6$
  Sodic–calcic
    Omphacite $(Ca,Na)(Mg,Fe^{2+},Fe^{3+},Al)Si_2O_6$
    Aegirine–augite $(Ca,Na)(Fe^{3+},Fe^{2+},Mg,Al)Si_2O_6$
  Sodic
    Jadeite $NaAlSi_2O_6$
    Acmite (aegirine) $NaFe^{3+}Si_2O_6$
  Lithium
    Spodumene $LiAlSi_2O_6$

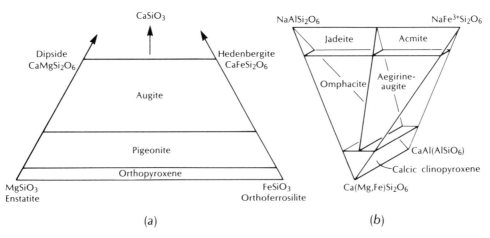

Figure 13.2 Pyroxene classification. (*a*) Calcic and iron–magnesium pyroxenes. (*b*) Sodic–calcic pyroxenes. After Clark and Papike (1968).

Orthopyroxene
(Enstatite–Orthoferrosilite)

$(Mg,Fe)_2Si_2O_6$
Orthorhombic
Biaxial $(+)$ or $(-)$
$n_\alpha = 1.649-1.768$
$n_\beta = 1.653-1.770$
$n_\gamma = 1.657-1.788$
$\delta = 0.007-0.020$
$2V_z = 50-132°$

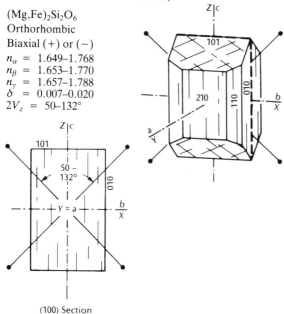

(100) Section

RELIEF IN THIN SECTION: Moderately high to high positive relief.

COMPOSITION: While the main chemical variation in the series is between Mg and Fe, up to 10 or 15 percent of other cations such as Al, Mn, Ti, Cr, and Ca may be present. The amount of Ca usually does not exceed 3 percent of the octahedral sites and appears to be related to the temperature of formation, with high temperature favoring more Ca.

PHYSICAL PROPERTIES: H = 5–6; G = 3.21–3.96; usually brown or greenish brown in hand sample, sometimes with a metallic cast (bronzite) but may range from white, tan, or light green to dark brown or greenish black; gray streak; vitreous to bronze submetallic luster.

COLOR AND PLEOCHROISM: Usually pale colored in thin section, with subtle pinkish to greenish pleochroism. Relatively pure enstatite can be colorless, and adding iron produces darker colors. The pleochroism is usually:

X = pink, brownish pink, pale yellow
Y = light brown, yellow, pinkish yellow, greenish yellow
Z = light green, gray green, bluish green

Longitudinal sections (parallel to c) which show one cleavage trace are pale greenish when the cleavage is parallel to the lower polar vibration and pinkish or yellowish in the 90° position.

FORM: Figure 13.3. Euhedral crystals are usually stubby prisms. Basal sections are four or eight sided and show the prismatic cleavages intersecting at nearly 90°. Longitudinal sections are usually roughly rectangular and show only one direction of cleavage. Orthopyroxene also forms anhedra, irregular grains occupying the space between other minerals, and poikiloblasts with numerous inclusions of associated minerals. Fibrous orthopyroxene may form reaction rims around other minerals such as olivine or garnet.

Orthopyroxene commonly contains exsolution lamellae of augite. The lamellae may be uniform and tabular, or they may pinch and swell and form rows of blebs. The lamellae form in orthopyroxene of the Bushveld type because Ca accommodated in the structure at high temperature is expelled with slow cooling to form lamellae of augite parallel to (100). In orthopyroxene of the Stillwater type, the pyroxene was originally pigeonite. On slow cooling from high temperature, it inverts to the orthopyroxene structure with associated exsolution of augite. The lamellae are oriented parallel to the (001) plane of the original pigeonite. Both sets of lamellae may be present. Lamellae of plagioclase in orthopyroxene are sometimes found in anorthositic rocks and apparently formed by exsolution. Schiller structure in bronzite and hypersthene is probably the result of exsolution of fine inclusions of ilmenite or other Ti minerals on the (010) and (001) planes.

CLEAVAGE: Like the other pyroxenes, there are two good cleavages parallel to the $\{210\}$ prism faces that intersect at about 88°. There also are partings on $\{100\}$ and $\{010\}$.

TWINNING: There may be very fine lamellar structure parallel to $\{100\}$ that may look like lamellar twinning. In most cases, it is very fine exsolution lamellae or the result of translation gliding. Orthopyroxene that has formed by inversion from pigeonite may preserve the position of the pigeonite twin planes.

OPTICAL ORIENTATION: $X = b$, $Y = a$, $Z = c$, optic plane = (100). Extinction is parallel in longitudinal

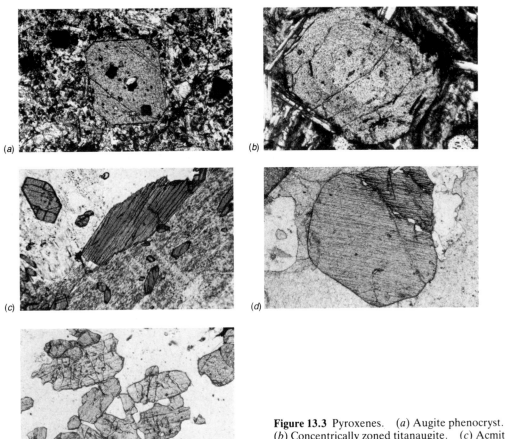

Figure 13.3 Pyroxenes. (*a*) Augite phenocryst. (*b*) Concentrically zoned titanaugite. (*c*) Acmite. (*d*) Rounded diopsidic augite in marble. (*e*) Orthopyroxene with clear plagioclase. Width of field of view for (*a*)–(*e*) is 1.5 mm.

sections (one cleavage direction visible) and the trace of the cleavage is length slow. Basal sections show symmetrical extinction. Fragments in grain mount tend to be elongate parallel to the *c* axis and are length slow with parallel extinction.

INDICES OF REFRACTION AND BIREFRINGENCE: The indices of refraction and birefringence increase systematically with increasing iron content (Figure 13.4). For enstatite, bronzite, and hypersthene (the compositions found in most common igneous and metamorphic rocks) interference colors in thin section are usually first-order yellow or below. More iron-rich compositions may have interference colors as high as lower second order. In grain mounts, fragments that lie on a cleavage surface are oriented to

allow measurement of n_γ, which can provide a fairly accurate estimate of composition.

INTERFERENCE FIGURE: The $2V$ angle varies systematically with iron content and can be used effectively to estimate composition to within ±5 or 10 percent enstatite in thin section. Enstatite and orthoferrosilite are biaxial positive with $2V_z$ between 50 and 90° and bronzite, hypersthene, ferro-hypersthene, and eulite are all negative. Acute bisectrix figures for enstatite and orthoferrosilite are seen in basal sections showing both cleavages. For the optically negative varieties, the acute bisectrix figure is seen in sections cut parallel to (010). Cleavage fragments yield off-center figures. Due to the large $2V$ angle usually encountered, optic axis figures are usually

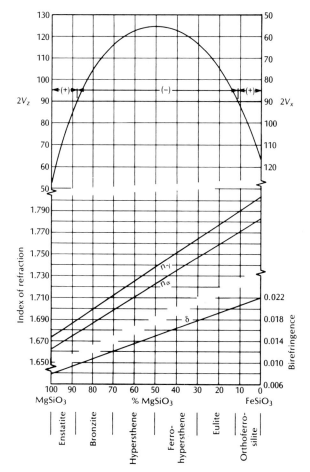

Figure 13.4 Indices of refraction, birefringence (δ), and $2V$ of orthopyroxene. Adapted from Deer and others (1978), Leake (1968), Jaffe and others (1975), and Jaffe and others (1978).

preferred to determine the sign and approximate $2V$ angle.

ALTERATION: Orthopyroxene may be altered to serpentine, talc, or fine-grained pale amphibole and other silicates.

DISTINGUISHING FEATURES: Orthopyroxene is distinguished from clinopyroxene by lower birefringence, parallel extinction, and the common pale pink to green pleochroism. Andalusite is similar to enstatite but does not have the pyroxene cleavage and is optically negative.

Monoclinic polymorphs of orthopyroxene (e.g.

clinoenstatite and clinohypersthene) are rare in terrestrial rocks. These polymorphs may form lamellae in calcic clinopyroxene or discrete crystals in clinker associated with burned coal beds. They may be distinguished by inclined extinction and smaller $2V_z$ (20–50°).

OCCURRENCE: Mg-rich orthopyroxene is common in mafic intrusive rocks of the gabbro and peridotite groups (e.g. gabbro, norite, anorthosite, peridotite, and pyroxenite) associated with plagioclase, calcic clinopyroxene, and olivine. More iron-rich orthopyroxene may be found in diorite, or even some syenite and granite. Volcanic rocks of basaltic to andesitic composition also commonly contain orthopyroxene as phenocrysts.

Very high-grade regional metamorphic rocks of the granulite facies commonly contain orthopyroxene. In charnockites, orthopyroxene may be associated with clinopyroxene, hornblende, biotite, and garnet. Very high-temperature contact metamorphic zones also may contain orthopyroxene.

Pigeonite

$(Mg,Fe^{2+},Ca)_2Si_2O_6$
Monoclinic
$\angle\beta = 108.5°$
Biaxial (+)
$n_\alpha = 1.682–1.732$
$n_\beta = 1.684–1.732$
$n_\gamma = 1.705–1.757$
$\delta = 0.023–0.029$
$2V_z = 0–32°$

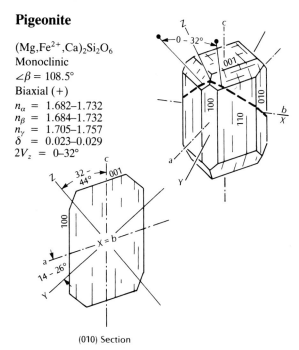

(010) Section

RELIEF IN THIN SECTION: High positive relief.

COMPOSITION: Pigeonite is a low calcium clinopyroxene with about 5 to 15 percent of the Fe-Mg replaced by Ca. Most pigeonites have an Fe:Mg ratio that ranges from 30:70 to 70:30. Minor amounts of Na, Ti, Cr, Al, and Fe^{3+} also may be present. High temperature of formation favors higher Ca contents, but slow cooling allows the Ca to exsolve from the structure to form lamellae of augite in the pigeonite.

PHYSICAL PROPERTIES: H = 6; G = 3.17–3.46; brown, greenish brown, or black in hand sample; light to dark gray streak; vitreous luster.

COLOR AND PLEOCHROISM: Colorless, pale brownish green, or pale yellowish green in thin section. More distinctly colored in the thicker fragments used in grain mounts. Pigeonite is generally not pleochroic, although weak pleochroism is sometimes found: X = colorless, pale greenish brown, or yellow; Y = pale brown or greenish brown; Z = colorless, pale yellow, or pale green.

FORM: Crystals are prismatic and elongate on the c axis. Basal cross sections are four or eight sided and show the two cleavages at about 87°. Longitudinal sections are usually roughly rectangular and show only one cleavage direction. In mafic to intermediate volcanic rocks, it forms anhedral to subhedral groundmass grains. Lamellae of exsolved augite parallel to {001} may be common in pigeonite from intrusive rocks.

Compositional zoning is fairly common and may be expressed by variation in the color, extinction angle, birefringence, or indices of refraction. In some cases, pigeonite is mantled with augite.

CLEAVAGE: Pigeonite, like the other pyroxenes, has two good cleavages at approximately 87° that are parallel to the {110} prism faces. There also are partings on {010} and {001}.

TWINNING: Single and lamellar twins with composition planes parallel to {100} and {001} are common. However, in many cases what appears to be twinning parallel to {001} is actually lamellae of augite that have exsolved as the pigeonite cooled. The combination of twinning on {100} and the lamellae may produce a herringbone pattern.

OPTICAL ORIENTATION: There are two orientations that can be found. In low-Ca pigeonite, $X = b$, $Y \wedge a$ = -14 to $-26°$, $Z \wedge c = +32$ to $+44°$, optic plane is normal to (010). This is the most common orientation. With increasing Ca content, the $2V$ angle closes on the Z axis to become uniaxial and then

opens in the (010) plane, so that the higher-Ca pigeonites are oriented $X \wedge a = -22$ to $-26°$, $Y = b$, $Z \wedge c = +40$ to $+44°$, optic plane = (010). Note that the orientations are related by a reversal in the position of the X and Y axes.

In basal sections showing both cleavages, extinction is symmetrical. In longitudinal sections showing only one cleavage direction, extinction ranges from parallel to inclined, depending on orientation. The $Z \wedge c$ extinction angle of 32 to 44° is seen in sections cut parallel to (010). These sections display maximum birefringence if the optic plane is (010) or close to maximum ($n_\gamma - n_\beta$) birefringence if the optic plane is perpendicular to (010). The trace of the cleavage is closer to the slow ray vibration direction (length slow).

INDICES OF REFRACTION AND BIREFRINGENCE: The indices of refraction and birefringence increase with increasing iron content (Figure 13.5). The curves shown here should be used with caution to estimate composition, because variation in the amount of Ca, Al, and Fe^{3+} can have a significant effect on indices, as can the presence of submicroscopic lamellae of calcic clinopyroxene. Interference colors in thin section range up to lower second order, with first-order yellow and red being common.

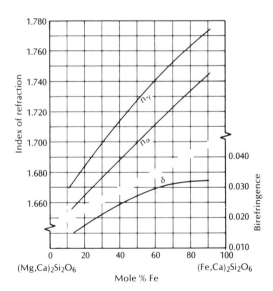

Figure 13.5 General variation of optical properties of pigeonite. After Hess (1949).

INTERFERENCE FIGURE: Most pigeonite is biaxial positive with a small 2V. Certain compositions may be uniaxial. Fragments lying on cleavage surfaces yield strongly off-center figures. Off-center acute bisectrix figures can be obtained from fragments lying on the {001} parting. Optic axis dispersion is usually weak, with either $r > v$ or $v > r$.

ALTERATION: Pigeonite may be uralitized (altered to fine-grained pale green amphibole), or altered to serpentine, talc, or chlorite.

DISTINGUISHING FEATURES: Pigeonite is distinguished from the other pyroxenes by its smaller optic angle. Orthopyroxene has lower birefringence and larger 2V. Olivine has no cleavage and higher birefringence.

OCCURRENCE: Pigeonite is common in dacitic, andesitic, and basaltic volcanic rocks as groundmass grains or, less commonly, as phenocrysts. It commonly crystallizes in mafic to intermediate intrusive rocks but often inverts to orthopyroxene on cooling, although it may be preserved in relatively shallow intrusives.

Pigeonite also can be formed in metamorphosed iron formations, but usually inverts to orthopyroxene, and may be found as exsolution lamellae in augite from granulite facies metamorphic rocks.

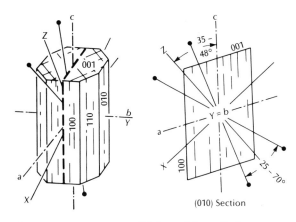

(010) Section

RELIEF IN THIN SECTION: High positive relief.

COMPOSITION: The term *augite* is used here to include calcium-rich monoclinic pyroxenes that contain approximately 20 to 50 percent wollastonite ($CaSiO_3$) molecule (Figure 13.2). These pyroxenes have been subdivided into a number of species based on their composition (Figure 13.6). However, all crystallize with the same structure, there is complete solid solution among all species, and they cannot be distinguished reliably based solely on their optical properties. An alternative is to use the term *calcic clinopyroxene* for pyroxenes falling in the augite field of Figure 13.2.

Calcic Clinopyroxene (Augite)

Monoclinic
$\angle \beta \cong 105°$
Diopside $CaMgSi_2O_6$
Hedenbergite $CaFeSi_2O_6$
Augite $(Ca,Mg,Fe,Al)_2(Si,Al)_2O_6$
Biaxial (+)
n_α = 1.664–1.745
n_β = 1.672–1.753
n_γ = 1.694–1.771
δ = 0.018–0.034
$2V_z$ = 25–70°

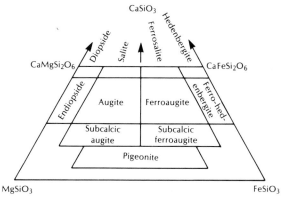

Figure 13.6 One commonly used classification scheme for the calcic clinopyroxenes and pigeonite. After Poldervaart and Hess (1951). In general, $2V_z$ increases from around 25° in the subcalcic augite and subcalcic ferroaugite fields to over 60° in the diopside, salite, ferrosalite, and hedenbergite fields, but there is substantial variability. The indices of refraction increase with increasing Fe content from right to left across the diagram (cf. Figure 13.7).

The primary compositional variation is expressed in the formula of augite given above. The majority of the eightfold M2 structural sites between the bases of the tetrahedra chains are occupied by Ca^{2+} with the remainder predominantly Fe^{2+} and Mg^{2+}. Significant amounts of Na^+ also may be present and the charge deficiency is usually made up by substituting Al^{3+} and Fe^{3+} for Mg^{2+} and Fe^{2+}, which usually occupy the octahedral M1 site. Substituting Ti, Cr, Mn, or other metal cations also is possible. Samples containing significant amounts of Al and Fe^{3+} are called *fassaite*, and Ti bearing samples are called *titanaugite*. In addition, there appears to be complete solid solution between hedenbergite-rich compositions and *johannsenite* ($CaMnSi_2O_6$). Slow cooling of plutonic igneous or metamorphic calcic clinopyroxene commonly results in the development of exsolution lamellae of pigeonite or orthopyroxene and enrichment of the remaining augite in Ca.

Clinopyroxenes with Ca contents intermediate between pigeonite and augite/ferroaugite (Figure 13.6) have been called subcalcic augite, or subcalcic ferroaugite if iron-rich. They appear to be stable only at high temperature and, with slow cooling, exsolve to form calcic clinopyroxene and either orthopyroxene or pigeonite.

PHYSICAL PROPERTIES: $H = 5\frac{1}{2}-6\frac{1}{2}$; $G = 3.19-3.56$ (increases with Fe); color ranges from white or pale green (Mg rich) to black (Fe rich) in hand sample, usually dark green, greenish brown or greenish black, less commonly brown or purplish brown; white to gray streak; vitreous luster.

COLOR AND PLEOCHROISM: Usually colorless, gray, pale green, pale brown, or brownish green in thin section; the darker colors are associated with iron-rich samples that may be weakly pleochroic: $X =$ pale green, bluish green; $Y =$ pale greenish brown, green, bluish green; $Z =$ pale brownish green, green, yellow-green. Titanaugite is more distinctly colored in shades of brown and violet. Color zoning is commonly found and, in some cases, forms a distinctive "hourglass" pattern consisting of four triangular segments radiating from the center of the crystal, particularly in titanaugite.

FORM: Figure 13.3. Crystals are usually stubby prisms elongate along the c axis. Basal cross sections are four or eight sided and show the two cleavages at approximately 87°. Longitudinal sections are roughly rectangular and show only one cleavage direction. Augite also forms anhedral grains, or irregular masses that may enclose associated minerals. Overgrowths of hornblende are relatively common. Lamellae of exsolved orthopyroxene are commonly found parallel to {100}, whereas pigeonite lamellae are inclined somewhat to either {100} or {001}. The lamellae may be inclined up to 22 and 17° from (100) and (001) respectively for Mg-rich augite. Smaller inclinations are found in Fe-rich augite.

CLEAVAGE: Calcic clinopyroxene has the typical pyroxene cleavages parallel to the {110} prism faces, which intersect at 87°. There also are partings parallel to {100} and {001}.

TWINNING: Simple and lamellar twins with {100} and {001} composition planes are common. In combination they may form a herringbone pattern.

OPTICAL ORIENTATION: $X \wedge a = -20$ to $-33°$, $Y = b$, $Z \wedge c = +35$ to $+48°$, optic plane = (010). Basal sections show both cleavages and have symmetrical extinction. Longitudinal sections cut parallel to (100) show parallel extinctions and are length slow. Longitudinal sections cut parallel to (010) show maximum birefringence, a single cleavage trace, and the $Z \wedge c$ extinction angle of 35 to 48° to the slow ray. Certain sections cut parallel to the c axis may display extinction angles greater than 35 to 48° as described in Chapter 7, but these sections always display low birefringence. The sign of elongation is ambiguous in sections showing a large extinction angle.

INDICES OF REFRACTION AND BIREFRINGENCE: Indices of refraction show a systematic increase with increasing iron content (Figure 13.7), but the amount of variability of the data precludes estimating composition from optical data. Interference colors in thin section are usually up to lower or middle second order. Birefringence generally increases with increasing iron content. Indices of refraction also show a fairly systematic variation toward johannsenite; $n_\alpha = 1.710$, $n_\beta = 1.719$ and $n_\gamma = 1.738$.

INTERFERENCE FIGURE: The $2V_z$ angle, which is between 25 and 70°, increases with increasing Ca content. Increasing iron content has a similar but less marked effect. The common range of $2V_z$ is $50 \pm 10°$. Fragments lying on cleavage surfaces yield strongly off-center figures, and basal sections yield off-center optic axis figures. Optic axis dispersion is

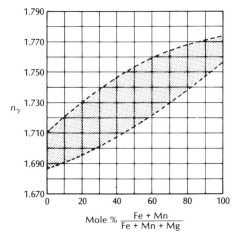

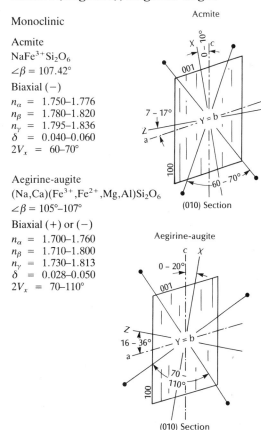

In metamorphic rocks, augite is found in amphibolite, hornblende gneiss, granulite, and related rocks. In general, higher temperatures of formation allow more Fe, Mg to replace Ca. Skarns, marble, and other metamorphosed carbonate-bearing rocks commonly contain calcic clinopyroxene falling in the diopside and salite fields of Figure 13.6. The associated minerals are commonly tremolite-actinolite, grossular, epidote, wollastonite, forsterite, monticellite, calcite, and dolomite.

Figure 13.7 Variation of n_γ with composition in augite. Data from Deer and others (1978), Hess (1949), and Jaffe and others (1975).

weak to strong, $r > v$; or strong, $v > r$, in titanaugite.

ALTERATION: The calcic clinopyroxenes are commonly altered to a material called uralite, which is predominantly fine-grained, light-colored amphibole, or may be altered to serpentine, chlorite, biotite, carbonates, or other silicates.

DISTINGUISHING FEATURES: Calcic clinopyroxenes are distinguished from orthopyroxene by inclined extinction and higher birefringence and from pigeonite by larger $2V$. The more sodium-rich clinopyroxenes (aegirine-augite, acmite, and omphacite) are usually more distinctly green and have larger $2V$. Acmite is optically negative. Some amphiboles may be similar but they have cleavage at 56 and 124°. Wollastonite is colorless, has lower indices and birefringence, and is optically negative. Olivine lacks cleavage and has higher birefringence.

OCCURRENCE: Augite is common in mafic igneous rocks such as gabbro, norite, anorthosite, peridotite, and pyroxenite. It may also be found in more silicic rocks, such as diorite or granodiorite, and hedenbergite-rich compositions are found in syenite and alkali granite. The commonly associated minerals are olivine, orthopyroxene, and plagioclase. Augite is also found in basaltic and andesitic volcanic rocks. Chrome-diopside is relatively common in ultramafic rocks such as kimberlite and peridotite.

Acmite (Aegirine), Aegirine-augite

Monoclinic

Acmite
$NaFe^{3+}Si_2O_6$
$\angle\beta = 107.42°$
Biaxial $(-)$
$n_\alpha = 1.750–1.776$
$n_\beta = 1.780–1.820$
$n_\gamma = 1.795–1.836$
$\delta = 0.040–0.060$
$2V_x = 60–70°$

Aegirine-augite
$(Na,Ca)(Fe^{3+},Fe^{2+},Mg,Al)Si_2O_6$
$\angle\beta = 105°–107°$
Biaxial $(+)$ or $(-)$
$n_\alpha = 1.700–1.760$
$n_\beta = 1.710–1.800$
$n_\gamma = 1.730–1.813$
$\delta = 0.028–0.050$
$2V_x = 70–110°$

RELIEF IN THIN SECTION: High positive relief.

COMPOSITION: The primary compositional variation in the acmite and aegirine-augite series is $NaFe^{3+} = Ca(Fe^{2+},Mg)$ and there appears to be a complete range to augite compositions. There is usually less than about 10 percent Al substituting in the octahedral M1 site, and almost none replacing Si in the tetrahedral sites. Minor amounts of Ti and Mn also may be present. There is no agreement on the division between acmite and aegirine-augite. The convention used here is to place the boundary at 80 percent Fe^{3+} in the octahedral M1 site (i.e., 80 percent acmite component). Others have placed the boundary at the change in optic sign, which occurs at about 40 percent Fe^{3+}. The latter definition is not recommended, because the optical data show significant scatter and the transition from optically positive to negative takes place over a range of about 20 mole percent Fe^{3+}. The boundary between aegirine-augite and augite is placed at 20 mole percent Na. Compositional zoning is common; the rims of grains are usually enriched in the acmite component.

PHYSICAL PROPERTIES: H = 6; G = 3.40–3.60; dark green to greenish black or reddish brown in hand sample; gray streak; vitreous luster. Some mineralogists prefer to restrict the term *acmite* to brownish samples, and use the term *aegirine* for green to black samples. The brown varieties are typically sharply pointed, whereas the green to black varieties are typically bluntly terminated.

COLOR AND PLEOCHROISM: Brown, yellowish brown, pale green, yellowish green, or dark green in thin section, darker with higher acmite content. Distinctly pleochroic with X = emerald green, dark green, bright green; Y = grass green, yellowish green, yellow; Z = brownish green, green, yellowish brown, yellow. Brown varieties are weakly pleochroic in shades of brown and yellow. Color zoning is common; the rim is usually darker than the core, and an hourglass pattern may be formed.

FORM: Figure 13.3. Acmite and aegirine-augite form stubby to quite elongate prisms. Eight-sided cross sections show both cleavages at about 87°. Longitudinal sections are elongate to roughly rectangular and show one direction of cleavage. The brown varieties may have acute or sharply pointed terminations. In the process of fenitization, fibrous acmite and aegirine-augite may form at the expense of biotite or hornblende. Spongy intergrowths of riebeckite and acmite may be found.

CLEAVAGE: There are two good {110} prismatic cleavages at about 87° and a parting on {100}.

TWINNING: Simple and lamellar twinning on {100} is common.

OPTICAL ORIENTATION: Acmite: $X \wedge c$ = +10 to 0°, Y = b, $Z \wedge a$ = +7 to +17°, optic plane = (010). Aegirine-augite: $X \wedge c$ = 0 to −20°, $Y = b$, $Z \wedge a$ = +16 to +36°, optic plane = (010). Basal sections yield symmetrical extinction. The $X \wedge c$ extinction angle is seen in sections with maximum birefringence cut parallel to (010) and varies systematically with composition. In acmite, the extinction angle is between about 0 and 10°, with the fast ray vibration direction lying in the obtuse angle between the c and a axes (Figure 13.8). In aegirine-augite, the extinction angle varies from about 0 to 20°, with the fast ray vibration direction usually lying in the acute angle between the c and a axes. The fast ray vibration direction is closer to the trace of cleavage in longitudinal sections (length fast).

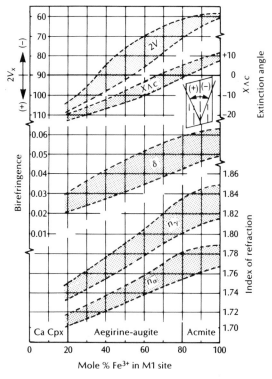

Figure 13.8 Variation of indices of refraction, birefringence (δ), extinction angle $X \wedge c$, and 2V in acmite and aegirine-augite. Data from Deer and others (1978), Larsen (1942), Sabine (1950), and Nolan (1969). See also Figure 13.9.

INDICES OF REFRACTION AND BIREFRINGENCE: The indices of refraction and birefringence show a systematic increase with increase in the amount of Fe^{3+} (Figure 13.8). The scatter in the data is probably due predominantly to variation in the $Mg:Fe^{2+}$ ratio. Maximum interference colors in thin section are third order for acmite and second to lower third order for aegirine-augite. The interference color may be masked by the mineral's color.

INTERFERENCE FIGURE: The $2V_x$ angle increases in a systematic manner from about 60° for relatively pure acmite, to 90° at 35 to 40 mole percent Fe^{3+}. More Fe^{3+}-rich compositions are optically positive with $2V_z$ decreasing to around 70° ($2V_x = 110°$). Fragments lying on a cleavage surface give highly off-center figures; those lying on the parting giving off-center optic axis figures. Optic axis dispersion is moderate to strong with $r > v$, with weak to moderate inclined bisectrix dispersion.

ALTERATION: The common alteration is to fine-grained amphibole (uralitization) or chlorite.

DISTINGUISHING FEATURES: Acmite and aegirine-augite are distinguished from the other pyroxenes by their color and pleochroism, small extinction angle, length-fast character, high indices, and birefringence. The optic sign is either negative or positive with a larger $2V$ than found with most other pyroxenes. The distinction between acmite and aegirine-augite can be made based on the extinction angle, $2V$ angle, indices, and birefringence. In addition, acmite is typically darker colored. Amphiboles have cleavage at 56 and 124° and are length slow. Epidote shows only one good cleavage.

OCCURRENCE: Acmite and aegirine-augite are common in alkaline igneous rocks such as alkali granite, syenite, nepheline syenite, and related rocks. They also are found in carbonatites, glaucophane- and riebeckite-bearing schists, or (rarely) in alkalic volcanic rocks.

Omphacite

$(Ca,Na)(Mg,Fe^{2+},Fe^{3+},Al)Si_2O_6$
Monoclinic
$\angle\beta = 105–108°$
Biaxial (+)
$n_\alpha = 1.662–1.701$
$n_\beta = 1.670–1.712$
$n_\gamma = 1.685–1.723$
$\delta = 0.012–0.028$
$2V_z = 56–84°$

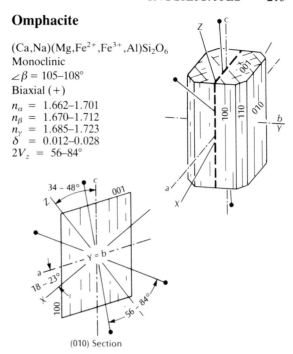

(010) Section

RELIEF IN THIN SECTION: High positive relief.

COMPOSITION: Omphacite differs from the calcic clinopyroxenes in that between 20 and 80 percent of the Ca^{2+} has been replaced by Na^+. The charge is balanced by substituting Fe^{3+} and Al^{3+} for Fe^{2+} and Mg^{2+} in the M1 site. Only very restricted amounts of Al^{3+} usually substitute for Si^{4+} in the tetrahedral sites. Aegirine-augite, which is the other sodium–calcium pyroxene, has more Fe^{3+} than Al in the octahedral site, whereas omphacite has more Al than Fe^{3+}.

PHYSICAL PROPERTIES: H = 5–6; G = 3.16–3.43; green to dark green in hand sample; gray streak; vitreous luster.

COLOR AND PLEOCHROISM: Colorless to pale green in thin section, darker in grain mount. Weakly pleochroic with X = colorless, Y = very pale green, and Z = very pale green to blue green.

FORM: Omphacite crystals are stubby prisms with four- or eight-sided cross sections. Basal sections show both cleavages at about 87°, and longitudinal sections are roughly rectangular and show only one cleavage direction. Also commonly anhedral granular.

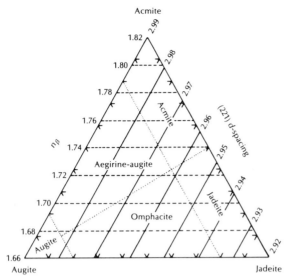

Figure 13.9 Diagram for estimating composition of sodic and sodic–calcic pyroxenes from glaucophane schist using the value of n_β (dashed lines) and the (221) d spacing (solid lines). After Essene and Fyfe (1967).

CLEAVAGE: Omphacite has the typical 87° pyroxene cleavages parallel to the {110} prism. There also is a parting on {100}.

TWINNING: Simple and lamellar twinning parallel to {100} is common.

OPTICAL ORIENTATION: $X \wedge a = -18$ to $-23°$, $Y = b$, $Z \wedge c = +34$ to $+48°$, optic plane = (010). Extinction in basal sections is symmetrical. In longitudinal sections, extinction ranges from parallel to a maximum of 34 to 48°, depending on orientation, Sections cut parallel to (010) show maximum birefringence and maximum extinction angle. Longitudinal sections showing a single cleavage direction and nearly parallel extinction are length slow. Cleavage fragments are elongate parallel to the c axis and are length slow.

INDICES OF REFRACTION AND BIREFRINGENCE: Although there is a substantial range of indices, the optical properties alone cannot be directly correlated with composition, However, the approximate composition ($\pm$ 5 percent) of omphacite from glaucophane schist can be determined from the (211) d spacing obtained from X-ray diffraction, and n_β (Figure 13.9). The birefringence is moderate, so maximum interference color in thin section is usually upper first or lower second order.

INTERFERENCE FIGURE: While the full range of $2V_z$ is 56 to 84°, it is usually between 60 and 75°. Fragments lying on the {100} parting yield off-center optic axis figures, and those on a cleavage surface yield strongly off-center figures. Optic axis dispersion is moderate with $r > v$ or $v > r$.

ALTERATION: Omphacite is commonly altered to fibrous green amphibole. In addition, omphacite may undergo a process of exsolution, which yields a symplectic intergrown plagioclase and diopside.

DISTINGUISHING FEATURES: Omphacite is distinguished from the calcic clinopyroxenes by its larger $2V$. Jadeite has lower indices, birefringence, and extinction angle and is always colorless in thin section. Acmite and aegirine-augite have higher indices, smaller extinction angles, and are usually more distinctly green and pleochroic in thin section.

OCCURRENCE: Omphacite and garnet are characteristic minerals in eclogite, which is a rock of roughly basaltic composition formed under very high-pressure igneous and metamorphic conditions. Associated minerals are plagioclase, quartz, orthopyroxene, kyanite, and Fe–Ti oxides. Omphacite is less commonly found in glaucophane schist, amphibolite, granulite, garnetiferous anorthosite, and granitic gneiss.

Jadeite

$NaAlSi_2O_6$
Monoclinic
$\angle\beta \cong 108°$
Biaxial (+) or (−)
$n_\alpha = 1.640–1.681$
$n_\beta = 1.645–1.684$
$n_\gamma = 1.652–1.692$
$\delta = 0.006–0.021$
$2V_z = 60–96°$

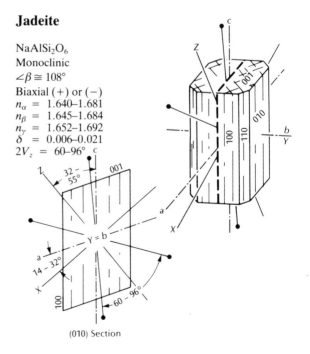

(010) Section

RELIEF IN THIN SECTION: Moderately high positive relief.

COMPOSITION: In jadeite, over 80 percent of the sites between the bases of the tetrahedral chains are occupied by Na^+, with the remainder occupied by Ca^{2+} or sometimes K^+. Typically, over 80 percent of the octahedral M1 sites are occupied by Al^{3+}, with the remainder occupied by Fe^{3+}, or to balance the Ca^{2+} in the M2 site, Mg^{2+} and Fe^{2+}. Small amounts of Ti and Mn also are usually present. There appears to be a miscibility gap between jadeite and the Na–Ca pyroxenes omphacite and aegirine-augite.

PHYSICAL PROPERTIES H = 6; G = 3.24–3.43; usually light to medium green in hand sample, also white, blue-green, blue, or brown; white streak; vitreous luster.

COLOR AND PLEOCHROISM: Colorless in thin section. Colorless or pale green with light yellow to green pleochroism in grain mount.

FORM: Crystals are stubby to elongate prisms with four- or eight-sided cross sections, Jadeite also forms anhedral granules and acicular to fibrous aggregates.

CLEAVAGE: Jadeite has the typical pyroxene cleavages on {110} that intersect at 87°. A parting on {100} is reported but is not common. Cleavage fragments are elongate parallel to the c axis.

TWINNING: Simple and lamellar twinning on {100} or {001} is sometimes found.

OPTICAL ORIENTATION: $X \wedge a = -14$ to $-32°$, $Y = b$, $Z \wedge c = +32$ to $+55°$, optic plane = (010). Basal sections show both cleavages intersecting at 87° and have symmetrical extinction. Longitudinal sections show only one cleavage direction and show parallel extinction in (100) sections and the $Z \wedge c$ extinction angle of 32 to 55° in (010) sections. Larger extinction angles are associated with higher contents of Fe^{3+}, Ca, and Mg. Sections cut parallel to (010) show maximum birefringence. Longitudinal sections showing nearly parallel extinction, and elongate cleavage fragments are length slow.

INDICES OF REFRACTION AND BIREFRINGENCE: The indices of relatively pure jadeite are around $n_\alpha = 1.654$, $n_\beta = 1.656$, and $n_\gamma = 1.666$. Increasing Fe^{3+}, Ca and Mg is accompanied by higher indices (Figure 13.9) but the variation may not be particularly regular, Lower indices may be caused by incipient alteration or by the presence of vacant sites in the structure. Birefringence is usually in the range 0.012 to 0.018, so maximum interference colors in thin section are first-order yellow or red. Samples with high Fe^{3+} content may show anomalous interference colors and lower birefringence.

INTERFERENCE FIGURE: The $2V_z$ angle for most jadeite averages around 70°, which is larger than for most other pyroxenes. Larger values have been reported from some locations. Substitution of Ca, Fe^{3+}, or other cations generally causes a decrease in $2V_z$. Fragments lying on cleavage surfaces yield highly off-center figures. Basal sections showing the two cleavages at about 87° yield off-center optic axis figures. Optic axis dispersion is weak to very strong with $r > v$.

ALTERATION: Jadeite may be replaced by amphibole or, less commonly, by analcime or nepheline.

DISTINGUISHING FEATURES: Jadeite has lower indices of refraction than any of the pyroxenes, except spodumene, which has a very different paragenesis. Omphacite has higher birefringence. Acmite and aegirine-augite have higher birefringence and are typically colored in thin section. Jadeite with anomalous interference colors resembles zoisite, but zoisite has parallel extinction and higher indices. Tremolite has a smaller extinction angle, higher birefringence, typical amphibole cleavage, and is optically negative.

OCCURRENCE: Jadeite has been found only in metamorphic rocks that have been subjected to relatively high pressure and low to moderate temperature, such as glaucophane schist, metagreywacke, serpentinite, and associated rocks. Jadeite occurs both as monomineralic pods and veins, and as disseminated grains. The commonly associated minerals are albite, glaucophane, lawsonite, quartz, chlorite, garnet, zoisite, micas, calcite, aragonite, and actinolite. In many cases, it appears that the jadeite has been derived from the albite component of plagioclase, and relict feldspar grains partially or completely replaced by fine radiating jadeite are not uncommon. Higher temperatures allow additional Ca to substitute for Na, so in equivalent higher-grade rocks, the pyroxene may be omphacite.

Spodumene

$LiAlSi_2O_6$
Monoclinic
$\angle \beta = 110°$
Biaxial (+)
$n_\alpha = 1.648–1.668$
$n_\beta = 1.655–1.671$
$n_\gamma = 1.662–1.682$
$\delta = 0.014–0.027$
$2V_z = 58–68°$

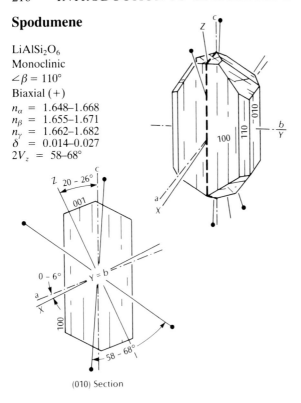

(010) Section

RELIEF IN THIN SECTION: Moderately high positive relief.

COMPOSITION: Spodumene shows little compositional variation. Minor amounts of Na may replace Li, and minor Fe^{3+} may replace Al. A wide variety of other elements may be present in trace amounts.

PHYSICAL PROPERTIES: H = $6\frac{1}{2}$–7; G = 3.03–3.23; usually white or grayish white in hand sample, also lightly colored in shades of blue, green, and yellow; white streak, vitreous luster.

COLOR AND PLEOCHROISM: Usually colorless in thin section and grain mount, The dark-colored gem varieties are pleochroic in shades of green or lilac.

FORM: Crystals are prismatic with roughly octagonal cross sections that show both cleavages intersecting at about 87°. Also as lathlike acicular crystals and cleavable masses. Because spodumene is usually found in pegmatites, its crystals may be far larger than normal thin sections.

CLEAVAGE: Spodumene shows typical pyroxene {110} cleavages that intersect at 87 and 93°. There also are {100} and {010} partings.

TWINNING: Twinning on {100} is not uncommon.

OPTICAL ORIENTATION: $X \wedge a = 0$ to $-6°$, $Y = b$, $Z \wedge c = +20$ to $+26°$, optic plane = (010). Extinction in basal sections showing both cleavages at about 87° is symmetrical. Longitudinal sections parallel to (010) show a single cleavage trace and the $Z \wedge c$ extinction angle of 20 to 26°, with the slow ray vibration direction closer to the cleavage trace (length slow).

INDICES OF REFRACTION AND BIREFRINGENCE: The indices of refraction and birefringence vary within a moderate range. The n_α index decreases and the birefringence increases with substitution of Na for Li, and substitution of Fe^{3+} for Al increases all indices and birefringence. Maximum interference color in thin section is upper first order to middle second order.

INTERFERENCE FIGURE: $2V_z$ is moderate (58–68°) with the melatopes near the edge of the field of view in acute bisectrix figures. Fragments lying on cleavage surfaces yield strongly off-center figures, those on the {100} parting yield off-center Bxo figures.

ALTERATION: Spodumene is readily altered to Li-mica, albite, and eucryptite ($LiAlSiO_4$), which is a uniaxial (+), colorless, granular, or prismatic mineral with $n_\omega = 1.572$, $n_\epsilon = 1.586$, and $\delta = 0.014$.

DISTINGUISHING FEATURES: Spodumene's occurrence in pegmatites and small extinction angle distinguish it from most of the other pyroxenes. Acmite has a similar extinction angle but is usually distinctly colored.

OCCURRENCE: Spodumene is a common mineral in Li-bearing pegmatites and is usually associated with quartz, albite, K-feldspar, beryl, lepidolite, tourmaline, and other Li-bearing minerals.

AMPHIBOLES

The amphiboles have a general formula $W_{0–1}X_2Y_5Z_8O_{22}(OH)_2$. They show considerable similarity to the pyroxenes both in structure and composition. The structure consists of double chains of silicon tetrahedra elongate parallel to the c axis and stacked in an alternating fashion so that the bases face each other (Figure 13.10). Cleavage between the chains yields the 56 and 124° angles characteristic of amphiboles. There are five struc-

tural sites between the bases and apices of the tetrahedra chains. In the formula given above, W represents large cations (Na, K) that occupy the $\sim$ tenfold coordinated A site. X cations (Ca, Mg, Fe, Na) occupy the six- or eight-fold coordinated M4 site, and Y cations (Mg, Fe, Al) occupy the octahed-

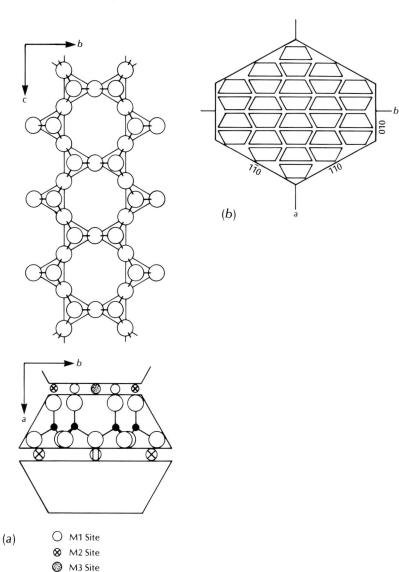

(b)

(a)

○ M1 Site
⊗ M2 Site
⊛ M3 Site
⊗ M4 Site
● Tetrahedral site
◍ A Site
○ Oxygen

Figure 13.10 Amphibole structure. (*a*) Idealized double chain of tetrahedra. The chains are stacked so that the M1, M2, and M3 sites are between the apices of the tetrahedra, and the M4 and A sites are between the bases.   (*b*) View down the c axis schematically showing the stacking of the chains and the typical amphibole cross section. Cleavage between the chains is parallel to the (110) and (1$\bar{1}$0) crystal faces at about 56° and 124°.

ral M1, M2 and M3 sites. Z represents Si or some Al that occupy the tetrahedra. There are four main groups of amphiboles, based primarily on the occupancy of the M4 site (X) (Leake, 1978), as follows:

Iron–magnesium amphiboles
  Anthophyllite
    $(Mg,Fe)_7Si_8O_{22}(OH)_2$
    (Orthorhombic)
  Gedrite
    $(Mg,Fe)_5Al_2(Al_2Si_6)O_{22}(OH_2)$
    (Orthorhombic)
  Cummingtonite–grunerite
    $(Fe,Mg)_7Si_8O_{22}(OH)_2$
    (Monoclinic)
Calcic amphiboles
  Tremolite–actinolite
    $Ca_2(Mg,Fe)_5Si_8O_{22}(OH)_2$
    (Monoclinic)
  Hornblende
    $(Na,K)_{0-1}Ca_2(Mg,Fe^{2+},Fe^{3+},Al)_5(Si,Al)_8O_{22}(OH)_2$
    (Monoclinic)
  Oxyhornblende
    $(Na,K)_{0-1}Ca_2(Mg,Fe^{2+},Fe^{3+},Al)_5(Si,Al)_8O_{22}(O,OH)_2$
    (Monoclinic)
  Kaersutite
    $NaCa_2(Mg,Fe^{2+})_4TiSi_6Al_2O_{22}(OH)_2$
    (Monoclinic)
Sodic–calcic amphiboles
  Katophorite
    $Na(Na,Ca)(Mg,Fe^{2+},Fe^{3+},Al)_5Si_7AlO_{22}(OH)_2$
    (Monoclinic)
  Richterite
    $Na(Na,Ca)(Mg,Fe)_5Si_8O_{22}(OH)_2$
    (Monoclinic)
Sodic amphiboles
  Glaucophane
    $Na_2(Mg,Fe)_3Al_2Si_8O_{22}(OH)_2$
    (Monoclinic)
  Riebeckite
    $Na_2(Mg,Fe^{2+})_3Fe^{3+}{}_2Si_8O_{22}(OH)_2$
    (Monoclinic)
  Arfvedsonite–eckermanite
    $NaNa_2(Mg,Fe^{2+})_4(Fe^{3+},Al)Si_8O_{22}(OH)_2$
    (Monoclinic)

As with the pyroxenes, there is substantial solid solution among most of the members of the group, and accurate determination of composition based on optical properties is usually not possible.

Anthophyllite

$(Mg,Fe)_7Si_8O_{22}(OH)_2$
Orthorhombic
Biaxial $(+)$ or $(-)$
$n_\alpha = 1.587-\sim1.660$
$n_\beta = 1.602-\sim1.670$
$n_\gamma = 1.613-\sim1.680$
$\delta = 0.013-0.028$
$2V_x = 65-90°$ (negative)
$2V_z = 90-58°$ (positive)

(010) Section

RELIEF IN THIN SECTION: Moderate to high positive relief.

COMPOSITION: The major compositional variation is in the relative amount of Fe^{2+} and Mg. Anthophyllite is usually restricted to no more than about 40 mole percent Fe^{2+}. At high temperatures, there appears to be continuous solid solution to gedrite, but there is a solvus at lower temperatures. With slow cooling, exsolution lamellae of gedrite may develop parallel to (010) in anthophyllite, but the lamellae are generally quite fine and are rarely visible. Mn can substitute for (Mg,Fe), as can small amounts of Ca.

PHYSICAL PROPERTIES: $H = 5\frac{1}{2}-6$; $G = 2.85-3.28$; usually medium brown in hand sample, less commonly green, gray, or light brown; gray streak; vitreous luster.

COLOR AND PLEOCHROISM: Colorless or pale brown in thin section, somewhat darker in grain mount. Weak pleochroism with $Z \geqslant Y \geqslant X$: X = colorless, pale yellow, pale gray-brown, pale greenish yellow; Y = colorless, pale yellow, pale gray-brown, yellow-green; Z = pale gray-brown, purplish brown, or gray-green. Grains are typically elongate parallel to

(a)

(b)

(c)

(d)

(e)

Figure 13.11 Amphiboles. (*a*) Anthophyllite. (*b*) Colorless tremolite in calcite which shows distinct cleavage. (*c*) Hornblende in amphibolite with both longitudinal and basal sections present. (*d*) Hornblende in granodiorite. Note the twin parallel to the length of the larger grain. (*e*) Glaucophane showing distinct amphibole cleavage. Widths of fields of view: (*a–c*) 1.7 mm; (*d*) 5 mm; (*e*) 0.9 mm.

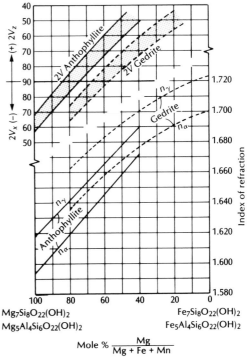

Figure 13.12 Indices of refraction and 2V for anthophyllite (solid curves) and gedrite (dashed curves). There is substantial scatter in the indices and only the trends are shown. Data from Deer and others (1963), Rabbitt (1948), and Winchell (1938).

the c axis and are darkest when the long dimension is parallel to the vibration direction of the lower polar. The darker colors are associated with higher iron content.

FORM: Figure 13.11. Commonly found as euhedral to subhedral columnar, bladed, or acicular crystals, with diamond-shaped cross section typical of amphiboles. Sometimes found as fine to coarse radiating aggregates. There also is an asbestiform variety.

CLEAVAGE: Perfect cleavages on $\{210\}$ intersect at about 55 and 125°, similar to the other amphiboles. Poor cleavages on $\{100\}$ and $\{010\}$ are generally not seen. A $\{001\}$ parting may be seen. Cleavage tends to control fragment shape.

TWINNING: None reported.

OPTICAL ORIENTATION: $X = a$, $Y = b$, $Z = c$, optic plane $= (010)$. Extinction is symmetrical in basal sections and parallel in longitudinal sections. Elongate grains are length slow. Cleavage fragments are elongate parallel to the c axis, show parallel extinction, and are length slow.

INDICES OF REFRACTION AND BIREFRINGENCE: Index of refraction increases with increasing iron content (Figure 13.12) and birefringence decreases moderately. Maximum interference colors in thin section are usually first-order red or second-order blue.

INTERFERENCE FIGURE: The $2V$ angle varies systematically with Fe:Mg ratio (Figure 13.12). Samples with more than about 80 percent Mg are optically negative; more iron-rich samples are usually optically positive. Cleavage fragments yield strongly off-center figures. Optic axis dispersion is weak to moderate, usually $r > v$ if optically negative, $v > r$ is positive.

ALTERATION: Fine-grained serpentine, talc, or other phyllosilicates are the usual alteration products.

DISTINGUISHING FEATURES: Parallel extinction in longitudinal sections distinguishes anthophyllite from the monoclinic amphiboles of which cummingtonite–grunerite is the most similar. Cummingtonite–grunerite has inclined extinction, somewhat higher birefringence, and common lamellar twinning. Sillimanite is usually colorless, has only one cleavage, and a smaller $2V$. Gedrite is quite similar and may be difficult to distinguish from anthophyllite with which it may be associated. However, gedrite has

higher indices of refraction, and optically negative gedrite is rare. Additionally, anthophyllite typically does not have visible exsolution lamellae, whereas gedrite may; and gedrite, being more iron-rich, may be somewhat darker colored and more strongly pleochroic. Chemical or X-ray tests may be required to distinguish anthophyllite from gedrite.

OCCURRENCE: Anthophyllite is restricted to medium- and high-grade mafic metamorphic rocks. The most common occurrences are in anthophyllite–cordierite gneiss and amphibolite, but anthophyllite may also be found in granitic gneiss, anthophyllite–talc schist, hornfels, or other metamorphic rocks. The commonly associated minerals are cordierite, hornblende, gedrite, plagioclase, quartz, garnet, sillimanite, talc, and other Fe–Mg–Al silicates.

Gedrite

$(Mg,Fe)_5Al_2(Al_2Si_6)O_{22}(OH)_2$
Orthorhombic
Biaxial $(+)$ or $(-)$
$n_\alpha = 1.627–1.694$
$n_\beta = 1.635–1.710$
$n_\gamma = 1.644–1.722$
$\delta = 0.017–0.027$
$2V_x = 72–90°$ (negative)
$2V_z = 90–47°$ (positive)

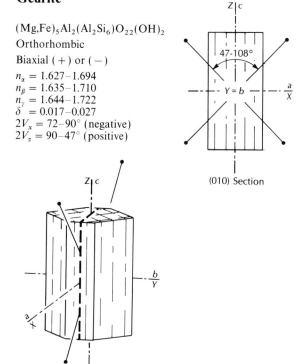

(010) Section

RELIEF IN THIN SECTION: Moderate to high positive relief.

COMPOSITION: Gedrite is closely related to anthophyllite and there appears to be continuous solid solution between the two minerals at high temperature. However, a solvus exists at lower temperature, so that with slow cooling, lamellae of anthophyllite may exsolve parallel to (010) in gedrite. The main compositional variation is in the relative amounts of Fe^{2+} and Mg. Gedrite is typically more iron-rich than associated anthophyllite and rarely has less than 20 mole percent Fe^{2+}. Mn may substitute for (Fe,Mg) as may small amounts of Ca.

PHYSICAL PROPERTIES: H = $5\frac{1}{2}$–6: G = ~3.0–3.57; usually medium brown in hand sample, less commonly green, gray, or light brown; gray streak, vitreous luster.

COLOR AND PLEOCHROISM: Pale shades of brown, yellow-brown, or greenish brown in thin section, somewhat darker in grain mount. Weak pleochroism with $Z > Y = X : X = Y =$ pale yellowish-, greenish-, or grayish-brown; $Z =$ yellowish brown, greenish gray, grayish brown, purplish brown. Grains are typically elongate parallel to the c axis and are darkest when the long dimension is parallel to the vibration direction of the lower polar. Darker colors are associated with higher iron content.

FORM: Commonly found as euhedral to subhedral columnar, bladed, or acicular crystals, with diamond-shaped cross section typical of the amphiboles. May be found as coarse or fine radiating aggregates and as an asbestiform variety.

CLEAVAGE: Perfect cleavages on $\{210\}$ intersect at about 55 and 125° similar to other amphiboles. Poor cleavages on $\{100\}$ and $\{010\}$ are generally not seen. A parting is reported on $\{001\}$. Cleavage tends to control fragment shape.

TWINNING: None reported.

OPTICAL ORIENTATION: $X = a$, $Y = b$, $Z = c$, optic plane = (010). Extinction is symmetrical in basal sections and parallel in longitudinal sections. Elongate grains are length slow. Cleavage fragments are elongate parallel to the c axis, show parallel extinction, and are length slow.

INDICES OF REFRACTION AND BIREFRINGENCE: Gedrite has somewhat higher indices of refraction than anthophyllite (Figure 13.12) and the indices of refraction increase with increasing iron content. Birefringence increases moderately with increasing iron content. Maximum interference colors in thin section are usually first-order red or second-order blue or green.

INTERFERENCE FIGURE: The $2V$ angle varies systematically with composition (Figure 13.12). Samples with more than about 65 percent Mg are optically negative; less Mg-rich samples are positive. Cleavage fragments yield strongly off-center figures. Optic axis dispersion is weak with $v > r$ for most optically positive varieties, $r > v$ for negative.

ALTERATION; Fine-grained serpentine, talc, or other phyllosilicates are the usual alteration products.

DISTINGUISHING FEATURES: Parallel extinction distinguishes gedrite from the monoclinic amphiboles of which cummingtonite–grunerite is the most similar. Cummingtonite–grunerite has inclined extinction, somewhat higher birefringence, and common lamellar twinning. Sillimanite is usually colorless, has only one cleavage, and a smaller $2V$. Anthophyllite is quite similar and may be difficult to distinguish from gedrite with which it may be associated. However, gedrite has higher indices of refraction, and optically negative gedrite is rare. Additionally, gedrite may have visible exsolution lamellae, whereas anthophyllite generally does not, and gedrite, being more iron-rich, may be somewhat darker colored and more strongly pleochroic. X-ray or chemical tests may be required to distinguish gedrite from anthophyllite.

OCCURRENCE: Gedrite is restricted to medium- and high-grade metamorphic rocks such as amphibolite, anthophyllite–cordierite gneiss, talc schist, hornfels, and granitic gneiss. It is commonly associated with garnet, cordierite, anthophyllite, plagioclase, hornblende, sillimanite, talc, and other Fe–Mg–Al silicates.

Cummingtonite–Grunerite

$(Mg,Fe)_7Si_8O_{22}(OH)_2$
Monoclinic
$\angle\beta \cong 102°$
Biaxial $(+)$ or $(-)$
$n_\alpha = 1.630–1.696$
$n_\beta = 1.638–1.709$
$n_\gamma = 1.655–1.729$
$\delta = 0.020–0.045$
$2V_z = 65–98°$

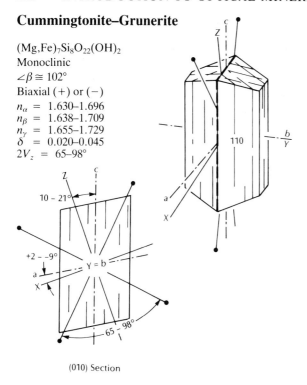

(010) Section

RELIEF IN THIN SECTION: Moderately high to high positive relief.

COMPOSITION: The primary compositional variation is in the relative amount of Mg and Fe^{2+}. The term grunerite is used for varieties having less than 30 percent Mg, and cummingtonite for between 30 and 70 percent Mg. Natural samples with more than about 70 mole percent Mg apparently do not occur. Some Mn also may be present, and there is apparently a solid solution series to *tirodite* $[Mn_2Mg_5Si_8O_{22}(OH)_2]$. Small amounts of Ca also are usually present.

PHYSICAL PROPERTIES: H = 5–6; G = 3.10–3.60; brown to dark green in hand sample; white or gray streak; vitreous luster.

COLOR AND PLEOCHROISM: Colorless to pale green in thin section; somewhat darker in grain mount. Darker colors are associated with higher iron content. Colored varieties are weakly pleochroic: $X =$ colorless, pale yellow; $Y =$ pale yellow, pale brown; $Z =$ pale green, pale brown. Cummingtonite commonly shows somewhat stronger pleochroism than grunerite.

FORM: Usually columnar, bladed, or acicular and elongate parallel to the c axis. Grains may be arranged in parallel or radiating aggregates. A fine asbestiform variety is called *amosite* or *montasite*. Basal sections show the typical diamond shape of the amphiboles.

CLEAVAGE: The good amphibole cleavages on {110} intersect at about 56 and 124°.

TWINNING: Simple and lamellar twinning on {100} is very common. Twin lamellae are parallel to the length of grains cut in longitudinal section. The lamellae are parallel to the long diagonal in basal sections.

OPTICAL ORIENTATION: $X \wedge a = +2$ to $-9°$, $Y = b$, $Z \wedge c = +10$ to $+21°$, optic plane = (010). Extinction is symmetrical in basal sections. Sections cut parallel to (010) have maximum birefringence and show the $Z \wedge c$ extinction angle of 10 to 15° (grunerite) and 15 to 21° (cummingtonite). Sections parallel to (100) show parallel extinction. The slow ray vibration is along the long diagonal between the cleavages in basal section, and in longitudinal section is closest to the cleavage traces (length slow).

INDICES OF REFRACTION AND BIREFRINGENCE: Indices of refraction and birefringence increase with increasing Fe^{2+} (Figure 13.13). Cummingtonite shows colors up to upper second order and grunerite may be as high as mid third order. Tirodite has optical properties $n_\alpha = 1.630–1.638$, $n_\beta = 1.644–1.651$, $n_\gamma = 1.652–1.665$; $2V_z = 73–89°$, $r > v$; $Z \wedge c = 20°$.

INTERFERENCE FIGURE: The $2V$ angle varies with Mg:Fe proportion. Samples with more than 30 to 40 percent Mg are positive ($2V_z = 65–90°$) and with less are negative ($2V_x = 82–90°$). Fragments lying on cleavages show strongly off-center figures. Weak optic axis dispersion is $r > v$ (cummingtonite) and $v > r$ (grunerite).

ALTERATION: Overgrowths of hornblende are not uncommon. Thin lamellae of hornblende may exsolve under conditions of slow cooling. Alteration to mafic phyllosilicates (e.g., chlorite and talc) is also possible.

DISTINGUISHING FEATURES: Cummingtonite and grunerite are distinguished from anthophyllite–gedrite by inclined extinction, higher birefringence, and prevalence of polysynthetic twinning. A positive

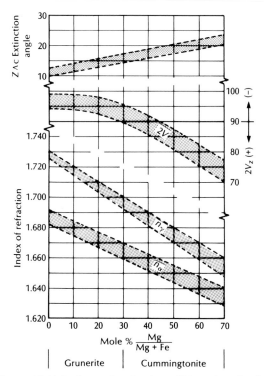

Figure 13.13 Indices of refraction, $2V$, and extinction angle $Z > c$ for cummingtonite and grunerite. Data from Klein (1964), Winchell (1938), and Deer and others (1963).

optic sign and higher indices distinguish most cummingtonite from tremolite–actinolite, which also has a different occurrence. The distinction between grunerite and cummingtonite is most easily made based on extinction angle and optic sign.

OCCURRENCE: Cummingtonite is usually found in fairly mafic metamorphic rocks such as amphibolite, cummingtonite–anthophyllite–cordierite gneiss, and hornfels, as well as granitic gneiss and granulite. It is commonly associated with anthophyllite, gedrite, garnet, cordierite, plagioclase, hornblende, and biotite. It is occasionally found in intermediate volcanic rocks, diorite, gabbro, norite, and in skarns. Grunerite is found in metamorphosed iron-rich sediments along with magnetite, quartz, and other iron oxides and silicates.

Tremolite–Actinolite

$Ca_2(Mg,Fe^{2+})_5Si_8O_{22}(OH)_2$
Monoclinic
$\angle\beta = 104.7°$
Biaxial ($-$)
$n_\alpha = 1.599–1.688$
$n_\beta = 1.612–1.697$
$n_\gamma = 1.622–1.705$
$\delta = 0.017–0.027$
$2V_x = 75–88°$

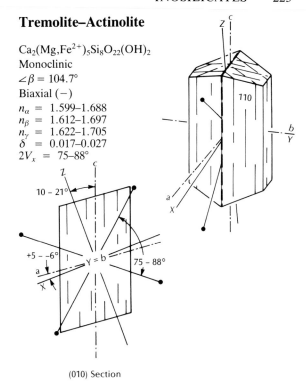

(010) Section

RELIEF IN THIN SECTION: Moderate to high positive relief.

COMPOSITION: The primary compositional variation is the relative proportion of Fe^{2+} and Mg. By convention the term tremolite is applied to samples having greater than 90 percent Mg, actinolite for compositions between 90 and 50 percent Mg, and ferro-actinolite for more iron-rich compositions. Ferro-actinolite compositions are uncommon. There appears to be substantial solid solution with the other calcic, sodic-calcic, and alkalic amphiboles. However, the solid solution is not continuous in all cases and miscibility gaps are present between tremolite–actinolite and certain hornblende compositions, riebeckite, and glaucophane and probably other compositions as well. The terms tremolite, actinolite, and ferro-actinolite are restricted to amphiboles having a tetrahedral Al content of less than $(Si_{7.5}Al_{0.5}O_{22})$, a high percentage of Ca in the M4 site, and little Fe^{3+} and Al in the M1, M2, and M3 sites. OH may be replaced by O, Cl, or F.

PHYSICAL PROPERTIES: H = 5–6; G = 2.96–3.46; tremolite is white or gray in hand sample, actinolite and ferro-actinolite are light to dark green or greenish black, darker with higher iron content; white or gray streak; vitreous luster.

COLOR AND PLEOCHROISM: Colorless, pale green to deep green in thin section or grain mount. Darker color and stronger pleochroism are associated with higher iron content. Colored varieties are pleochroic with $Z > Y \geqslant X$: X = colorless, pale yellow-green; Y = pale yellow-green, pale blue-green, green; Z = pale green, green, blue-green, dark green.

FORM: Figure 13.11. Commonly as columnar, bladed, or acicular crystals elongate parallel to the c axis. Sometimes fibrous or asbestiform. Basal sections are diamond shaped and show the typical amphibole cleavage. Longitudinal sections (parallel to the c axis) are roughly rectangular and show only one cleavage trace.

CLEAVAGE: Typical amphibole cleavage on {110}, intersects at 56 and 124°.

TWINNING: Simple and lamellar twins with {100} as the composition plane are common. Fine lamellar twins on {001} are rare. The {100} twin lamellae are parallel to the long diagonal in basal sections.

OPTICAL ORIENTATION: $X \wedge a = +5$ to $-6°$, $Y = b$, $Z \wedge c = +10$ to $+21°$, optic plane = (010). Basal sections show symmetrical extinction with the slow ray parallel to the long diagonal. The $Z \wedge c$ extinction angle of 10 to 21° is seen in sections parallel to (010). Sections parallel to (100) show parallel extinction. The elongation is length slow. Extinction angles show a general decrease with increasing iron content (Figure 13.14).

INDICES OF REFRACTION AND BIREFRINGENCE: Indices of refraction increase with increasing iron content (Figure 13.14). Maximum interference colors in thin section are upper first order to mid second order. The (010) section that shows the $Z \wedge c$ extinction angle also shows maximum birefringence.

INTERFERENCE FIGURE: The entire series is optically negative with the $2V_x$ angle decreasing from about 88 to below 75° with increasing iron content. Cleavage fragments yield highly off-center figures. Because the acute bisectrix is nearly parallel to the a axis, moderately elongate sections showing darker colors and parallel extinction yield acute bisectrix or optic axis sections.

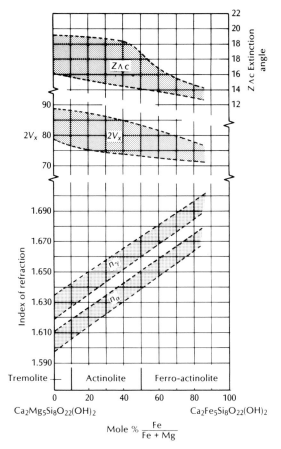

Figure 13.14 Indices of refraction, $2V$, and $Z \wedge c$ extinction angle for the tremolite–actinolite series. Data from Deer and others (1963), and Winchell (1945).

ALTERATION: The more common alteration products are chlorite, talc, and carbonates.

DISTINGUISHING FEATURES: Tremolite is distinguished from anthophyllite by inclined extinction and from cummingtonite by the negative sign and lower indices. Actinolite and ferro-actinolite closely resemble members of the hornblende group; however, hornblende usually has a smaller $2V$ angle and may have a larger extinction angle. If the identity is in doubt, the term *calcic clinoamphibole* is appropriate.

OCCURRENCE: The common occurrence is in contact and regionally metamorphosed limestone, dolomite, and other calcareous sediments. The commonly associated minerals are forsterite, garnet, diopside, wollastonite, calcite, dolomite, talc, and members of the epidote group. Both actinolite and hornblende may be present in some rocks. Tremolite and actinolite also are found in metamorphosed mafic and ultramafic rocks and in glaucophane schist and associated rocks. Uralite is a common fine-grained alteration product of pyroxenes that is usually composed primarily of tremolite or actinolite.

Hornblende

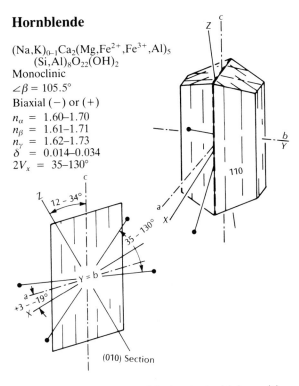

$(Na,K)_{0-1}Ca_2(Mg,Fe^{2+},Fe^{3+},Al)_5$
$\quad (Si,Al)_8O_{22}(OH)_2$
Monoclinic
$\angle\beta = 105.5°$
Biaxial $(-)$ or $(+)$
$n_\alpha = 1.60-1.70$
$n_\beta = 1.61-1.71$
$n_\gamma = 1.62-1.73$
$\delta = 0.014-0.034$
$2V_x = 35-130°$

RELIEF IN THIN SECTION: Moderate to high positive relief.

COMPOSITION: Hornblende has a wide range of compositions. In comparison with tremolite–actinolite, hornblende contains significant amounts of Na and K in the A structural site, and Fe^{3+} and Al^{3+} in the M1, M2, and M3 sites. The charge is balanced by substituting Al^{3+} for Si^{4+} in the tetrahedral sites up to about $Al_{2.25}Si_{5.75}$. The material collectively called hornblende here includes members of the following series:

Magnesio-hornblende—ferro-hornblende
$\quad Ca_2(Mg,Fe^{2+})_4Al(Si_7AlO_{22})(OH)_2$

Tschermakite—ferro-tschermakite
$\quad Ca_2(Mg,Fe^{2+})_3Fe^{3+}{}_2Si_6Al_2O_{22}(OH)_2$
Edenite—ferro-edenite
$\quad NaCa_2(Mg,Fe^{2+})_5Si_7AlO_{22}(OH)_2$
Pargasite—ferro-pargasite
$\quad NaCa_2(Mg,Fe^{2+})_4AlSi_6Al_2O_{22}(OH)_2$
Hastingsite—magnesio-hastingsite
$\quad NaCa_2(Mg,Fe^{2+})_4Fe^{3+}Si_6Al_2O_{22}(OH)_2$

Other elements such as Ti, Mn, and Cr, may also be present and F, Cl, and O may replace (OH). The compositional complexity makes it impossible to use optical data alone to determine the composition.

PHYSICAL PROPERTIES: H = 5–6; G = 3.02–3.45; greenish black in hand sample, less commonly brown; greenish gray to gray streak; vitreous luster.

COLOR AND PLEOCHROISM: Distinctly colored and pleochroic in both thin section and grain mount, usually in shades of green, yellow-green, blue-green, and brown. The green varieties usually have X = light yellow, light yellow-green, light blue-green; Y = green, yellow-green, gray-green; Z = dark green, dark blue-green, dark gray-green. The brownish varieties usually show X = yellow, greenish yellow, light greenish brown; Y = yellow-brown, brown, reddish brown; Z = gray-brown, dark brown, reddish brown. Concentric or patchy color zoning is common. Sections are usually darker colored when the long dimension is roughly parallel to the vibration direction of the lower polar.

FORM: Figure 13.11. Commonly found as slender prismatic to bladed crystals with a diamond-shaped cross section showing the amphibole cleavage at 56 and 124°. Also found as anhedral to strongly irregular grains, which in some cases may poikilitically enclose associated minerals. Large grains or fine fibrous masses of hornblende may mantle or replace pyroxene.

CLEAVAGE: The typical amphibole cleavages on {110} intersect at 56 and 124°. Fragment shape is controlled by cleavage and is usually elongate parallel to the c axis. There also are partings on {100} and {001}.

TWINNING: Simple and lamellar twins on {100} are not uncommon.

OPTICAL ORIENTATION: $X \wedge a = +3$ to $-19°$, $Y = b$, $Z \wedge c = +12$ to $+34°$, optic plane = (010). Basal sections show symmetrical extinction with the slow ray parallel to the long diagonal between cleavages. Longitudinal sections are length slow. The $Z \wedge c$ extinction angle of 12 to 34° is seen in sections parallel to (010), which show one cleavage direction and maximum birefringence. In most cases, the extinction angle is between 14 and 25°.

INDICES OF REFRACTION AND BIREFRINGENCE: The indices of refraction show an increase with increasing iron content (Figure 13.15). Most hornblende has indices in the range $n_\alpha = 1.655 \pm 0.010$, $n_\beta = 1.665 \pm 0.010$, and $n_\gamma = 1.675 \pm 0.015$, with birefringence in the range 0.018 to 0.028. The highest interference colors in thin section are usually upper first or lower second order but are often masked by the mineral color.

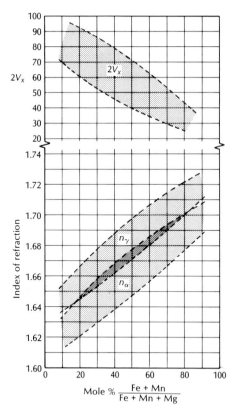

Figure 13.15 Indices of refraction and $2V$ for hornblende. Data from Deer and others (1963), and other sources. The bands cover the common range of indices and $2V$ but do not include the full range reported in the literature.

INTERFERENCE FIGURE: Common hornblende is biaxial negative, although much pargasite and some tschermakite and edenite are optically positive with large $2V$. The $2V$ angle shows a general decrease with increasing iron content (Figure 13.15). For common hornblende, the $2V_x$ angle falls in the range $75 \pm 10°$. Elongate sections that show darker colors and parallel or symmetrical extinction are cut perpendicular to the (010) optic plane and are most likely to yield acute bisectrix and optic axis figures. Optic axis dispersion is moderate with $r > v$ or $v > r$.

ALTERATION: Hornblende may be altered to biotite, chlorite, or other Fe–Mg silicates.

DISTINGUISHING FEATURES: Cleavage and crystal habit distinguish hornblende from the dark-colored pyroxenes. Tremolite is light colored and has lower indices. Actinolite has lower indices and a smaller extinction angle. Cummingtonite is usually optically positive, lighter colored, and characteristically shows fine lamellar twinning. Anthophyllite shows parallel extinction, is lighter colored, and may be optically positive. Oxyhornblende is dark brown, strongly pleochroic, and has higher birefringence. The sodic amphiboles are blue. Biotite shows only one cleavage, parallel extinction, and has a smaller $2V$. The optical properties of hornblende merge with those of actinolite, and distinguishing between the two may, in some cases, be impractical without chemical or other information. The term *calcic clinoamphibole* is appropriate for samples of uncertain identity.

OCCURRENCE: Hornblende is a common mineral found in a variety of geological environments. It can be found in almost any igneous rock but is typical of plutonic and extrusive rocks of intermediate composition such as diorite, granodiorite, trondhjemite, and andesite. It is not uncommon in more silica-rich granitic and rhyolitic rocks and crystallizes as a late magmatic mineral in gabbro, norite, and related mafic rocks.

Hornblende is a common mineral in medium- and high-grade regional metamorphic terranes in amphibolite, hornblende gneiss, or other schists and gneisses. It is also found in marble, skarns, or other metamorphosed carbonate rocks, but tremolite–actinolite is more common in these rocks.

Pyroxene commonly alters to a fine-grained aggregate of amphibole called uralite. The amphibole is more commonly tremolite–actinolite but may be hornblende. The alteration usually begins along grain margins or cracks and progresses inward. The fine-grained amphibole may recrystallize to larger optically continuous grains that mantle or replace the pyroxenes.

Oxyhornblende

$(Na,K)_{0-1}Ca_2(Mg,Fe^{2+},Fe^{3+}Al)_5(Si,Al)_8O_{22}(O,OH)_2$
Monoclinic
$\angle \beta \cong 106°$
Biaxial $(-)$
$n_\alpha = 1.65–1.70$
$n_\beta = 1.67–1.78$
$n_\gamma = 1.68–1.80$
$\delta = 0.018–0.083$
$2V_x = 56–88°$

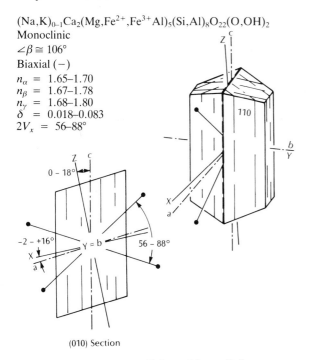

(010) Section

RELIEF IN THIN SECTION: High positive relief.

COMPOSITION: Oxyhornblende is not a distinct mineral species in the conventional sense, but rather appears to be hornblende in which a substantial amount of the iron has been oxidized to the Fe^{3+} state, and $(OH)^-$ is replaced by O^{2-} to balance the charge. There is a continuous range of compositions and properties between members of the hornblende group and oxyhornblende, but the relatively restricted occurrence in volcanic rocks warrants a separate description. Conventional hornblende can be converted to oxyhornblende by heating in an oxidizing environment, and the process may be reversed by heating in a reducing environment. The term *basaltic hornblende* has been used for oxyhornblende but is now discredited. Zoning is common, with the rim darker colored and more oxidized than the core.

PHYSICAL PROPERTIES: H = 5–6; G = 3.2–3.3; brown to black in hand sample; gray streak; vitreous luster.

COLOR AND PLEOCHROISM: Oxyhornblende is distinctly brown or reddish brown in thin section and grain mount and has strong pleochroism with $Z > Y > X$: X = pale yellow, yellow, pale brown, greenish yellow; Y = brown, reddish brown, brownish green; Z = dark brown, dark reddish brown, dark green-brown.

FORM: Oxyhornblende usually forms euhedral or subhedral phenocrysts in volcanic and shallow intrusive rocks. Crystals have the typical amphibole cross section with cleavages at 56 and 124°, and longitudinal sections are roughly rectangular with a single cleavage trace. Grain margins may be irregular or corroded due to reaction with the magma. Pleochroic halos around radioactive mineral inclusions may be prominent.

CLEAVAGE: Typical amphibole cleavages on {110} intersect at 56 and 124°.

TWINNING: Simple and lamellar twins on {100} are not uncommon.

OPTICAL ORIENTATION: $X \wedge a = +16$ to $-2°$, $Y = b$, $Z \wedge c = 0$ to $+18°$, optic plane = (010). Extinction in basal section is symmetrical with the slow ray parallel to the long diagonal between the cleavages. Longitudinal sections show a single cleavage trace and are length slow. The $Z \wedge c$ extinction angle of 0 to 18° is seen in sections parallel to (010), which also show maximum birefringence and the strongest pleochroism. The extinction angle decreases with increasing degree of oxidation.

INDICES OF REFRACTION AND BIREFRINGENCE: Oxyhornblende has higher indices and birefringence than comparable hornblende, and there appears to be continuous variation between the two, depending on the degree of oxidation. The highest interference colors in thin section are between upper second and fourth order, but the strong mineral color masks the interference color.

INTERFERENCE FIGURE: $2V_x$ is fairly large (56–88°) and falls in the same range as hornblende. Elongate sections showing darker colors and parallel or symmetrical extinction can be used to obtain acute bisectrix or optic axis figures. Optic axis dispersion is weak to moderate with $v > r$.

ALTERATION: The periphery, or sometimes entire grains, may be replaced by fine-grained aggregates of pyroxene, biotite, plagioclase, magnetite, and hematite.

DISTINGUISHING FEATURES: The darker brown color, smaller extinction angle, and higher birefringence and indices distinguish oxyhornblende from hornblende and the other clinoamphiboles. Biotite may have similar strong pleochroism and color but has a different crystal habit, only one cleavage, essentially parallel extinction, and a smaller $2V$. Kaersutite is very similar but has somewhat different occurrences. The pleochroism, however, may be different with the colors associated with Y and Z being less different in kaersutite than in oxyhornblende.

   Katophorite (also spelled kataphorite) $[Na(Na,Ca)(Mg,Fe^{2+},Fe^{3+}Al)_5Si_7AlO_{22}(OH)_2]$ is a sodic-calcic amphibole with similar color and pleochroism that is found in relatively mafic alkalic intrusives associated with acmite, aegirine-augite, riebeckite, or arfvedsonite. Katophorite has lower indices ($n_\alpha = 1.639–1.681$, $n_\beta = 1.658–1.688$, $n_\gamma = 1.660–1.690$), $2V_x = 0–50°$, $X \wedge c$ extinction angle of 36–70°, optic plane either $\perp$ or $\parallel$ to (010).

OCCURRENCE: Oxyhornblende is usually restricted to volcanic and shallow intrusive rocks such as trachytes, basalt, andesite, latite, and other intermediate compositions.

Kaersutite

$NaCa_2(Mg,Fe^{2+})_4TiSi_6Al_2O_{22}(OH)_2$
Monoclinic
$\angle \beta \cong 106°$
Biaxial $(-)$
$n_\alpha = 1.670–1.707$
$n_\beta = 1.690–1.741$
$n_\gamma = 1.700–1.772$
$\delta = 0.019–0.083$
$2V_x = 66–84°$

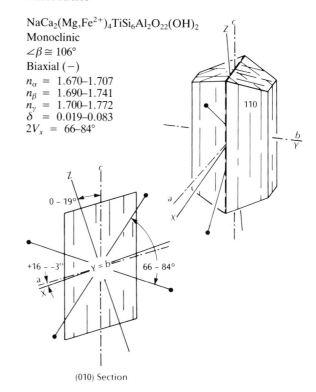

(010) Section

RELIEF IN THIN SECTION: High positive relief.

COMPOSITION: Kaersutite is a titaniferous amphibole analogous to titanaugite in the pyroxenes. There is probably a continuous range of compositions and optical properties between kaersutite and members of the hornblende group.

PHYSICAL PROPERTIES: H = 5–6; G = 3.2–3.3; dark brown or black in hand sample; gray streak; vitreous luster.

COLOR AND PLEOCHROISM: Usually yellowish brown, brown, reddish brown or greenish brown in thin section or grain mount. Pleochroic with X = yellow to brownish yellow, Y = reddish brown, Z = dark reddish brown or greenish brown.

FORM: Kaersutite forms typical amphibole crystals. It may also be intergrown with titanaugite, or titanian biotite, or may form overgrowths on titanaugite. Many samples are color zoned and it commonly is rimmed by fine grained Fe–Ti oxides.

CLEAVAGE: Typical amphibole cleavages on {110} intersect at 56 and 124°. There may also be partings on {100} and {001}.

TWINNING: Simple or lamellar twins on {100} are common.

OPTICAL ORIENTATION: $X \wedge a = +16$ to $-3°$, $Y = b$, $Z \wedge c = 0$ to $19°$, optic plane = (010). Extinction in basal sections is symmetrical. Longitudinal sections show a single cleavage trace and are length slow. The $Z \wedge c$ extinction angle is seen in (010) sections, which also display maximum birefringence.

INDICES OF REFRACTION AND BIREFRINGENCE: Indices of refraction and birefringence increase rapidly with increasing Ti content and increase moderately with increasing ratio of ferric to ferrous iron. Maximum interference colors in thin section range from second order to as high as fourth or fifth order, but the strong mineral color may mask the color.

INTERFERENCE FIGURE: Kaersutite is biaxial negative with $2V_x$ between 66 and 84°. Basal sections yield moderately off-center obtuse bisectrix figures. Acute bisectrix and optic axis figures are obtained from sections cut perpendicular to (010). Optic axis dispersion is strong, $r > v$.

ALTERATION: Kaersutite may alter to dark amphiboles, titanaugite, Fe–Ti oxides, and chlorite.

DISTINGUISHING FEATURES: Oxyhornblende may be difficult to distinguish from kaersutite, because the optical properties are essentially the same. However, at least for fairly magnesian kaersutite, the pleochroism between Y and Z seen in grains oriented to yield acute bisectrix figures is significantly less strong than for oxyhornblende. Katophorite has lower indices of refraction, smaller $2V$, and a large $X \wedge c$ extinction angle.

OCCURRENCE: Kaersutite is not a common amphibole but may be found in alkalic volcanic rocks such as trachybasalt, trachyandesite, and trachyte, and in intrusive rocks such as monzonite, syenite, camptonite, and in lamphrophyre dikes. Kaersutite is commonly associated with titanaugite, titanian biotite, Fe–Ti oxides, and titanite.

Richterite

$NaCaNa(Mg,Fe)_5Si_8O_{22}(OH)_2$
Monoclinic
$\angle\beta = 104.3°$
Biaxial $(-)$
$n_\alpha = 1.605–1.685$
$n_\beta = 1.615–1.700$
$n_\gamma = 1.622–1.712$
$\delta = 0.015–0.029$
$2V_x = 64–87°$

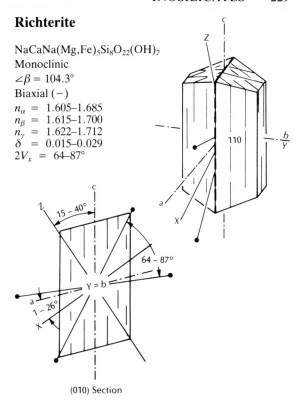

(010) Section

RELIEF IN THIN SECTION: Moderate to high positive relief.

COMPOSITIONS: The primary composition variation is in the relative amounts of Mg and Fe^{2+}. Most richterite is fairly Mg rich. There also may be significant solid solution toward more calcic amphiboles (e.g., tremolite–actinolite) and more sodic amphiboles (e.g., arfvedsonite–eckermanite), and significant amounts of Fe^{3+}, Al, or Mn may substitute for (Mg, Fe).

PHYSICAL PROPERTIES: H = 5–6; G = 2.97–3.45; brown, yellow, brownish red, or light to dark green in hand sample; white to gray streak; vitreous luster.

COLOR AND PLEOCHROISM: Colorless to yellow or violet in thin section and grain mount. Pleochroism variable in shades of pale yellow or green with orange or red tints, or sometimes violet or blue. The absorption is usually $Y > Z > X$, so some samples may be darker when oriented perpendicular to the lower polar vibration direction.

FORM: Crystals are bladed with the typical amphibole cross section; also fibrous or asbestiform.

CLEAVAGE: Typical amphibole cleavages on {110} intersect at 56 and 124°.

TWINNING: Simple or lamellar twinning on {100} may be present.

OPTICAL ORIENTATION: $X \wedge a = -1$ to $-26°$, $Y = b$, $Z \wedge c = +15$ to $+40°$, optic plane $= (010)$. Basal sections show symmetrical extinction with the slow ray parallel to the long diagonal. Sections parallel to (010) are length slow and show the $Z \wedge c$ extinction angle of 15 to 40° to the single cleavage trace. The lower extinction angles are for the more common Mg-rich varieties.

INDICES OF REFRACTION AND BIREFRINGENCE: Magnesium-rich varieties have indices at the lower end of the range given above. Synthetic Mg-richterite has indices of $n_\alpha = 1.604$ and $n_\gamma = 1.622$. Interference colors in thin section range up to upper first order or mid second order. Iron-rich varieties may show anomalous interference colors.

INTERFERENCE FIGURE: Richterite is biaxial negative with a large $2V_x$ (64–87°). Optic axis dispersion is slight to strong, $r > v$, with stronger dispersion associated with iron-rich samples.

ALTERATION: Alteration to pyroxene has been reported.

DISTINGUISHING FEATURES: Colorless varieties resemble tremolite and may be difficult to distinguish from it. The color in pleochroic varieties of richterite is distinctive.

OCCURRENCE: Richterite is found in metamorphosed limestone and skarn deposits. It is possible that amphibole identified as tremolite from these rocks is actually richterite. It is sometimes found associated with carbonatites, in hydrothermal deposits, and has been reported in pegmatites.

Arfvedsonite and Eckermanite

Monoclinic
$\angle\beta = 105°$

Arfvedsonite
$NaNa_2Fe^{2+}_4Fe^{3+}Si_8O_{22}(OH)_2$
Biaxial (+) or (−)
$n_\alpha = 1.638{-}1.700$
$n_\beta = 1.643{-}1.709$
$n_\gamma = 1.650{-}1.710$
$\delta = 0.005{-}0.016$
$2V_x = $ variable (0–100°?)

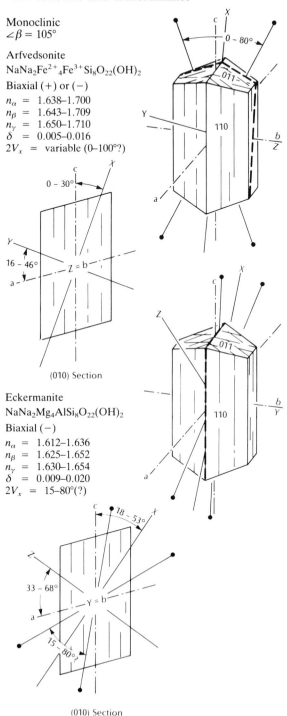

(010) Section

Eckermanite
$NaNa_2Mg_4AlSi_8O_{22}(OH)_2$
Biaxial (−)
$n_\alpha = 1.612{-}1.636$
$n_\beta = 1.625{-}1.652$
$n_\gamma = 1.630{-}1.654$
$\delta = 0.009{-}0.020$
$2V_x = 15{-}80°(?)$

(010) Section

RELIEF IN THIN SECTION: Arfvedsonite has high positive relief and eckermanite moderately high positive relief.

COMPOSITION: There appears to be complete solid solution between arfvedsonite and eckermanite. The division between them is placed at $\dfrac{Fe^{3+}}{Fe^{3+} + Al} = 0.5$. Arfvedsonite with more Mg than Fe^{2+} is called magnesio-arfvedsonite, and eckermanite with more Fe^{2+} than Mg is called ferro-eckermanite. Ca usually is present in amounts up to $NaNa_{1.5}Ca_{0.5}$, some K may replace Na, and F may replace (OH). In addition, there appears to be some solid solution toward members of the glaucophane–riebeckite group and to richterite. Compositional zoning is common and is expressed by variation in color and optical properties. Exsolution lamellae of cummingtonite may occur in some magnesio-arfvedsonite.

PHYSICAL PROPERTIES: H = 5–6; G = 3.0–3.5; green-black to black in hand sample, less commonly blue-green or brownish for Mg-rich compositions; gray streak; vitreous luster.

COLOR AND PLEOCHROISM: Strongly colored in shades of blue-green, brown, or violet with distinct pleochroism. Arfvedsonite has variable absorption with X = blue-green; Y = light blue-green, brown-yellow, gray-violet; Z = yellow-green, blue-gray, brownish green, black. Eckermanite usually has absorption of $X > Y > Z$, with X = blue-green, Y = pale blue-green, and Z = pale yellowish green. Arfvedsonite may be nearly opaque in some orientations.

FORM: Commonly found as elongate lath shaped to prismatic crystals with typical diamond shaped amphibole cross sections. Also found as fibrous aggregates and as poikiloblasts enclosing grains of associated minerals.

CLEAVAGE: Typical amphibole cleavage on {110} intersects at 56 and 124°. Also has a parting on {010}.

TWINNING: Simple and lamellar twins on {100} may be present.

OPTICAL ORIENTATION: Arfvedsonite: optic plane is perpendicular to (010) with $X \wedge c$ = 0 to $-30°$, $Y \wedge a$ = $+16$ to $+46°$, $Z = b$. Basal sections show symmetrical extinction, and longitudinal sections show parallel or inclined extinction with the maximum extinction angle of 0 to 30° seen in sections parallel to (010). Longitudinal sections are length fast. Extinction may not be complete due to strong dispersion.

Eckermanite and some intermediate compositions are oriented $X \wedge c = -18$ to $-53°$, $Y = b$, $Z \wedge a = +33$ to $+68°$, optic plane = (010). Longitudinal sections may be either length fast or length slow. Extinction may be irregular or flamy, and the maximum extinction angle to the fast ray ($X \wedge c$) in (010) sections is usually greater than 40°.

INDICES OF REFRACTION AND BIREFRINGENCE: Indices of refraction decrease in a fairly systematic manner from arfvedsonite to eckermanite (i.e., with increasing Mg and Al content), but there is substantial scatter. Most arfvedsonite has low birefringence (0.005–0.012), yielding lower to middle first-order colors in thin section. Eckermanite may have birefringence up to about 0.020. Interference colors are usually masked by the mineral color and may be anomalous.

INTERFERENCE FIGURE: Although biaxial positive arfvedsonite has been reported, most appears to be biaxial negative. Acute bisectrix figures are seen from near-basal sections. The $2V_x$ angle is quite variable and values anywhere between 0 and 100° appear to be possible. Eckermanite has a similar large range of $2V_x$ angles (~15–80°). Interpreting interference figures is often difficult due to strong coloring from the mineral. Fragments lying on cleavage surfaces yield strongly off-center figures.

ALTERATION: Arfvedsonite and eckermanite may be uralitized (converted to fine-grained, light-colored amphibole) or altered to iron oxides and carbonates. They also may be replaced by biotite or acmite.

DISTINGUISHING FEATURES: Eckermanite resembles actinolite and hornblende but is distinguished by its usually larger extinction angle and more deeply colored fast ray vibration direction. Arfvedsonite is distinguished from most amphiboles because the optic plane is oriented perpendicular to (010). To check, obtain an acute bisectrix figure, place the optic plane E–W and return to orthoscopic illumination. If the long diagonal between the cleavages is E–W, the optic plane is normal to (010). Riebeckite, crossite, and katophorite may have the same orientation but have higher birefringence and distinctly different colors.

OCCURRENCE: Arfvedsonite and eckermanite are usually found in alkalic igneous rocks, such as alkalic granite, syenite, nepheline syenite, shonkinite, trachyte, phonolite, and associated pegmatites. The associated minerals are often acmite, aegirine-augite, biotite, K-feldspar, plagioclase, or nepheline. The sodic pyroxenes may form parallel intergrowths.

RELIEF IN THIN SECTION: Moderate to moderately high positive relief.

COMPOSITION: Glaucophane and riebeckite display a continuous solid solution series. The principal substitutions are Mg for Fe^{2+}, and Al for Fe^{3+}. Members of the glaucophane group have $Fe^{3+}/(Fe^{3+}+Al) < 0.3$, and are called glaucophane if $Mg > Fe^{2+}$ and ferro-glaucophane if $Fe^{2+} > Mg$. Members of the riebeckite group have $Fe^{3+}/(Fe^{3+}+Al) > 0.7$ and are called riebeckite if $Fe^{2+} > Mg$, and magnesio-riebeckite if $Mg > Fe^{2+}$. Crossite occupies the intermediate compositions (Figure 13.16). In addition, there is significant solid solution

Glaucophane–Riebeckite Series

Monoclinic
$\angle\beta = 103–104°$
Biaxial $(-)$ or $(+)$
$n_\alpha = 1.606–1.701$
$n_\beta = 1.622–1.711$
$n_\gamma = 1.627–1.717$
$\delta = 0.006–0.029$

Glaucophane
$Na_2Mg_3Al_2Si_8O_{22}(OH)_2$
Biaxial $(-)$
$2V_x = 10–45°$

Crossite
$Na_2(Mg,Fe^{2+})_3(Al,Fe^{3+})_2Si_8O_{22}(OH)_2$
Biaxial $(-)$ or $(+)$
$2V_x = 0–180°(?)$

Riebeckite
$Na_2Fe^{2+}_3Fe^{3+}_2Si_8O_{22}(OH)_2$
Biaxial $(-)$ or $(+)$
$2V_z = 0–135°$

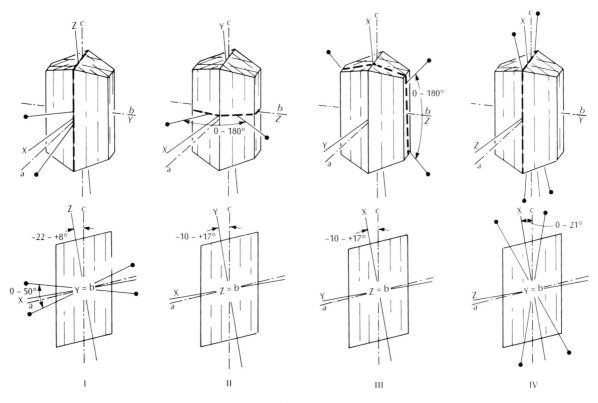

(010) Sections

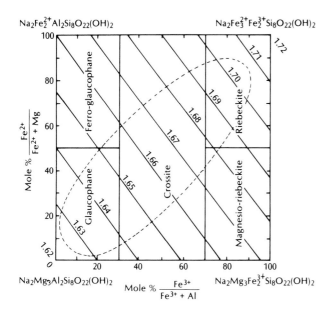

$Na_2Fe_2^{2+}Al_2Si_8O_{22}(OH)_2$

$Na_2Fe_3^{2+}Fe_2^{3+}Si_8O_{22}(OH)_2$

Mole % $\dfrac{Fe^{2+}}{Fe^{2+} + Mg}$

$Na_2Mg_3Al_2Si_8O_{22}(OH)_2$

Mole % $\dfrac{Fe^{3+}}{Fe^{3+} + Al}$

$Na_2Mg_3Fe_2^{3+}Si_8O_{22}(OH)_2$

Figure 13.16 Approximate variation of n_γ with composition in the sodic amphiboles. There is substantial variability. Compositions found in blueschists and related rocks fall in the field trending from glaucophane to riebeckite. Compositions found in igneous rocks and metamorphosed iron formations fall in the riebeckite and magnesio-riebeckite fields. Data from Deer and others (1963), Borg (1967), Ernst and Wai (1970), and Coleman and Papike (1968).

toward arfvedsonite and the sodic-calcic amphiboles. F may substitute for (OH), and Ti, Mn, and other metal cations may be present.

PHYSICAL PROPERTIES: H = 5–6; G = 3.02–3.42 (increasing with iron); lavender-blue, blue, dark blue, black, or gray in hand sample (darker with increasing iron); white to gray streak; vitreous luster.

COLOR AND PLEOCHROISM: Glaucophane and crossite are colorless to medium blue in thin section with distinct pleochroism: X = colorless to pale blue or yellow; Y = lavender-blue, bluish green; Z = blue, greenish blue, violet. Absorption is usually $Z > Y > X$ but may be variable, especially for intermediate compositions. Color is usually darkest when the long dimension of elongate sections is parallel to the vibration direction of the lower polar.

Riebeckite is usually dark blue in thin section or grain mount, and strongly pleochroic: X = dark blue; Y = indigo blue or gray-blue; Z = blue, yellow-green, or yellow-brown. Magnesio-riebeckite tends to be lighter colored.

FORM: Figure 13.11. Glaucophane commonly forms bladed to slender prismatic crystals with the typical diamond-shaped amphibole cross section showing the prismatic cleavage; less commonly fibrous or granular. Because glaucophane is common in metamorphic rocks, it is often foliated.

Riebeckite commonly forms slender bladed or prismatic crystals, usually without terminations. Cross sections show the typical diamond shape of amphiboles and the trace of the two cleavages. Grains may be irregular and poikilitic, enclosing associated minerals. The fibrous to asbestiform variety *crocidolite* is common in metamorphosed iron formations.

CLEAVAGE: Typical amphibole cleavage on {110} intersects at about 55 and 125°.

TWINNING: Simple or lamellar twins with a {100} composition plane are not common.

OPTICAL ORIENTATION: The position of the indicatrix axes remains relatively constant for all compositions, with one nearly parallel to c, one parallel to b, and one near a. However, the positions of the optic plane and the relative lengths of the axes change with increasing iron content, as shown in the orientation diagrams. Most glaucophane shows orientation I, crossite shows orientation II (or III), and riebeckite shows III. Orientation IV is found in samples intermediate between riebeckite and arfvedsonite. All samples show symmetrical extinction in basal sections and inclined extinction in (010) sections. The extinction angle is less than ~21° for all compositions and usually less than 10°. In glaucophane the Z axis is close to c, so it is length slow; and in riebeckite the X axis is close to c, so it is length fast. In crossite, either the X or Y axis may be closer to c, so it is either length fast or length slow, depending on which indicatrix orientation is present and how the section is cut.

INDICES OF REFRACTION AND BIREFRINGENCE: Indices of refraction increase with substitution of Fe^{2+} for Mg, and Fe^{3+} for Al, as shown in Figure 13.16. Most glaucophane has indices less than $n_\alpha = 1.65$, $n_\beta = 1.66$, $n_\gamma = 1.66$, and most riebeckite has indices greater than $n_\alpha = 1.69$, $n_\beta = 1.70$, $n_\gamma = 1.70$. Crossite and magnesio-riebeckite have the intermediate indices. The reader is cautioned that Figure 13.16 is schematic and that substantial variability exists. Birefringence shows a general decrease with increasing iron content but with considerable variability. Birefringence of glaucophane is usually 0.008–0.029, crossite 0.010–0.016, and riebeckite 0.006–0.017. Interference colors in thin section are usually middle to upper first order but may be masked by the mineral color.

INTERFERENCE FIGURE: Glaucophane is biaxial negative, with $2V_x$ usually in the range 10 to 45°. Crossite may be either biaxial negative or positive, with $2V$ anywhere between 0 (uniaxial) to 90°. Riebeckite is biaxial negative, although some Mg-rich samples may be biaxial positive; the $2V_x$ angle is typically greater than about 45°. In dark-colored samples, interference figures may be difficult to interpret. Glaucophane shows weak optic axis dispersion ($v > r$) and weak inclined bisectrix dispersion. Riebeckite and crossite show strong optic axis dispersion ($r > v$ or $v > r$) with horizontal or inclined bisectrix dispersion.

ALTERATION: Glaucophane and crossite may be altered to other amphiboles, and riebeckite is sometimes altered to quartz, iron oxides, and carbonates.

DISTINGUISHING FEATURES: Distinguished from the other amphiboles by the distinct blue colors. Riebeckite is usually darker blue than glaucophane and may be biaxial positive, is length fast, has a larger $2V$, and occurs in different types of rocks. Glaucophane and crossite are distinguished by the change of optic orientation. In glaucophane, the optic plane is (010) and in crossite the optic plane is roughly parallel to (001) (there are exceptions, however). To check, obtain an acute bisectrix figure and put the trace of the optic plane north–south. Return to orth-

oscopic illumination. If the cleavage trace is N–S the mineral is probably glaucophane and if E–W, crossite. Blue tourmaline is darkest when the long dimension is perpendicular to the vibration direction of the lower polar.

OCCURRENCE: Glaucophane and crossite are characteristic minerals of high-pressure–low-temperature regional metamorphic rocks, which, because of their presence, are called blueschists. Glaucophane and crossite are usually associated with lawsonite, garnet, pumpellyite, chlorite, albite, quartz, and epidote group minerals. They are sometimes found in eclogites with garnet, hornblende, and omphacite.

While some of the sodic amphibole in blueschist metamorphic terranes fall into the riebeckite range, riebeckite is most commonly found in alkali granite, syenite, nepheline syenite, trachyte, and related alkalic rocks. It also is found as a constituent of metamorphosed iron formations, usually in the form of crocidolite. Authigenic riebeckite/magnesio-riebeckite also has been reported in sedimentary rocks.

OTHER INOSILICATES

Sapphirine

$(Mg,Fe,Al)_8O_2(Al,Si)_6O_{18}$
Monoclinic
$\angle\beta = 125°$ (variable)
Biaxial (−) or (+)
$n_\alpha = 1.701–1.731$
$n_\beta = 1.703–1.741$
$n_\gamma = 1.705–1.743$
$\delta = 0.005–0.012$
$2V_x = 47–114°$

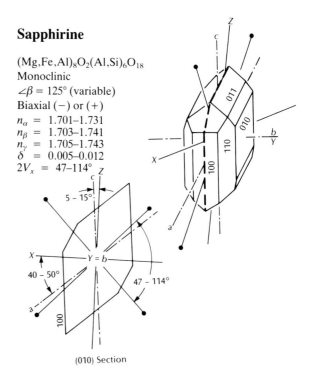

(010) Section

RELIEF IN THIN SECTION: High positive relief.

COMPOSITION AND STRUCTURE: The structure consists of essentially cubic close-packed oxygen with layers parallel to (100). The cations occupy the tetrahedral and octahedral sites between the oxygen layers so that there are layers of octahedrally coordinated Mg, Fe, and Al alternating with layers of tetrahedrally coordinated Si and Al. The octahedrally coordinated cations are arranged so that they form bands of octahedra parallel to the c axis, and the tetrahedrally coordinated Si and Al are arranged to form irregular single chains parallel to the c axis. Most sapphirine is quite aluminous, and around one-half of the octahedral and two-thirds to three-quarters of the tetrahedral sites are occupied by aluminum.

PHYSICAL PROPERTIES: H = $7\frac{1}{2}$; G = 3.40–3.58; light blue or green in hand sample, sometimes gray to pale red; white streak; vitreous luster.

COLOR AND PLEOCHROISM: Usually colorless to shades of blue in thin section and grain mount, sometimes pinkish. Pleochroic with $Z > Y > X$: X = colorless, pale pink, pale yellow, yellowish green, greenish blue, greenish gray; Y = sky blue, pale blue, lavender-blue, gray-blue, green; Z = blue, indigo blue, blue-green, pale green.

FORM: Often as anhedral granules or granular aggregates. Crystals are usually tabular and somewhat elongate parallel to the c axis.

CLEAVAGE: There is a poor to fair cleavage on {010}, which is generally not displayed in thin section, and poor cleavages on {100} and {001}.

TWINNING: Lamellar twinning on {010} is reported but may actually be subparallel lamellar intergrowths.

OPTICAL ORIENTATION: $X \wedge a$ = +40 to +50°, $Y = b$, $Z \wedge c$ = −5 to −15°, optic plane = (010). If crystals are elongate parallel to c, the maximum extinction angle to the elongation is 5 to 15°, and the elongation is length slow.

INDICES OF REFRACTION AND BIREFRINGENCE: Indices of refraction increase with increasing proportion of iron to magnesium, but there is considerable variability. Birefringence is relatively uniform and yields first-order gray colors in thin sections. Anomalous interference colors are sometimes found.

INTERFERENCE FIGURE: Most sapphirine is biaxial negative with $2V_x$ between 50 and 70° but there is considerable variability and some is positive with a large $2V_z$. $2V_x$ generally decreases with increasing proportion of iron to magnesium. Optic axis dispersion is strong with $v > r$, and there is usually distinct inclined bisectrix dispersion.

ALTERATION: Sapphirine may alter to a mixture of corundum and biotite or other phyllosilicates.

DISTINGUISHING FEATURES: Sapphirine is recognized by its high relief, blue color, and low birefringence. It is similar to corundum, but corundum commonly forms euhedral hexagonal crystals and is uniaxial. Kyanite has a bladed habit and distinct cleavage. The blue amphiboles have distinctly different habit and good cleavage.

OCCURRENCE: Sapphirine is usually found in high-grade aluminous contact or regional metamorphic rocks associated with other aluminous minerals such as cordierite, corundum, kyanite, sillimanite, spinel, calcic plagioclase, and almandine, as well as ortho-pyroxene, anthophyllite, and hornblende. It usually is not associated with quartz.

Wollastonite

CaSiO$_3$
Triclinic
$\angle \alpha$ = 90°02′
$\angle \beta$ = 95°22′
$\angle \gamma$ = 103°26′
Biaxial (−)
n_α = 1.616–1.645
n_β = 1.628–1.652
n_γ = 1.631–1.656
δ = 0.013–0.017
$2V_x$ = 36–60°

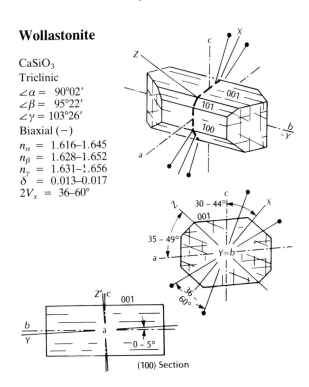

(100) Section

RELIEF IN THIN SECTION: Moderate to moderately high positive relief.

COMPOSITION AND STRUCTURE: The structure is similar to the pyroxenes, with the principal difference being that the single chains of silicon tetrahedra are kinked or twisted. The Ca occupies octahedral sites between the chains. The main compositional variation is substitution of up to about 20 mole percent Fe^{2+} and 20 percent Mn^{2+}. Small amounts of Mg also may be present. Most wollastonite is relatively pure $CaSiO_3$, however. Wollastonite-2M and pseudowollastonite are polymorphs with similar structures but neither is at all common.

PHYSICAL PROPERTIES: H = $4\frac{1}{2}$–5; G = 2.86–3.09; white, gray, or very pale green in hand sample; white streak; vitreous to pearly luster.

COLOR AND PLEOCHROISM: Colorless in thin section and grain mount.

FORM: Usually bladed, columnar, or fibrous; elongate parallel to the b axis, and flattened parallel to {100} or {001}.

CLEAVAGE: There is a single perfect cleavage on {100} and good cleavages on {001} and {$\bar{1}$02}. The {100} and {001} cleavages intersect at $84\frac{1}{2}°$, and the {$\bar{1}$02} cleavage intersects the {100} cleavage at 70°. Fragment orientation tends to be controlled by cleavages.

TWINNING: Twins with composition plane {100} are common.

OPTICAL ORIENTATION: The optic plane is nearly parallel to (010) with $X \wedge c = -30$ to $-44°$, $Y \wedge b \cong 0$ to 5°, $Z \wedge a = +35$ to $+49°$. Iron-rich wollastonite has been reported with $X \wedge c$ as small as 3°. Sections cut parallel to the b axis are elongate, show a single cleavage direction, and have nearly parallel extinction ($\pm5°$), as do elongate cleavage fragments. Elongation may be either length fast or length slow, depending on orientation. All other sections show inclined extinction.

INDICES OF REFRACTION AND BIREFRINGENCE: Pure wollastonite has indices of about $n_\alpha = 1.618$, $n_\beta = 1.630$, and $n_\gamma = 1.632$. Adding iron or manganese increases the indices by about 0.0012 per mole percent Fe + Mn. Maximum interference color in thin section is first-order yellow.

INTERFERENCE FIGURE: Most wollastonite has a $2V_x$ angle of about 40°, although $2V_x$ in iron-rich samples may be larger. Fragments lying on cleavages yield off-center optic axis or acute bisectrix figures. Optic axis dispersion is weak to moderate with $r > v$.

ALTERATION: Pectolite, calcite, and apophyllite may replace wollastonite.

DISTINGUISHING FEATURES: Tremolite and pectolite have higher birefringence and are consistently length slow. Calcic clinopyroxene has higher relief, larger 2V, is optically positive, and has typical pyroxene cleavage. Zoisite and clinozoisite have higher relief and 2V.

OCCURRENCE: Wollastonite is a common mineral in metamorphosed siliceous limestone and dolomite. It is commonly associated with calcite, dolomite, tremolite, grossular, epidote, augite, monticellite, forsterite, and other calc-silicate minerals. Wollastonite is rarely found in carbonatites and alkalic igneous rocks.

Rhodonite

$MnSiO_3$
Triclinic
$\angle \alpha = 108.6°$
$\angle \beta = 102.9°$
$\angle \gamma = 82.5°$
Biaxial (+)
$n_\alpha = 1.711$–1.739
$n_\beta = 1.715$–1.748
$n_\gamma = 1.724$–1.760
$\delta = 0.011$–0.021
$2V_z = 63$–87°

(001) Section   (110) Section

RELIEF IN THIN SECTION: High positive relief.

COMPOSITION AND STRUCTURE: The structure is similar to wollastonite and consists of distorted chains of tetrahedra tied laterally through Mn in octahedral coordination. Up to 15 or 20 percent of the Mn may be replaced by Ca, and a similar amount may be replaced by (Fe, Mg). Average rhodonite shows a ratio of Mn:Ca:(Fe, Mg) = 80:10:10. There is not continuous solid solution between rhodonite and wollastonite. *Bustamite* [(Mn,Ca)SiO$_3$] is a separate mineral with intermediate compositions.

PHYSICAL PROPERTIES: H = $5\frac{1}{2}$–$6\frac{1}{2}$; G = 3.54–3.76; pink, rose red, or red-brown in hand sample; white streak; vitreous luster.

COLOR AND PLEOCHROISM: Colorless to pale pink in thin section, somewhat darker in grain mount. Pleochroism is weak with X = yellowish red, Y = pinkish red, and Z = pale yellowish red.

FORM: Crystals are tabular parallel with (001). Also found as anhedral grains and granular aggregates.

CLEAVAGE: There are two perfect cleavages on {110} and {1$\bar{1}$0}, and a fair cleavage on {001}. The angle between the perfect cleavages is $92\frac{1}{2}°$.

TWINNING: Lamellar twins with {010} composition planes are not common.

OPTICAL ORIENTATION: The optic plane is nearly perpendicular to (100) with $X \wedge a \cong 5°$, $Y \wedge b \cong 20°$, $Z \wedge c \cong 25°$. Extinction is inclined in all sections.

INDICES OF REFRACTION AND BIREFRINGENCE: Indices of refraction decrease in an essentially linear manner for increasing amounts of (Ca,Mg)SiO$_3$, and the birefringence shows a small decrease (Figure 13.17). However, there is substantial variability in the data, so the indices cannot be used to reliably estimate composition. The maximum interference color in thin section is first-order yellow.

INTERFERENCE FIGURE: $2V_z$ shows a small increase with increasing amounts of (Ca, Mg) SiO$_3$, but the scatter in the data is too large to allow $2V$ to be used to estimate composition. Basal sections showing low birefringence (~ 0.005) and two cleavages at about 93° yield off-center acute bisectrix figures. Optic axis dispersion is weak, $v > r$, and crossed bisectrix dispersion may be distinct.

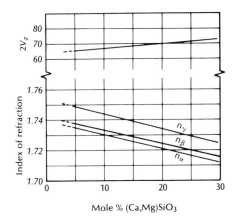

Figure 13.17 Variation of indices of refraction and $2V_z$ of rhodonite with composition. After Sundias (1931) and Momoi (1964).

ALTERATION: Rhodonite may alter to rhodochrosite or dark-colored Mn oxides and hyroxides (e.g. pyrolusite).

DISTINGUISHING FEATURES: Pink color in hand sample, and sometimes in thin section, along with inclined extinction and association with other Mn minerals is distinctive. *Pyroxmangite* [(Mn,Fe)SiO$_3$] is a closely related mineral with higher birefringence and smaller $2V$. Bustamite is optically negative with $2V_x$ less than 60°.

OCCURRENCE: The usual occurrence for rhodonite is in Mn-bearing hydrothermal ore deposits and skarns. It is usually associated with other Mn minerals, such as rhodochrosite and pyrolusite, as well as quartz, other carbonates, and a variety of sulfide minerals. Rhodonite has been reported in pegmatite.

Pectolite

Ca$_2$NaH(SiO$_3$)$_3$ Biaxial (+)
Triclinic n_α = 1.592–1.610
∠α = 90°31′ n_β = 1.603–1.615
∠β = 95°11′ n_γ = 1.630–1.645
∠γ = 102°28′ δ = 0.026–0.039
                      2V_z = 35–63°

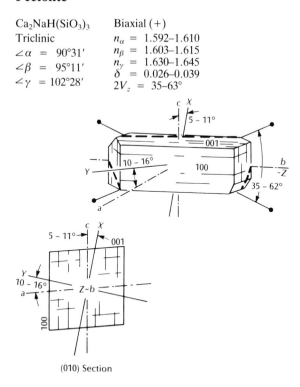

(010) Section

RELIEF IN THIN SECTION: Moderate positive relief.

COMPOSITION AND STRUCTURE: Pectolite has a structure similar to wollastonite, with distorted single chains of tetrahedra bonded laterally through Ca in octahedral coordination and Na in distorted octahedral coordination. The hydrogen is probably situated between two oxygens on adjacent tetrahedra, which are brought into proximity by the distortion of the chain. There is continuous solid solution to *serandite* [Mn$_2$NaH(SiO$_3$)$_3$], but most pectolite is reasonably close to its ideal composition. Minor amounts of Mg and Fe also may substitute in the structure.

PHYSICAL PROPERTIES: H = $4\frac{1}{2}$–5; G = 2.84–2.90; colorless or white in hand sample; white streak; vitreous luster.

COLOR AND PLEOCHROISM: Colorless in thin section and grain mount.

FORM: Pectolite usually forms radiating groups of acicular crystals that are elongate parallel to the *b* axis.

CLEAVAGE: There are two perfect cleavages, one on {100}, the other on {001}, which intersect at 95°.

TWINNING: Rare twinning is found with a composition plane nearly parallel to {100}. Twin lamellae are parallel to the length of elongate grains.

OPTICAL ORIENTATION: The optic plane is nearly parallel to (100) with $X \wedge c$ = −5 to −11°, $Y \wedge a$ = +10 to +16°, and $Z \wedge b \cong$ 2°. Elongate sections cut parallel to the *b* axis and cleavage fragments show essentially parallel extinction (±2°) and are length slow. Sections cut parallel to (010) are nearly square and show inclined extinction of up to about 16° to the cleavage traces. However, small grain size may make this section difficult to work with.

INDICES OF REFRACTION AND BIREFRINGENCE: The indices of essentially pure pectolite are n_α = 1.592, n_β = 1.603, and n_γ = 1.630. Substituting Mn for Ca produces a roughly linear increase in indices to n_α = 1.680, n_β = 1.682, and n_γ = 1.705 for serandite and a decrease in birefringence from 0.038 to 0.025. Most pectolite has little Mn and shows indices near those of pure pectolite. Interference colors in thin section are up to upper second order.

INTERFERENCE FIGURE: 2V_z for relatively pure pectolite is about 62° and most pectolite falls in the range 50 to 60°, although 2V_z as small as 35° has been reported. Substituting Mn causes a decrease in 2V; serandite has 2V_z = 33°. Sections parallel to (010) yield nearly centered acute bisectrix figures. Fragments lying on cleavage surfaces yield off-center flash or obtuse bisectrix figures. Optic axis dispersion is $v > r$ or $r > v$.

ALTERATION: Pectolite may be replaced by pale pink *stevensite* [Mg$_3$Si$_4$O$_{10}$(OH)$_2$].

DISTINGUISHING FEATURES: Pectolite is distinguished from the zeolites, with which it is often associated, by its higher birefringence. Wollastonite has a different occurrence, lower birefringence, and shows both length slow and length fast elongation.

OCCURRENCE: Pectolite is usually found in amygdules and other cavities in basalt flows, diabase dikes, and related rock types. It is commonly associated with zeolites, calcite, prehnite, and datolite. Pectolite is also found in serpentinites, as a primary mineral in alkalic igneous rocks such as nepheline syenite and phonolite, and in skarns and other Ca-rich metamorphic rocks.

REFERENCES

Borg, I. Y., 1967, Optical properties and cell parameters in the glaucophane–riebeckite series: Contributions to Mineralogy and Petrology, v. 15, p. 67–92.

Clark, J. R., and Papike, J. J., 1968, Crystal–chemical characterization of omphacites: American Mineralogist, v. 53, p. 840–868.

Coleman, R. G., and Papike, J. J., 1968, Alkali amphiboles of the Cazadero, California: Journal of Petrology, v. 9, p. 105–122.

Deer, W. A., Howie, R. A., and Zussman, J., 1963, Rock forming minerals, Volume 2, Chain silicates: Longman, London, 379 p.

Deer, W. A., Howie, R. A., and Zussman, J., 1978, Rock forming minerals, Volume 2A, Single chain silicates: Wiley, New York, 668 p.

Ernst, W. G., and Wai, C. M., 1970, Mossbauer, infrared, X-ray and optical study of cation ordering and dehydrogenation in natural and heat-treated sodic amphiboles: American Mineralogist, v. 55, p. 1226–1258.

Essene, E. J., and Fyfe, W. S., 1967, Omphacite in Californian metamorphic rocks: Contributions to Mineralogy and Petrology, v. 15, p. 1–23.

Hess, H. H., 1949. Chemical composition and optical properties of common clinopyroxenes, Part I: American Mineralogist, v. 34, p. 621–666.

Jaffe, H. W., Robinson, P., and Tracy, R. J., 1975, Orientation of pigeonite exsolution lamellae in metamorphic augite: Correlation with composition and calculated optimal phase boundaries: American Mineralogist, v. 60, p. 9–28.

Jaffe, H. W., Robinson, P., and Tracy, R. J., 1978, Orthoferrosilite and other iron-rich pyroxenes in microperthite gneiss of the Mount Marcy area, Adirondack Mountains: American Mineralogist, v. 63, p. 1116–1136.

Klein, C., 1964, Cummingtonite-grunerite series: A chemical, optical, and X-ray study: American Mineralogist, v. 49, p. 963–982.

Larsen, E. S., 1942, Alkali rocks of Iron Hill, Gunnison County, Colorado: U. S. Geological Survey Professional Paper 197A.

Leake, B. E., 1968, Optical properties and composition in the orthopyroxene series: Mineralogical Magazine, v. 36, p. 745–747.

Leake, B. E., 1978, Nomenclature of amphiboles: American Mineralogist, v. 63, p. 1023–1052.

Momoi, H., 1964, Mineralogical study of rhodonites in Japan, with special reference to contact metamorphism: Kyushu University Faculty of Science Memoirs, Series D, Geology, 15, p. 39–63.

Nolan, J., 1969, Physical properties of synthetic and natural pyroxenes in the system diopside-hedenbergite-acmite: Mineralogical Magazine, v. 37, p. 216–229.

Poldervaart, A., and Hess, H. H., 1951, Pyroxenes in the crystallization of basaltic magmas: Journal of Geology, v. 59, p. 472–489.

Rabbitt, J. C., 1948, A new study of the anthophyllite series: American Mineralogist, v. 33, p. 263–323.

Sabine, P. A., 1950, The optical properties and composition of the acmitic pyroxenes: Mineralogical Magazine, v. 29, p. 113–125.

Sundias, N., 1931, On the triclinic manganiferous pyroxenes: American Mineralogist, v. 16, p. 411–518.

Winchell, A. N., 1938, The anthophyllite and cummingtonite-grunerite series: American Mineralogist, v. 23, p. 329–333.

Winchell, A. N., 1945, Variations in composition and optical properties of the calciferous amphiboles: American Mineralogist, v. 28, p. 27–50.

14

Phyllosilicates

The structural feature common to all members of the phyllosilicates is a continuous network of silicon tetrahedra called a T sheet (Figure 14.1), which in effect is composed of two planes of oxygen and hydroxyl with Si in the tetrahedral sites between the anion planes. The hydroxyl is situated in the top plane in the center of the rings formed by "points" of the tetrahedra. Each tetrahedra shares three of its oxygen with adjacent tetrahedra along the base of the sheet, and the fourth, or apical, oxygen bonds to other cations in octahedral sites formed with the addition of another plane of hydroxyl. This forms a 1:1 T-O or tetrahedra-octahedral sandwich or layer. A 2:1 T-O-T layer also can be formed by sandwiching the octahedral sheet between two tetrahedral sheets.

In the dioctahedral phyllosilicates, two out of every three octahedral sites are occupied by a cation, usually Al, as in the mineral gibbsite [$Al_2(OH)_6$]. In the trioctahedral phyllosilicates, three out of three of the octahedral sites in the octahedral sheet are occupied by a cation—usually Fe or Mg—as in the mineral brucite [$Mg_3(OH)_6$].

The T-O and T-O-T sandwiches or layers are bonded together either through weak electrostatic bonds or through large interlayer cations, such as K, Na, and Ca. The charge of the cations between the layers is balanced by replacing Si^{4+} with Al^{3+} in the tetrahedral sites in most cases. In the chlorite group of minerals, an octahedral layer [$Mg_2(OH)_6$] is inserted between the T-O-T layers. The structures of some common phyllosilicates are shown in Figure 14.1. The weak bonds between the layers are responsible for the perfect cleavage of the phyllosilicates.

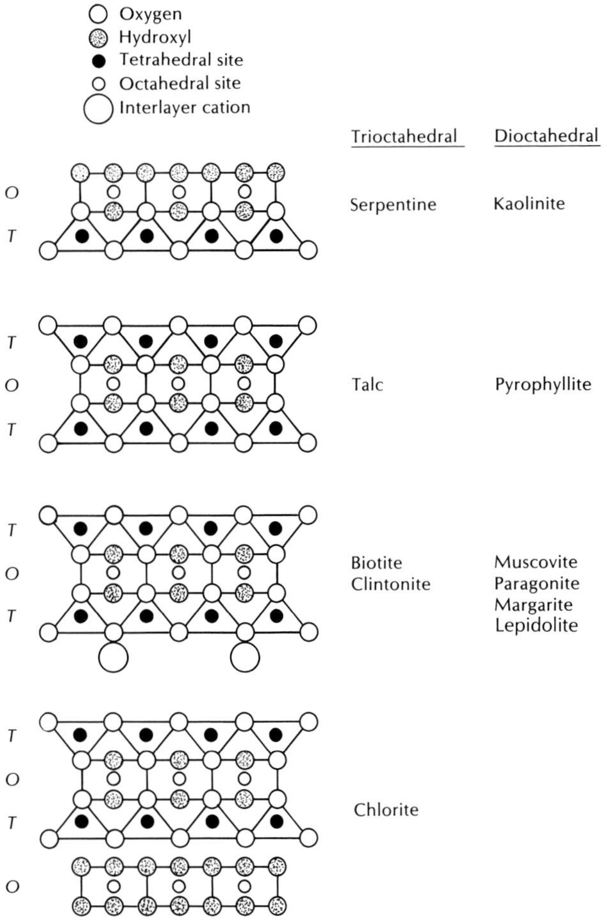

Figure 14.1 Idealized structure of the di- and trioctahedral phyllosilicates. T and O indicate the tetrahedral and octahedral sheets, respectively.

241

Clay Minerals

The clay minerals are a large group of complex phyllosilicates constructed by combining brucite, gibbsite, talc, pyrophyllite, kaolinite, and serpentine layers in various repeating patterns. While optical properties of many different clay minerals have been described, it is almost impossible to distinguish individual species or even broad groups of clays from each other based on optical means, even when the individual grains are large enough to examine with the microscope. It is strongly recommended that the petrographer use the term clay for all members of the group and employ X ray or other techniques to identify the various species.

The common groups of clay minerals include the kaolin (kandite), montmorillonite, illite, vermiculite, and mixed-layer groups. The palygorskite group is also briefly described here.

The members of the kaolin group [$\sim$ Al$_2$Si$_2$O$_5$(OH)$_4$] are dioctahedral clays constructed of T-O layers bonded together with weak electrostatic bonds. Some varieties contain H_2O molecules between the layers. All varieties are monoclinic with $\angle \beta = 90$–$104°$ and have optical properties in the range $n_\alpha = 1.55$–1.57, $n_\beta = 1.55$–1.57, $n_\gamma = 1.56$–1.57; $\delta = 0.000$–0.008; $2V$ moderate to large, either positive or negative. They are oriented so that the X axis is roughly perpendicular to (001), Y is inclined less than $20°$ to a, and $Z = b$. Maximum extinction angle to the trace of cleavage is therefore less than about $20°$ and cleavage is length slow.

The montmorillonite group [$\frac{1}{2}$(Ca,Na)$_{0.67}$(Al,Mg,Fe)$_{4-6}$(Si,Al)$_8$O$_{20}$(OH)$_4$·nH$_2$O] has a structure similar to talc and pyrophyllite and is constructed of T-O-T layers. The octahedral sites may contain either Al, Fe, or Mg. Some of the tetrahedral Si is replaced by Al; the charge is balanced by inserting interlayer cations (usually Ca or Na). Montmorillonite is often expansive; it swells when it absorbs water. All members of the group are monoclinic ($\angle \beta \cong 90°$) and have optical properties that fall in the range $n_\alpha = 1.48$–1.61, $n_\beta = 1.49$–1.63, $n_\gamma = 1.50$–1.64; $\delta = 0.01$–0.04; $2V_x =$ small to moderately large, all optically negative. The indicatrix is oriented $X \cong c$, $Y = b$, $Z \cong a$, so the extinction angle to the trace of cleavage is close to parallel and cleavage is length slow.

The members of the illite group [(K,Na,H$_3$O)$_{1-2}$Al$_4$(Si$_{7-6}$Al$_{1-2}$)O$_{20}$(OH)$_4$] have structures similar to muscovite. This group is constructed of dioctahedral T-O-T layers, containing predominantly Al in the octahedral sites, although some are trioctahedral and contain Mg or Fe. Some Si is replaced by Al in the tetrahedral sites, with the charge balanced by inserting interlayer cations (usually K, Na, or H$_3$O) between the layers. Members of the illite group are monoclinic with $\angle \beta \cong 90°$ and have optical properties in the range $n_\alpha = 1.53$–1.57, $n_\beta = 1.56$–1.61, $n_\gamma = 1.56$–1.61; $\delta \cong 0.03$; $2V_x < 10°$, optically negative. $X \cong c$ so the trace of cleavage is length slow and extinction is close to parallel. Glauconite is an illite group mineral described in a following section.

Vermiculite [(Mg,Ca) (Mg, Fe^{2+})$_5$ (Fe^{3+}, Al) (Si$_5$Al$_3$)O$_{20}$(OH)$_4$ · 8H$_2$O] is a trioctahedral clay mineral constructed of T-O-T layers with Mg or Ca interlayer cations sandwiched between two layers of water molecules. It is monoclinic with $\angle \beta \cong 97°$ and has optical properties in the range $n_\alpha = 1.52$–1.56, $n_\beta = 1.54$–1.58, $n_\gamma = 1.54$–1.58; $\delta = 0.02$–0.03; $2V_x < 8°$, optically negative. The indicatrix is oriented so that $X \wedge c = 3$–$6°$, $Y = b$, $Z \wedge a = 1$–$4°$. The extinction angle is less than $4°$ and the trace of cleavage is length slow.

There are also mixed-layer clays which consist of alternating layers of several clay types. Clay may also have the chlorite structure. Chlorite is described in a later section.

Palygorskite [(Mg,Al)$_4$Si$_8$O$_{20}$(OH)$_2$·8H$_2$O] is a complex clay mineral constructed of amphibolelike double chains of tetrahedra aligned parallel to the c axis. Because of its structure, palygorskite is typically fibrous, rather than scaly like the other clay minerals, and may form flexible matted sheets that are sometimes called mountain leather. It is monoclinic with; $\beta \cong 96°$, or orthorhombic, and the indicatrix is oriented so that $Z \cong c$; thus, fibers are length slow. Optical properties usually fall in the range $n_\alpha = 1.50$–1.52, n_β 1.54–1.56; $\delta = 0.020$–0.035; $2V_x =$ small to large. Palygorskite may be mistaken for serpentine. Sepiolite is a similar mineral constructed of triple chains of tetrahedra.

DISTINGUISHING FEATURES: Clay is recognized in thin section as fine-grained gray, brown, or white earthy-looking aggregates. Clay may or may not display interference colors, depending on grain size and which clay minerals are present. Members of the kaolin group have low birefringence (< 0.008), whereas members of other groups (illite, montmorillonite, palygorskite, mixed layer, etc.) may have birefringence as high as 0.04. Most clay minerals are monoclinic, and are biaxial negative or, less commonly, positive. A few are hexagonal, orthorhombic, or triclinic. If clay minerals are sufficiently large, cleavage fragments yield nearly centered bisectrix figures. Indices are usually in the range 1.5–1.6, so relief is typically low. Individual grains are thin plates or scales, and all have a perfect cleavage parallel to (001), like the micas.

Other minerals, notably the zeolites, may occur as fine-grained material in sedimentary and altered rocks, and are all too easily dismissed as clay.

OCCURRENCE: Clays are abundant in soil and in a wide variety of sedimentary rocks. They are produced by the weathering of feldspars, micas, and other silicates. Hydrothermal activity can also form clays in areas adjacent to ore deposits and in the vicinity of hot springs. In many rocks, incipient weathering or other alteration produces clays that may cause primary minerals to appear turbid or cloudy, or which partially replace them along fractures, cleavages, and grain boundaries. Vermiculite is commonly formed by the weathering or hydrothermal alteration of biotite, and pseudomorphs after biotite are not uncommon. Palygorskite and sepiolite are formed in veins and cavities in altered or weathered carbonate rocks and mafic igneous rocks. They may also be found in deposits from saline lakes.

Glauconite

$(K,Na)(Fe^{3+},Al,Fe^{2+},Mg)_2(Si,Al)_4O_{10}(OH)_2$
Monoclinic
$\angle \beta \cong 100°$
Biaxial $(-)$
$n_\alpha = 1.56–1.61$
$n_\beta = 1.58–1.65$
$n_\gamma = 1.58–1.65$
$\delta = 0.014–0.032$
$2V_x = 0–20°$

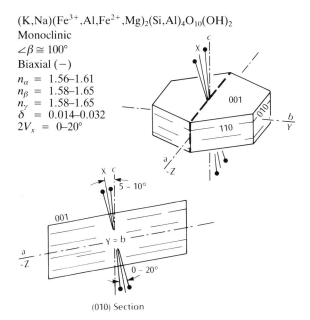

(010) Section

RELIEF IN THIN SECTION: Moderate positive relief.

COMPOSITION AND STRUCTURE: The structure of glauconite is essentially the same as muscovite and consists of dioctahedral T-O-T layers. Adjacent T-O-T layers are bonded together through K and Na in 12-fold coordination with oxygen. The octahedral sites usually contain more Fe^{3+} than Al^{3+} and significant amounts of Mg^{2+} and Fe^{2+} are common. The charge deficiency resulting from the presence of divalent rather than trivalent cations in octahedral sites is typically balanced by substituting Si^{4+} for Al^{3+} in tetrahedral sites. The structure usually also contains layers of expandable-type clay in variable proportions.

PHYSICAL PROPERTIES: H = 2; G = 2.4–3.0; usually green or blue-green in hand sample, sometimes with limonite stains; green streak; earthy or dull luster.

COLOR AND PLEOCHROISM: Usually green in thin section or grain mount; if Al rich, glauconite is light green or colorless. Pleochroism is distinct with $X =$ yellowish green to green; $Y = Z =$ green, bluish green, darker yellowish green.

FORM: Figure 14.2. Glauconite forms small pellets or granules in sedimentary rocks, or casts of foraminifera or other microfossils. The pellets and casts usually consist of aggregates of fine, irregular flakes. A variety called celadonite is essentially the same but forms small radiating masses in vesicles in volcanic rocks.

CLEAVAGE: Glauconite has perfect {001} cleavage, as do all the phyllosilicates. Fine grain size may preclude seeing it, however.

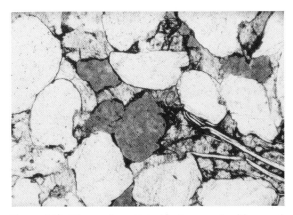

Figure 14.2 Glauconite pellets (darker gray) with round quartz grains and elongate detrital muscovite. The cement between the quartz grains is calcite and the opaque material is hematite. Field of view is 1.2 mm wide.

OPTICAL ORIENTATION: $X \wedge c = +5$ to $+10°$, $Y = b$, $Z \cong a$, optic plane = (010). Extinction to cleavage is essentially parallel, and the trace of cleavage is length slow.

INDICES OF REFRACTION AND BIREFRINGENCE: Indices increase with increasing iron content and decrease with increasing percentage of expanding clay structure. Maximum interference colors in thin section are upper first to mid second order but are usually masked by the mineral color. Basal sections are nearly isotropic since $n_\beta \cong n_\gamma$.

INTERFERENCE FIGURE: Figures usually cannot be obtained due to fine grain size. Basal sections and cleavage fragments yield biaxial negative acute bisectrix figures with a small $2V_x$ (0–20°). Optic axis dispersion is $r > v$.

ALTERATION: Limonitic staining due to weathering is common.

DISTINGUISHING FEATURES: The distinctive occurrence and green color are characteristic. Some varieties of chlorite may form similar aggregates but have lower birefringence.

OCCURRENCE: Glauconite characteristically forms small rounded pellets in clastic sediments deposited in marine conditions. "Greensands" owe their green color to glauconite. It also may be found in limestones and marls. Celadonite is found as a vesicle filling in basalts and related volcanic rocks.

Serpentine

$Mg_3Si_2O_5(OH)_4$
Monoclinic (also triclinic or hexagonal)
$\angle \beta \cong 90–93°$
Biaxial (−)
$n_\alpha = 1.529–1.595$
$n_\beta = 1.530–1.603$
$n_\gamma = 1.537–1.604$
$\delta = 0.001–0.010$ (rarely higher)
$2V_x = $ highly variable, may be sensibly uniaxial

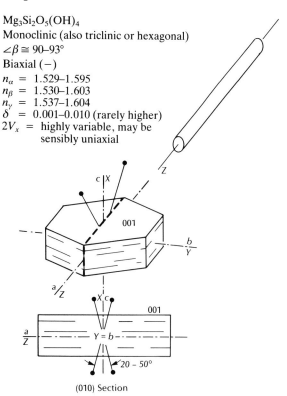

(010) Section

RELIEF IN THIN SECTION: Low to moderately low negative or positive relief.

COMPOSITION AND STRUCTURE: Serpentine is a trioctahedral phyllosilicate consisting of T-O layers bonded together with weak electrostatic bonds. The ideal T and O sheets have slightly different dimensions. The mechanisms that accommodate the dimensional mismatch account for the different varieties of serpentine. In *chrysotile*, the different dimensions are accommodated by allowing the T-O layers to be curved so that they roll up into slender, hollow fibers elongate parallel to the *a* axis. In *antigorite*, the tetrahedral sheets are continuous, but periodically there is a reversal in the direction that the apical oxygens face and the side on which the octahedral sheet is mated. This yields a periodic reversal in the sense of curvature of the T-O layers, so that they remain essentially flat over relatively long dimensions rather than curling up. In *lizardite*, the pattern of tetrahedra in the T sheet is distorted to allow a match to be achieved with the O sheet. Most serpentine has a composition close to the ideal formula given above, but Fe^{2+} may replace Mg. *Berthierine* [$(Fe_2Al)SiAlO_5(OH)_4$] and *amesite* [$(Mg_2Al)SiAlO_5(OH)_2$] are related species produced by substituting Al^{3+} for Mg^{2+} or Fe^{2+} with associated replacement of Si^{4+} by Al^{3+} to balance the charges. *Greenalite* [$Fe_3Si_2O_5(OH)_4$] is a related mineral with a somewhat more complex structure, which occurs in weakly metamorphosed iron formations.

PHYSICAL PROPERTIES: H = $2\frac{1}{2}$–$3\frac{1}{2}$; G = 2.5–2.6; pale yellow-green to dark green in hand sample, less commonly white, gray, or blue-green; white or gray streak. Antigorite and lizardite have a pearly or waxy luster, chrysotile is distinctly silky due to its fibrous nature.

COLOR AND PLEOCHROISM: Colorless to pale green in thin section; slightly darker in fragments. Pleochroism is weak, from colorless to pale yellow-green or pale green. Chrysotile fibers are usually darker when aligned parallel to the vibration direction of the lower polar, and folia are darker when the cleavage trace is aligned with the vibration direction of the lower polar. Fine-grained material may not be discernibly pleochroic, due to random grain orientation.

FORM: Figure 14.3. Fibrous serpentine is typically chrysotile and it commonly forms veins with fibers across the width or matted masses. Lizardite is commonly very fine-grained and may form an irregular net-like pattern with uneven or undulatory extinction. Antigorite is often more or less micaceous and may form foliated or scaly masses. However, different varieties may be intergrown, and habit is not always a reliable guide to variety. Serpentine is typically produced as a consequence of alteration of minerals such as olivine and pyroxene, and remnants of those minerals are commonly found within masses of fine-grained serpentine. Greenalite, berthierine, and amesite typically form very fine-grained aggregates that are colorless to light green and are less common as folia resembling chlorite.

CLEAVAGE: There is perfect {001} cleavage, but this is visible only in antigorite. Chrysotile fibers are easily separated.

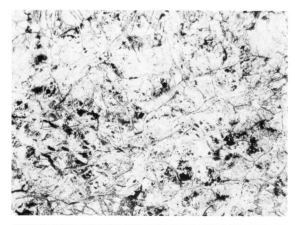

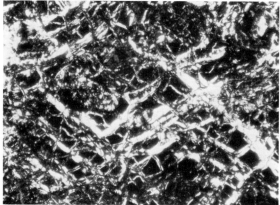

Figure 14.3 Serpentine with abundant opaque inclusions. (*Top*) Plane light. (*Bottom*) Crossed polars. Field of view is 1.7 mm.

TWINNING: Identifiable twins are rare.

OPTICAL ORIENTATION: The slow ray vibration direction is typically parallel to the length of fibers in chrysotile, which show parallel extinction. In antigorite, the optic plane is either parallel or perpendicular to (010), with $X \cong c$, $Y = b$, and $Z \cong a$, or $X \cong c$, $Y \cong a$, and $Z = b$. Extinction is, therefore, essentially parallel to the trace of cleavage, which is length slow.

INDICES OF REFRACTION AND BIREFRINGENCE: Indices of refraction increase rapidly with substitution of Fe for Mg, and less rapidly with substitution of Al. Determining indices is difficult and often only a mean index can be estimated in grain mount. Birefringence is low and maximum interference colors are rarely above first-order gray or white. Anomalous interference colors may be seen in some samples.

The mean indices of greenalite, berthierine, and amesite are 1.65–1.68, 1.62–1.67 and 1.59–1.62, respectively. Birefringence of all three is weak, and the typical fine-grained aggregates are essentially isotropic.

INTERFERENCE FIGURE: Interference figures are usually difficult to obtain due to fine grain size. Cleavage fragments and basal sections of antigorite yield biaxial negative figures with $2V_x$ usually between 20 and 50°, although smaller or larger $2V_x$ is not uncommon. Biaxial positive or negative figures can sometimes be obtained from chrysotile. Optic axis dispersion is $r > v$ or $r < v$.

ALTERATION: Oxidation of associated magnetite may stain the serpentine brown or yellow. The serpentine may be altered to chlorite or replaced by quartz.

DISTINGUISHING FEATURES: Except for cross-fibers of chrysotile in veins, the varieties of serpentine cannot reliably be distinguished without X-ray diffraction or related techniques. The fibrous amphiboles (riebeckite, anthophyllite, tremolite, etc.) are distinguished because they have higher indices of refraction and birefringence. Chlorite is more strongly pleochroic than serpentine and brucite is uniaxial and has abnormal interference colors.

OCCURRENCE: Serpentine is commonly formed by hydrothermal alteration of mafic and ultramafic rocks such as peridotite and pyroxenite, which contain olivine and pyroxene. Replacement of the olivine and pyroxene is usually associated with a release of iron which forms fine-grained disseminated magnetite. Associated minerals are talc, calcite, brucite, chlorite, and chromite. Serpentine also may be found in contact metamorphosed carbonate rocks in association with calcite, forsterite, dolomite, magnesite, and calc-silicate minerals. Greenalite and berthierine are found in weakly metamorphosed iron formations associated with stilpnomelane, minnesotaite, chlorite, and riebeckite. Berthierine also may be found in sedimentary rocks and may occur in oolitic forms. Amesite is rather rare, and has been found in emery deposits, and in the contact zone of mafic intrusives associated with vesuvianite and chlorite.

Pyrophyllite

$Al_2Si_4O_{10}(OH)_2$
Monoclinic and Triclinic
$\angle\beta \cong 100°$
Biaxial $(-)$
$n_\alpha = 1.534–1.556$
$n_\beta = 1.586–1.589$
$n_\gamma = 1.596–1.601$
$\delta = 0.046–0.062$
$2V_x = 53–62°$

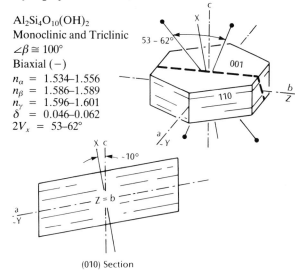

(010) Section

RELIEF IN THIN SECTION: Low to moderate positive relief.

COMPOSITION AND STRUCTURE: Pyrophyllite is a dioctahedral phyllosilicate with T-O-T layers held together with weak electrostatic bonds. There are no interlayer cations. The composition of pyrophyllite is usually close to the ideal formula, although minor subsition of Fe and Mg for Al, and of Al for Si is possible. There are triclinic and disordered polytypes, but the optical properties are essentially the same.

PHYSICAL PROPERTIES: H = 1–2; G = 2.65–2.90; white, gray, pale green, yellow, or brown in hand sample; white streak; pearly to greasy luster.

COLOR AND PLEOCHROISM: Colorless in thin section and grain mount.

FORM: Commonly found as foliated, radial, columnar, or massive aggregates of micalike flakes; also fibrous. Individual grains may be strongly bent.

CLEAVAGE: Perfect {001} cleavage, like the micas, controls fragment orientation.

TWINNING: Twinning generally is not observed.

OPTICAL ORIENTATION: $X \wedge c \cong 10°$, $Y \wedge a \cong 0°$, $Z \cong b$, optic plane is perpendicular to (010) and nearly parallel to (100). The trace of cleavage is length slow and shows essentially parallel extinction.

INDICES OF REFRACTION AND BIREFRINGENCE: The birefringence is high (0.046–0.062) and interference colors in thin section may be as high as third or fourth order.

INTERFERENCE FIGURE: Cleavage fragments and basal sections yield nearly centered acute bisectrix figures with $2V_x = 53$ to $62°$. Optic axis dispersion is weak, with $r > v$.

ALTERATION: Not readily altered.

DISTINGUISHING FEATURES: Resembles talc and muscovite, but both have smaller $2V$. The clay minerals typically have lower birefringence. Gibbsite and brucite are biaxial and uniaxial positive, respectively, and have lower birefringence.

OCCURRENCE: Pyrophyllite is produced by the hydrothermal alteration of aluminous minerals such as feldspars, andalusite, kyanite, corundum, and topaz. It may form irregular veins cutting through altered rocks. Hydrothermally altered silicic volcanic rocks commonly contain fine-grained pyrophyllite associated with quartz, clay, sericite, zoisite, and other aluminous minerals.

Talc

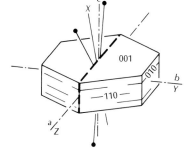

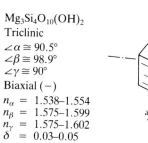

$Mg_3Si_4O_{10}(OH)_2$
Triclinic
$\angle\alpha \cong 90.5°$
$\angle\beta \cong 98.9°$
$\angle\gamma \cong 90°$
Biaxial $(-)$
$n_\alpha = 1.538–1.554$
$n_\beta = 1.575–1.599$
$n_\gamma = 1.575–1.602$
$\delta = 0.03–0.05$
$2V_x = 0–30°$

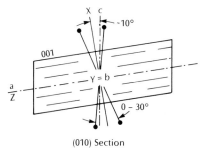

(010) Section

RELIEF IN THIN SECTION: Low to moderate positive relief.

COMPOSITION AND STRUCTURE: Talc is a trioctahedral phyllosilicate composed of T-O-T layers held together with weak electrostatic bonds because there are no interlayer cations. Most talc is relatively close to the ideal formula $Mg_3Si_4O_{10}(OH)_2$, with minor substitution of Fe, Mn, or other cations for Mg, and minor substitution of Al for Si in the tetrahedral site. *Minnesotaite* $[Fe_3Si_4O_{10}(OH)_2]$ is a related mineral with a somewhat more complex structure found in metamorphic iron formations.

PHYSICAL PROPERTIES: Talc: H = 1; G = 2.58–2.83; white, gray, or green in hand sample; white streak; pearly to somewhat greasy luster.

COLOR AND PLEOCHROISM: Talc is colorless in thin section and grain mount.

FORM: Talc is common as foliated, radiating, or randomly oriented aggregates of irregular flakes or fibers resembling the micas. Individual grains may be bent.

CLEAVAGE: Perfect cleavage on {001}, like the micas.

TWINNING: Generally not observed.

OPTICAL ORIENTATION: $X \wedge c \cong +10°$, $Y \cong b$, $Z \wedge a \cong 0°$, optic plane = (010). The trace of cleavage is length slow and shows essentially parallel extinction.

INDICES OF REFRACTION AND BIREFRINGENCE: The indices of refraction increase with increasing iron content. Birefringence is strong and the maximum interference colors seen in thin section are usually third order. Basal sections and fragments lying on the cleavage show low birefringence, because n_β and n_γ are nearly the same. Iron-rich minnesotaite had indices in the vicinity of $n_\alpha = 1.592$, $n_\beta = 1.622$, $n_\gamma = 1.623$ ($\delta = 0.030–0.040$).

INTERFERENCE FIGURE: Cleavage flakes and basal sections yield essentially centered acute bisectrix figures with a small $2V_x$ (0–30°). Optic axis dispersion is weak to moderate with $r > v$.

ALTERATION: Talc may alter to chlorite.

DISTINGUISHING FEATURES: Talc may be mistaken for muscovite and pyrophyllite but has a smaller $2V$. Phlogopite has higher indices, stronger relief in thin section, and is usually pale brown. Brucite is uniaxial positive and often shows anomalous interference colors with lower birefringence. Minnesotaite has higher indices and is weakly colored and pleochroic in shades of colorless, pale yellow and pale green, has small $2V_x$ and typically is fibrous and may form rosettes of needles.

OCCURRENCE: Talc is found in hydrothermally altered mafic and ultramafic rocks in association with serpentine, magnesite, and relict grains of olivine and pyroxene. It is commonly formed in the metamorphism of siliceous dolomite in association with calcite, dolomite, tremolite, and related calc-silicate minerals. Talc is a major constituent of talc schist that also may contain magnetite, tremolite, chlorite, anthophyllite, or serpentine.

Minnesotaite is found in metamorphosed iron formations associated with magnetite and other iron oxides, riebeckite, greenalite, stilpnomelane, siderite, quartz, and chlorite.

Muscovite

$KAl_2(AlSi_3O_{10})(OH)_2$
Monoclinic
$\angle\beta = 95.5°$
Biaxial (−)
$n_\alpha = 1.552–1.580$
$n_\beta = 1.582–1.620$
$n_\gamma = 1.587–1.623$
$\delta = 0.036–0.049$
$2V_x = 30–47°$

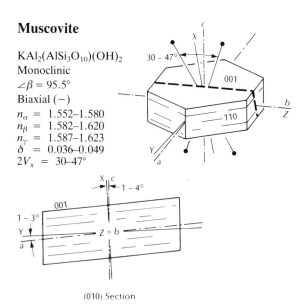

(010) Section

RELIEF IN THIN SECTION: Moderate positive relief; modest change with rotation.

COMPOSITION: Muscovite is a dioctahedral mica consisting of T-O-T layers bonded together through K^+ in 12-fold coordination. The charge of the K^+ is balanced by replacing a Si^{4+} with an Al^{3+} in a tetrahedral site. Some Na may replace K, but there is not

continuous solid solution to *paragonite*, which is the sodium analog. Other large cations such as Ba, Rb, Cs, and Ca also may replace K in minor amounts, as may H_3O groups in the variety *hydromuscovite*. In the octahedral sites, Fe^{2+}, Fe^{3+}, Mg, Mn, Li, Cr, and V may replace minor amounts of Al. *Phengite* is a name given to varieties with significant Mg and Fe^{2+}. In most cases, some of the OH is replaced by F, or, to a lesser extent, Cl.

PHYSICAL PROPERTIES: H = $2\frac{1}{2}$–3; G = 2.77–2.88; colorless, white, or pale brown in hand sample, also shades of pale pink or pale green; white streak; vitreous to pearly luster.

COLOR AND PLEOCHROISM: Typically colorless in thin section and grain mount, rarely very pale pink or green.

FORM: Figure 14.4. Well-formed tabular crystals with a roughly hexagonal outline are rare. Usually found as micaceous flakes or tablets with irregular outlines, tabular parallel to {001}. *Sericite* is a name given to very fine, ragged grains and aggregates of white mica—usually muscovite or phengite—produced by the alteration of feldspars or other minerals. Included grains of zircon or other radioactive minerals may form pale pleochroic halos in surrounding muscovite.

CLEAVAGE: Perfect cleavage on {001} is well displayed in thin section and controls fragment orientation.

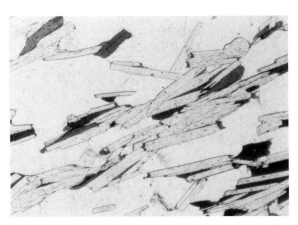

Figure 14.4 Muscovite laths with dark biotite and clear quartz. Note the single rounded apatite grain to the upper right. Field of view is 1.7 mm wide.

TWINNING: Twinning with a composition plane on {001} is reported but is rarely seen.

OPTICAL ORIENTATION: $X \wedge c = +1$ to $+4°$, $Y \wedge a = +1$ to $+3°$, $Z = b$, optic plane is perpendicular to (010). Extinction is essentially parallel to cleavage in all orientations, because the maximum extinction angle is less than $3°$. Grains are commonly bent and may show wavy or undulose extinction. The trace of the cleavage is always length slow. The orientation of paragonite is essentially the same.

INDICES OF REFRACTION AND BIREFRINGENCE: Relatively pure muscovite has indices at the lower end of the range given above. Substituting Fe, Mg, Mn, Ti, Cr, and V causes the indices to increase. Substituting Li causes a decrease in indices. Birefringence is high (0.036–0.049) and interference colors in thin sections may be as high as third order, and vivid colors of the second order are typical.

Paragonite has indices in the range $n_\alpha = 1.564–1.580$, $n_\beta = 1.594–1.609$, $n_\gamma = 1.600–1.609$; $\delta = 0.028–0.038$.

INTERFERENCE FIGURE: The $2V_x$ angle is usually between 30 and $47°$. Pure muscovite has the larger $2V_x$, and substituting Fe, Mn, and Cr usually causes a decrease. Phengite may have $2V_x$ in the range 0 to $20°$. Cleavage fragments give excellent, nearly centered, acute bisectrix figures. Good figures in thin section are yielded by grains that do not show the cleavage and that have lower to middle first-order colors. Optic axis dispersion may be noticeable with $r > v$. Paragonite is biaxial $(-)$ with $2V_x = 0–40°$.

ALTERATION: Muscovite is not generally altered but may be converted to clay minerals by weathering.

DISTINGUISHING FEATURES: Muscovite is distinguished from most biotite by its lack of color. Mg-rich biotite (phlogopite) may be nearly colorless but has a smaller $2V$. A distinctive character of muscovite and the other micas is a pebbly surface texture seen near extinction called "birds-eye" extinction, alluding to the pattern seen in birds-eye maple wood. Talc is colorless like muscovite but has a smaller $2V$. Pale-colored chlorite shows lower birefringence, often with anomalous interference colors. Pyrophyllite has high birefringence like muscovite but has a larger $2V$. Paragonite has very similar optical properties, and X-ray or staining techniques (Laduron, 1971) are needed to distinguish them.

OCCURRENCE: Muscovite is very common in a wide variety of metamorphic rocks, including slate, schist, phyllite, gneiss, hornfels, and quartzite. In igneous rocks, it is commonly found in granite, granodiorite, aplite, pegmatite, and related felsic rocks and, less commonly, in felsic volcanics. Sericite, which may recrystallize to form coarser-grained muscovite, is formed by hydrothermal or deuteric alteration of feldspars or other minerals. The alteration may be selective, replacing only the cores of plagioclase grains or only certain twin lamellae.

Muscovite is commonly found as detrital grains in immature clastic sediments. Phengite and paragonite are usually found in relatively low-grade regional metamorphic rocks. Paragonite is probably more abundant than has often been supposed and has commonly been misidentified as muscovite.

Biotite

$K_2(Mg,Fe)_3AlSi_3O_{10}(OH,O,F)_2$
Monoclinic
$\angle\beta = 99.3°$
Biaxial $(-)$
$n_\alpha = 1.522–1.625$
$n_\beta = 1.548–1.672$
$n_\gamma = 1.549–1.696$
$\delta = 0.03–0.07$
$2V_x = 0–25°$

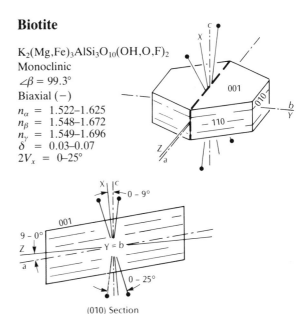

(010) Section

RELIEF IN THIN SECTION: Moderate to moderately high positive relief; low if Mg rich.

COMPOSITION: Biotite shows substantial variability in composition and, for that reason, correlation of optical properties with composition has not been particularly successful. The primary compositional variation in biotite is in the occupancy of the octahedral sites, and most biotite falls between *phlogopite* [$K_2Mg_3AlSi_3O_{10}(OH,F)_2$] and *annite*

$[K_2Fe^{2+}{}_3AlSi_3O_{10}(OH,F)_2]$. *Siderophyllite* contains up to 10 or 15 percent Al^{3+} which is balanced by replacing additional Si^{4+} by Al^{3+} in the tetrahedral sites or by replacing $(OH,F)^-$ with O^{2-}. Fe^{3+} also is common and is usually balanced by replacing $(OH,F)^-$ with O^{2-}. Samples with substantial amounts of ferric iron, and oxygen in the hydroxyl sites are called *oxybiotite*. Small amounts of Ti and Mn are common, up to about 10 percent of the K may be replaced by Na, and F usually occupies a substantial percentage of the hydroxyl sites. *Ferriannite* is a variety found in metamorphosed iron formations that has substantial Fe^{3+} in tetrahedral sites.

PHYSICAL PROPERTIES: $H = 2\frac{1}{2}-3$; $G = 2.7-3.3$ (increases with iron content); brown to black in hand sample, sometimes greenish, darker with increasing iron content; white or gray streak; vitreous luster.

COLOR AND PLEOCHROISM: Typically brown, brownish green, or reddish brown, and distinctly pleochroic. Phlogopite may be nearly colorless or pale brown. With rare exceptions found in phlogopite, the absorption is $Z \cong Y > X$, so grains are darker when the trace of cleavage is parallel to the vibration direction of the lower polar. The common colors are X = colorless, light tan, pale greenish brown, pale green; $Y \cong Z$ = brown, olive-brown, dark green, dark red-brown. The intensity of the color generally increases with increasing iron content. Cleavage flakes and sections cut parallel to (001) yield darker colors with little pleochroism. Pleochroic halos around radioactive minerals such as zircon or allanite are common.

FORM: Figure 14.5. Euhedral crystals are not uncommon, are usually tabular parallel to (001), and have a roughly hexagonal cross section. Also found as micaceous or tabular grains, or grains with irregular outline. Grains may be bent, particularly in metamorphic rocks.

CLEAVAGE: The perfect cleavage on {001} is easily seen in thin section, and controls fragment orientation and shape.

TWINNING: Twins may be present with composition planes {001} and {110} but are rarely visible.

OPTICAL ORIENTATION: $X \wedge c = +9$ to $0°$, $Y = b$, $Z \wedge a = 0$ to $+9°$, optic plane = (010). Extinction is parallel or close to parallel with a maximum angle of

Figure 14.5 Biotite. (*Top*) Biotite with dark pleochroic halos around radioactive inclusions. (*Bottom*) Light colored laths of phlogopite (center) in marble. Width of fields of view: (*top*) 1.7 mm; (*bottom*) 2.3 mm.

a few degrees, rarely up to 9°. The trace of cleavage is length slow. Most samples are the 1M polytype described here. A few are the $2M_1$ polytype with the optic plane perpendicular to (010) as shown in muscovite, but they are still length slow and show near-parallel extinction. Bent grains show wavy extinction.

INDICES OF REFRACTION AND BIREFRINGENCE: The indices of refraction increase with increasing iron content (Figure 14.6), but the compositional diversity makes determination of composition from optical properties impossible. Oxybiotite may have indices above the range given above. Birefringence is strong and yields maximum interference colors in the third, or occasionally fourth, order. The strong mineral color usually masks the color, however. Cleavage flakes and sections cut parallel to (001)

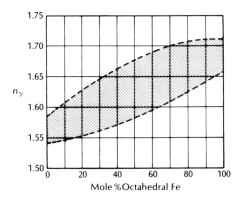

Figure 14.6 Approximate range of n_γ with increasing proportion of octahedral iron in biotite.

show low birefringence. Ferriannite may have indices as high as $n_\alpha = 1.677$ and $n_\gamma = 1.721$.

INTERFERENCE FIGURE: Biotite is biaxial negative with $2V_x$ usually between 0 and 10°, and no more than about 25°. Some biotite is sensibly uniaxial and shows no isogyre separation with rotation. Cleavage flakes and sections cut parallel to (001) yield acute bisectrix figures with several orders of isochromes, the color of which is usually strongly influenced by the inherent mineral color. The (001) sections can be recognized because they are dark with little pleochroism and do not show cleavage traces. Optic axis dispersion is weak with $v > r$ or, less commonly, $r > v$ for Mg-rich varieties.

ALTERATION: Biotite commonly alters to chlorite, which may mantle the biotite or be interleaved along cleavage traces (cf. Figures 12.4 and 14.9). Biotite also may alter to clay minerals that have lower birefringence and lighter color. Other products of the breakdown of biotite are sericite, iron–titanium oxides, epidote, calcite, and various sulfides.

DISTINGUISHING FEATURES: Light-colored phlogopite resembles muscovite but has smaller $2V$. Chlorite has lower birefringence and often displays anomalous interference colors. Stilpnomelane is very similar but the basal cleavage is less well developed than in biotite, and there is a second cleavage at right angles to the basal cleavage. The pebbly or finely mottled "birds-eye" extinction that is characteristic of the micas is not as well developed in stilpnomelane. Green or brown tourmaline resembles biotite but is darker when its long dimension is oriented at

right angles to the vibration direction of the lower polar. Amphiboles have two cleavages with the distinct amphibole cross section and show inclined extinction without the "birds-eye" characteristic.

OCCURRENCE: Biotite is a common mineral in a wide variety of igneous and metamorphic rocks. In igneous rocks, it is characteristic of silicic and alkalic rocks such as granite, granodiorite, trondjhemite, aplite, pegmatite, syenite, nepheline syenite, and the volcanic equivalents. It also is found as a late-stage magmatic product in diorite, gabbro, norite, anorthosite, and related rocks. Phlogopite is found in periodotite and other ultramafic varieties.

In metamorphic rocks, biotite is common in a wide variety of phyllites, schists, and gneisses and may persist from greenschist facies through strongly migmatitic rocks. Biotite is common in hornfels, and phlogopite is often found in contact metamorphosed carbonates.

Biotite also is a relatively common detrital mineral, particularly in immature sediments, but usually yields to clay minerals with extended weathering.

Lepidolite

$K(Li,Al)_3(Si,Al)_4O_{10}(F,OH)_2$
Monoclinic
$\angle\beta \cong 100°$
Biaxial (−)
$n_\alpha = 1.525–1.548$
$n_\beta = 1.548–1.585$
$n_\gamma = 1.554–1.587$
$\delta = 0.018–0.038$
$2V_x = 0–58°$, usually 30–50°

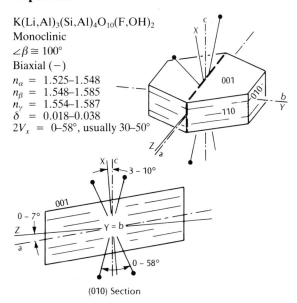

(010) Section

RELIEF IN THIN SECTION: Low to moderate positive relief.

COMPOSITION: Lepidolite is a mica with a T-O-T structure essentially the same as muscovite. The principal chemical variation is in substitution of Al^{3+} for Li^+ in the octahedral sites, with associated exchange of Si^{4+} and Al^{3+} in the tetrahedral sites to balance the charge. There is substantial solid solution toward muscovite [$KAl_2AlSi_3O_{10}(OH,F)_2$], and variable amounts of Fe^{2+}, Fe^{3+}, Mn, and Mg may be present in the octahedral sites. Significant amounts of Na, Rb, or Cs may substitute for K in the interlayer site, and most lepidolite contains a substantial amount of F in the hydroxyl site.

PHYSICAL PROPERTIES: H = $2\frac{1}{2}$–4; G = 2.8–2.9; white, pink, or lavender in hand sample; white streak; vitreous to pearly luster.

COLOR AND PLEOCHROISM: Colorless in thin section. Large fragments in grain mount may show pale colors with pleochroism in shades of pink, violet, or green.

FORM: Crystals are tabular parallel to (001) and have a roughly hexagonal outline. Lepidolite is commonly found in aggregates of small scaly flakes or as felted to cryptocrystalline masses.

CLEAVAGE: Perfect cleavage on {001}, like the other micas. The cleavage controls fragment orientation.

TWINNING: Twins with a {001} composition plane have been reported but are rather rare.

OPTICAL ORIENTATION: $X \wedge c$ = +10 to +3°, Y = b, $Z \wedge a$ = 0 to +7°, optic plane = (010). Extinction is parallel or nearly parallel to cleavage with a maximum extinction angle of 0 to 7°. The trace of cleavage is length slow.

INDICES OF REFRACTION AND BIREFRINGENCE: The normal range for the indices is n_α = 1.533 ± 0.004, n_β = 1.559 ± 0.007, n_γ = 1.562 ± 0.007. The indices do not appear to be strongly influenced by the relative amounts of Li and Al, but do increase with increasing Fe and Mn content, and decrease with increasing substitution of F for OH. The birefringence is usually around 0.03, so interference colors in thin section may be as high as middle second order.

INTERFERENCE FIGURE: Lepidolite is biaxial negative with $2V_x$ usually in the range 30 to 50°, although the full range is 0 to 58°, and some may be essentially uniaxial. Cleavage fragments and sections cut parallel to (001) produce nearly centered acute

bisectrix figures. The (001) sections can be recognized because they show low birefringence and do not show the cleavage. Optic axis dispersion is weak with $r > v$.

ALTERATION: Lepidolite may be replaced by muscovite.

DISTINGUISHING FEATURES: Lepidolite closely resembles muscovite in thin section, but has lower relief, and somewhat lower birefringence. The pink to lavender color in hand sample is characteristic. In some cases, it may be necessary to use X-ray techniques to distinguish pink lithian muscovite from lepidolite.

Zinnwaldite [$K(Fe^{2+}_{1-0.5}Li_{1-1.5}Al)(Si_{3-3.5}Al_{1-0.5})O_{10}(OH,F)_2$] is a closely related mineral that is colorless to light brown in thin section and resembles phlogopite but is found in Li-bearing pegmatites which almost never contain phlogopite.

OCCURRENCE: Lepidolite is restricted to granitic pegmatites where it is associated with quartz, K-feldspar, albite, beryl, tourmaline, and other lithium minerals such as spodumene.

Stilpnomelane

~$K_{0.6}(Fe,Mg)_6(Si_8Al)(O,OH)_{27}\cdot 2$–$4H_2O$
Triclinic (pseudomonoclinic)
$\angle\alpha = 124°$
$\angle\beta = 96°$
$\angle\gamma = 120°$

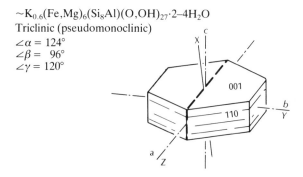

Biaxial (−), essentially uniaxial (−)
n_α = 1.543–1.634
n_β = 1.576–1.745
n_γ = 1.576–1.745
δ = 0.030–0.110
$2V_x \cong 0°$

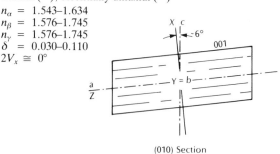

(010) Section

RELIEF IN THIN SECTION: High positive relief; may show significant change with rotation.

COMPOSITION AND STRUCTURE: Stilpnomelane is an unusual trioctahedral phyllosilicate. The basic structure consists of T-O-T layers like biotite, but some of the tetrahedra in the tetrahedral sheets are arranged so that apices point outward, away from the central octahedral sheet. The oxygen on the apices of the outward-pointing tetrahedra are shared with similar tetrahedra on adjacent layers. The interlayer cation is usually K, although Na or Ca also may be present and the total number can be quite variable. Water molecules are also located between the layers. The cations in the octahedral site are predominantly Fe^{3+}, Fe^{2+}, and Mg, although Mn, Ti, and Al also may be present.

PHYSICAL PROPERTIES: H = 3–4; G = 2.59–2.96; golden brown, reddish brown, dark green, or black in hand sample; white to tan streak; pearly to submetallic luster.

COLOR AND PLEOCHROISM: Pale yellow, dark brown, or green in thin section or grain mount (like biotite). Distinct pleochroism with $Z \cong Y > X$: X = pale yellow to golden yellow, $Y \cong Z$ = brown, reddish brown, greenish brown, black. Sections are darker when the cleavage trace is parallel to the vibration direction of the lower polar. Basal sections or cleavage flakes show negligible pleochroism and are dark. In grain mount or abnormally thick section, basal sections may be sensibly opaque.

FORM: Figure 14.7. Stilpnomelane has a typical micaceous habit. Grains may be arranged in radiating, plumose, or sheaflike aggregates and may be interlayered with chlorite or other micaceous minerals.

CLEAVAGE: Cleavage on {001} is not quite as perfect as in the micas. There also is a fair cleavage on {010} that intersects the {001} cleavage at 90°. The {001} cleavage controls fragment orientation.

TWINNING: None reported.

OPTICAL ORIENTATION: Relative to axes assigned for the monoclinic habit, the orientation is: $X \wedge c \cong$ +6°, $Y = b$, $Z \wedge a \cong 0°$, optic plane = (010). The trace of the basal {001} cleavage is length slow and shows essentially parallel extinction.

INDICES OF REFRACTION AND BIREFRINGENCE: Indices of refraction and birefringence increase rapidly with increasing Fe^{3+} (Figure 14.8). An increase in Mg

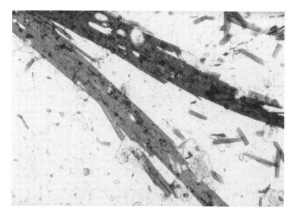

Figure 14.7 Elongate stilpnomelane grains. Field of view is 1.2 mm wide.

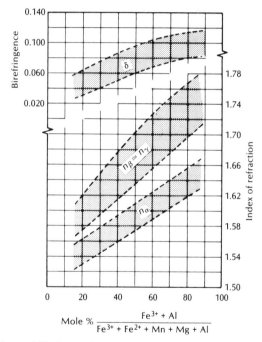

Figure 14.8 Approximate range of indices of refraction and birefringence (δ) in stilpnomelane. Data from Deer and others (1962), Blake (1964), Hutton (1956), and Eggelton and Chappell (1978).

content causes a substantial reduction in indices. Maximum interference colors in thin section are from upper second order, to creamy fifth or sixth order, but the strong mineral color tends to mask the interference color. Cleavage fragments and sections cut parallel to (001) are essentially isotropic.

INTERFERENCE FIGURE: Despite the fact that it is triclinic, stilpnomelane usually yields essentially uniaxial negative figures from cleavage plates and basal sections. $2V_x$ as large as $40°$ has been reported. The colors of the isochromes in the figure are strongly influenced by mineral color.

ALTERATION: Stilpnomelane may alter to chlorite, clay, and iron oxides.

DISTINGUISHING FEATURES: Stilpnomelane closely resembles biotite. The following characteristics serve to distinguish them:

1. The basal cleavage of stilpnomelane is less perfect than in biotite and is intersected by a second cleavage.
2. The pebbly or "birds-eye" texture seen near extinction with biotite is subdued or absent in stilpnomelane.
3. Birefringence of Fe^{3+}-rich varieties is higher than in biotite.

Chlorite, chloritoid, and clintonite have much lower birefringence.

OCCURRENCE: Stilpnomelane is fairly common in low-grade regional metamorphic rocks derived from graywacke and related sediments. The commonly associated minerals are chlorite, muscovite, garnet, actinolitic amphibole, glaucophane, calcite, epidote group, and pumpellyite. Stilpnomelane is also common in metamorphosed iron formations associated with riebeckite, siderite, chlorite, and various iron oxides and silicates.

Margarite

$CaAl_2(Al_2Si_2)O_{10}(OH)_2$
Monoclinic
$\angle \beta = 95.4°$
Biaxial $(-)$
$n_\alpha = 1.630–1.638$
$n_\beta = 1.642–1.648$
$n_\gamma = 1.644–1.650$
$\delta = 0.010–0.014$
$2V_x = 26–67°$

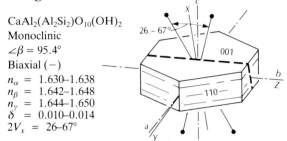

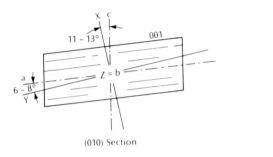

(010) Section

RELIEF IN THIN SECTION: Moderately high positive relief.

COMPOSITION: Margarite is a dioctahedral brittle mica analogous to muscovite. Ca occupies the interlayer sites and half of the tetrahedral sites are occupied by Al. Up to about 30 percent of the Ca^{2+} may be replaced by Na^+, with the charge balanced either by exchanging Si^{4+} for Al^{3+}, filling additional octahedral sites, or exchanging OH^- for O^{2-}. Minor amounts of Fe, Mg, Mn, or Ti may substitute in the octahedral sites.

PHYSICAL PROPERTIES: $H = 3\frac{1}{2}–4\frac{1}{2}$; $G = 3.0–3.1$; grayish pink, pale yellow, or pale green in hand sample; white streak; pearly to vitreous luster.

COLOR AND PLEOCHROISM: Colorless in thin section and grain mount.

FORM: Margarite has the same form as the micas, forming platy or scaly aggregates as well as granular masses. Crystals are uncommon and are tabular parallel to (001) with a roughly hexagonal cross section. Margarite may be intergrown with paragonite and muscovite.

CLEAVAGE: Perfect {001} cleavage, like the micas.

TWINNING: Twins with composition planes parallel to (001) are not particularly common.

OPTICAL ORIENTATION: $X \wedge c = +11$ to $+13°$, $Y \wedge a = -6$ to $-8°$, $Z = b$, optic plane is perpendicular to (010) and nearly parallel to (100). The maximum extinction angle to cleavage is 6 to 8°, and the trace of cleavage is length slow.

INDICES OF REFRACTION AND BIREFRINGENCE: Indices decrease with increased substitution of Na for Ca, and birefringence increases. Interference colors in thin section are usually no higher than first-order yellow.

INTERFERENCE FIGURE: Margarite is biaxial negative with $2V_x$ normally in the range of 40 to 50°, although higher and lower values have been reported. Cleavage fragments and basal sections (which show low birefringence and no cleavage traces) yield nearly centered acute bisectrix figures. Optic axis dispersion is $v > r$.

ALTERATION: Margarite may alter to clay minerals.

DISTINGUISHING FEATURES: Margarite resembles muscovite, paragonite, and talc, but has higher indices, a slightly larger extinction angle, and lower birefringence. Chlorite and chloritoid are green and pleochroic.

OCCURRENCE: Margarite is probably more common than was once thought and is found in graphite-bearing aluminous metamorphic rocks. It occurs with other micas or replaces such aluminous minerals as andalusite and corundum.

Clintonite

$Ca_2(Mg_{4.6}Al_{1.4})(Si_{2.5}Al_{5.5})O_{20}(OH)_4$
Monoclinic
$\angle\beta = 95–100°$
Biaxial $(-)$
$n_\alpha = 1.643–1.648$
$n_\beta = 1.655–1.662$
$n_\gamma = 1.655–1.663$
$\delta = 0.012–0.015$
$2V_x = 0–33°$

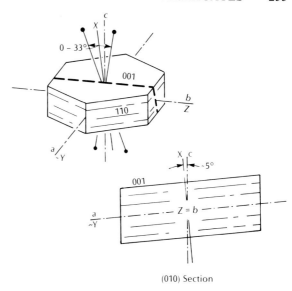

(010) Section

RELIEF IN THIN SECTION: Moderately high positive relief.

COMPOSITION AND STRUCTURE: Clintonite is the trioctahedral brittle mica analog of phlogopite. The interlayer cation is predominantly Ca, and about two-thirds of the tetrahedral sites are occupied by Al, and the remainder by Si. About one quarter of the octahedral sites contain Al^{3+}. The remaining octahedral sites contain predominantly Mg^{2+}, with minor amounts of Fe, Mn, or Ti. The term *xanthophyllite* has been used for certain varieties of trioctahedral brittle micas but the distinction between clintonite and xanthophyllite is poorly defined. The term clintonite is preferred for all species.

PHYSICAL PROPERTIES: $H = 3\frac{1}{2}–4\frac{1}{2}$; $G = 3.0–3.1$; yellow, reddish brown, or green in hand sample, rarely colorless; white or gray streak; vitreous luster.

COLOR AND PLEOCHROISM: Usually colorless or pale brown in thin section, darker in grain mount. Pleochroism is X = colorless to light orange, yellow, or reddish brown; $Y \cong Z$ = pale brown, pale green; $X < Y \cong Z$. Sections are darker when the long dimension is parallel to the vibration direction of the lower polar.

FORM: Clintonite has a typical micaceous habit, with irregular platy grains or sheaves of nearly parallel grains.

CLEAVAGE: The perfect {001} cleavage is not quite as good as in the micas.

TWINNING: Twins are reported with {001} composition planes, but are not common.

OPTICAL ORIENTATION: $X \wedge c \cong +5°$, $Y \cong a$, $Z = b$, optic plane is perpendicular to (010) and nearly parallel to (100). Also, $X \wedge c \cong +10°$, $Y = b$, $Z \cong a$, optic plane = (010). In both orientations extinction is sensibly parallel to cleavage with a maximum extinction angle of no more than a few degrees. As a practical matter, the two optical orientations are nearly impossible to distinguish optically. The trace of cleavage is length slow.

INDICES OF REFRACTION AND BIREFRINGENCE: The range of indices is relatively small. Samples with higher iron content have the higher indices. The maximum interference colors in thin section are first-order yellow.

INTERFERENCE FIGURE: Cleavage fragments and sections cut parallel to (001) show low birefringence and no cleavage traces, and yield nearly centered acute bisectrix figures with small $2V_x$ (0–33°). Optic axis dispersion is weak with $v > r$.

ALTERATION: Clintonite may alter to clay minerals.

DISTINGUISHING FEATURES: Clintonite is distinguished from muscovite and biotite by lower birefringence and from margarite by smaller $2V$. The common varieties of chlorite are more distantly green, have lower relief in thin section, lower birefringence, and, in some cases, are biaxial positive. Chloritoid is more distinctly green, and the biaxial negative variety has larger $2V$.

OCCURRENCE: Clintonite is found in chlorite and talc schists and in metamorphosed dolomite. In the latter occurrence, clintonite usually is associated with forsterite, calcite, diopside, phlogopite, and other calc-silicate minerals.

Chlorite

$(Mg,Al,Fe)_3(Si,Al)_4O_{10}(OH)_2 \cdot (Mg,Al,Fe)_3(OH)_6$
Monoclinic and Triclinic
$\angle \beta \cong 97°$
Biaxial $(+)$ or $(-)$
n_α = 1.55–1.67
n_β = 1.55–1.69
n_γ = 1.55–1.69
δ = 0.0–0.015
$2V_z$ = 0–60° (positive)
$2V_x$ = 0–40° (negative)

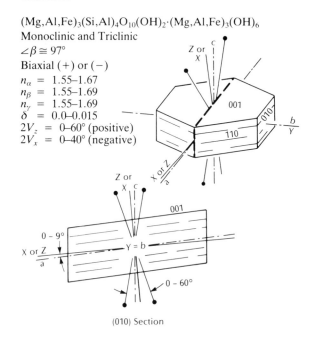

(010) Section

RELIEF IN THIN SECTION: Moderate to moderately high positive relief.

COMPOSITION AND STRUCTURE: The term *chlorite* is applied to a large group of minerals whose structure consists of alternating T-O-T layers equivalent to talc layers $[Mg_3Si_4O_{10}(OH)_2]$ and octahedral interlayers equivalent to brucite sheets $[Mg_3(OH)_6]$. The layers are held together with weak electrostatic forces and there are no interlayer cations (Figure 14.1). There is a wide range of compositional variation involving substitution of Mg, Al, Fe^{2+}, and Fe^{3+} in both octahedral sheets and substitution of Al^{3+} for Si^{4+} in the tetrahedral sheets. The common chlorite compositions can be expressed as $(Mg,Fe)_5Al(Si_3Al)O_{10}(OH)_8$. The term *clinochlore* is often used for the Mg end of this series, and *chamosite* for the Fe end. As with the other phyllosilicates, there are a number of polytypes related by differences in the stacking of successive sheets, but these can only be distinguished with X ray techniques.

Dozens of different names have been applied to different species of chlorite but because of the compositional and structural diversity, a widely accepted classification scheme has not been devised (e.g.

Hey, 1954; Phillips, 1964; Müller, 1966; Carroll, 1970; Bailey, 1988). The commonly encountered names include clinochlore, pennine, prochlorite, and chamosite. Others include sheridanite, ripidolite, brunsvigite, diabantite, thuringite, corundophilite, daphnite, diabantite, strigovite, klementite, delessite, pseudothuringite, aphrosiderite, bavalite, and helminthe. The optical properties cannot reliably be used as an indicator of composition or structure, so it is strongly recommended that petrographers use the term *chlorite* for all varieties and defer to chemical and structural studies for more detailed classification.

PHYSICAL PROPERTIES: H = 2–3; G = 2.6–3.3; green in hand sample, less commonly white, gray, or brown; white, gray or pale green streak; vitreous to greasy luster.

COLOR AND PLEOCHROISM: Usually light to medium green in thin section and pleochroic. The pleochroism is usually expressed in shades of colorless, pale green, yellowish green, green, or brownish green. Grains are darker when the trace of cleavage is parallel with the vibration direction of the lower polar. For optically positive varieties, the pleochroic formula is $X \cong Y > Z$, with $X = Y$ = green, pale green, brownish green; Z = colorless, pale green, pale yellowish green. Optically negative varieties show $X < Y \cong Z$, with X = colorless, pale green, pale yellowish green; $Y = Z$ = pale green, green, olive-green. The darker varieties usually have higher iron content.

FORM: Figure 14.9. Rare crystals are tabular parallel to (001) with a roughly hexagonal outline. Commonly found as plates or scales similar to the micas. In sediments, chlorite is a common constituent of the clay fraction and may form oolitic or spherulitic balls similar to glauconite.

CLEAVAGE: Perfect cleavage on {001} controls fragment orientation.

TWINNING: Twins with composition plane on (001) are common but are sometimes difficult to recognize.

OPTICAL ORIENTATION: The optic plane is usually parallel to (010) and the acute bisectrix, either Z or X, is approximately perpendicular to the {001} cleavage. The obtuse bisectrix, either X or Z, is within a few degrees of the a crystal axis, and $Y = b$. The

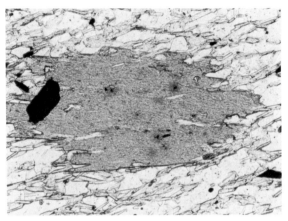

Figure 14.9 Chlorite. (*Top*) Chlorite porphyroblast with fine scaly chlorite in the groundmass. (*Bottom*) Chlorite interleaved with dark biotite. The dark high relief grains are titanite. Field of view is 0.66 mm wide for both photographs.

maximum extinction angle to the trace of cleavage is usually no more than a few degrees and rarely more than 9°. The trace of cleavage in optically positive varieties is length fast and in optically negative varieties is length slow.

INDICES OF REFRACTION AND BIREFRINGENCE: Indices of refraction increase with increasing Fe + Mn + Cr content (Figure 14.10). Birefringence is usually low, so maximum interference colors are rarely above first-order white or yellow. Anomalous brown, bluish, or purplish interference colors are common; brownish colors are more common in optically positive varieties and bluish or purplish colors are more common in optically negative varieties. Some varieties may be essentially isotropic.

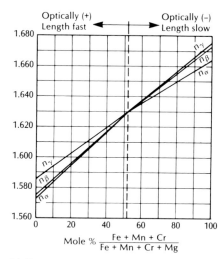

Figure 14.10 Variation of indices of refraction of common chlorite. The optic sign changes from positive to negative at about 52% Fe + Mn + Cr. There is significant scatter about each curve, and some low Fe + Mn + Cr samples may be optically negative and some higher samples may be positive. After Albee (1962).

INTERFERENCE FIGURE: Cleavage fragments and basal sections yield approximately centered biaxial acute bisectrix figures. Low Fe + Mn + Cr chlorite is usually optically positive, while high Fe + Mn + Cr samples are usually negative; there are, however, numerous exceptions, and some samples may yield essentially uniaxial figures. Because birefringence is low, the isogyres are usually broad and diffuse. Optic axis dispersion is usually $v > r$, and less commonly, $r > v$. The sign of elongation may be used to confirm the optic sign in fine-grained material. Positive varieties are length fast, and negative varieties are length slow.

ALTERATION: Oxidation may produce iron stains, and weathering produces clay minerals, but otherwise, chlorite is fairly resistant to alteration.

DISTINGUISHING FEATURES: Chlorite is distinguished from muscovite and biotite by its green color, pleochroism, and weak birefringence. Serpentine usually has lower refractive indices and is less pleochroic. In soils and sediments, chlorite is usually indistinguishable from the other clay minerals and must be identified by X-ray or other techniques. The various members of the chlorite group can only be

reliably distinguished with chemical and X-ray techniques.

OCCURRENCE: Chlorite is a widespread mineral in contact and regional metamorphic rocks of low to medium grade. In pelitic rocks, it is commonly found with biotite, garnet, staurolite, andalusite, muscovite, chloritoid, and cordierite. In more mafic rocks, it is found with talc, serpentine, actinolite, hornblende, epidote and garnet. It also commonly forms by the alteration of other mafic minerals such as pyroxenes, amphiboles, biotite, staurolite, cordierite, garnet, and chloritoid. In igneous rocks, chlorite is formed by hydrothermal or deuteric alteration of biotite, amphibole, pyroxene, or other mafic minerals. It is also found in amygdules and fractures in altered volcanic rocks.

The hydrothermal alteration of almost any rock type may result in the production of chlorite, either from recrystallization of clay minerals, or from the alteration of mafic minerals such as biotite and hornblende. Chlorite also is found in hydrothermal vein deposits. Chlorite is formed by weathering of mafic minerals, so it is a common constituent of soils and the clay fraction of sedimentary rocks. It also may form small oolites or spherulites similar to glauconite.

Prehnite

$Ca_2Al(AlSi_3O_{10})(OH)_2$
Orthorhombic
Biaxial (+)
$n_\alpha = 1.610–1.637$
$n_\beta = 1.615–1.647$
$n_\gamma = 1.632–1.670$
$\delta = 0.020–0.035$
$2V_z = 64–70°$

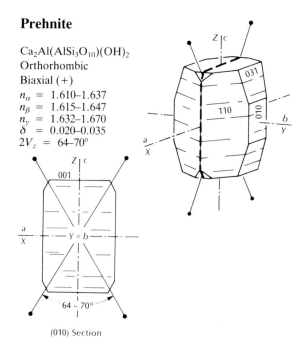

(010) Section

RELIEF IN THIN SECTION: Moderately high positive relief.

COMPOSITION AND STRUCTURE: Prehnite is a phyllosilicate with a complex structure. Most prehnite has a composition close to the ideal formula given above, but some may have up to 60 percent of the octahedral Al replaced by Fe^{3+}, and additional Al may replace Si in small amounts in the tetrahedral sites.

PHYSICAL PROPERTIES: $H = 6-6\frac{1}{2}$; $G = 2.9$; pale green in hand sample, less commonly white or gray; white streak; vitreous luster.

COLOR: Colorless in thin section and grain mount.

FORM: Commonly as radiating, fanlike, or sheaflike aggregates. A pattern resembling a bow tie is common. The individual grains in the aggregates are often joined nearly parallel to (001). Also as lamellar or columnar masses. Crystals are uncommon but tend to be either tabular parallel to (001) or prismatic parallel to the c axis. A pattern of symmetrical intergrowths may show a radial or "hourglass" pattern.

CLEAVAGE: The basal {001} cleavage is good, and two prismatic cleavages on {110} are poor and not usually seen. The basal cleavage tends to control fragment orientation.

TWINNING: Fine lamellar twinning has been reported but is not common.

OPTICAL ORIENTATION: $X = a$, $Y = b$, $Z = c$, optic plane (010). The trace of {001} cleavage and the contacts between grains in radiating and sheaflike bundles show parallel or nearly parallel extinction. Extinction may be wavy or incomplete.

INDICES OF REFRACTION AND BIREFRINGENCE: Indices of refraction and birefringence show a general increase with increasing iron content. Aluminous varieties are more common and have indices near the lower end of the range given above. Interference colors in thin section may be as high as lower to middle second order, and they may be anomalous.

INTERFERENCE FIGURE: Prehnite is biaxial positive with a fairly large $2V_z$ (64–70°). The $2V_z$ angle may vary from place to place on a single grain, and anomalously small angles may be found in sectors with anomalous interference colors and incomplete extinction. Optic axis dispersion is weak with $r > v$, but may be strong with $v > r$ and with crossed bisectrix dispersion in anomalous sectors.

ALTERATION: Prehnite may alter to chlorite or zeolites.

DISTINGUISHING FEATURES: The "bow tie" structure and anomalous optical properties are distinctive. Lawsonite, pumpellyite, and epidote have higher indices and lower birefringence. Datolite has poor cleavage, higher birefringence, and is optically negative.

OCCURRENCE: Prehnite is common in amygdules, veins, or other cavities in mafic to intermediate volcanic rocks in association with zeolites, datolite, calcite, pectolite, chlorite, and epidote group minerals. It also is found in contact metamorphosed carbonates associated with calcite, wollastonite, garnet, tremolite, and other calc-silicate minerals. Its most abundant occurrence, however, is in fairly low-grade regionally metamorphosed graywacke, basalt, and other mafic to intermediate rocks in association with pumpellyite, lawsonite, albite, epidote group, chlorite, and zeolites. Sausserite, which is an alteration product after plagioclase, or occasionally amphiboles and pyroxenes, may contain prehnite as well as albite and an epidote group mineral. Small masses of prehnite also have been found along the cleavage in biotite from granitic rocks.

Apophyllite

$KCa_4Si_8O_{20}(F,OH)\cdot 8H_2O$
Tetragonal
Uniaxial (+) or (−)
$n_\omega = 1.531-1.542$
$n_\epsilon = 1.533-1.543$
$\delta = 0.000-0.003$

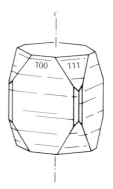

RELIEF IN THIN SECTION: Low negative or positive relief.

COMPOSITION AND STRUCTURE: The structure consists of fourfold rings of silicon tetrahedra linked together to form kinked sheets parallel to (001). The OH, F, H_2O, K, and Ca are located between the sheets.

While most apophyllite is relatively fluorine rich, there is complete solid solution between fluorapophyllite and hydroxyapophyllite, which are rich in F and OH, respectively. Minor amounts of Na may replace K and small amounts of Al may replace Si. Natroapophyllite is an orthorhombic Na analog of apophyllite.

PHYSICAL PROPERTIES: H = $4-4\frac{1}{2}$; G = 2.33–2.37; white, pale pink, yellow, or green in hand sample; white streak; pearly to vitreous luster, sometimes earthy.

COLOR: Colorless in thin section and grain mount.

FORM: Crystals are equant, or platy parallel to (001). Apophyllite also may be anhedral granular.

CLEAVAGE: There is a single perfect cleavage parallel to {001} that controls fragment orientation. Two poor prismatic cleavages parallel to {110} intersect at 90° but are not usually apparent.

TWINNING: Twins on {111} are uncommon.

OPTICAL ORIENTATION: Apophyllite is uniaxial, either positive or negative, so the trace of the basal {001} cleavage always shows parallel extinction and is usually length fast (optically positive), although in some samples it may be length slow (optically negative). Because elongate crystals have their long dimension parallel to the c axis, they are length slow (optically positive) or length fast (optically negative).

INDICES OF REFRACTION AND BIREFRINGENCE: Hyroxyapophyllite has indices of $n_\omega = 1.542$ and $n_\epsilon = 1.543$, and it appears that indices decrease with increasing substitution of F for OH. Birefringence is very low (0.0 to 0.003), so maximum interference colors in thin section are no higher than first-order gray. The interference colors are often anomalous.

INTERFERENCE FIGURE: Apophyllite is commonly uniaxial positive, less commonly uniaxial negative. Biaxial varieties with $2V$ up to 60° and crossed axial plane dispersion have been reported. The optic sign and, if biaxial, the position of the optic plane, may vary from place to place on a single grain. Basal sections and cleavage fragments yield centered optic axis figures.

ALTERATION: Apophyllite may alter to clay minerals, calcite, opal, or quartz.

DISTINGUISHING FEATURES: Apophyllite has higher indices than most zeolites with which it may be associated, and it has a perfect basal cleavage, which the low birefringent zeolites lack. It also commonly has anomalous interference colors. Natroapophyllite is biaxial (+) with $2V_z = 32°$, $n_\alpha = 1.536$, $n_\beta = 1.538$, and $n_\gamma = 1.544$, $\delta = 0.008$.

OCCURRENCE: Apophyllite is usually found in amygdules, cavities, and veins in basaltic volcanics in association with prehnite, zeolites, pectolite, datolite, and calcite. It is also found in cavities in granite, in the contact aureole around granitic intrusions, and in metamorphosed limestone or other calc-silicate rock. It may form by the alteration of wollastonite. Natroapophyllite is found in skarns.

REFERENCES

Albee, A. L., 1962, Relationships between the mineral association, chemical composition and physical properties of the chlorite series: American Mineralogist, v. 47, p. 851–870.

Bailey, S. W., 1988, Chlorites: structures and chrystal chemistry: in Bailey, S. W. (ed.), Hydrous phyllosilicates: Mineralogical Society of America, Washington, D.C., p. 347–403.

Blake, R. L., 1964, Some iron phyllosilicates of the Cayuna and Mesabi districts of Minnesota: U. S. Bureau of Mines Report, Investigation 3694.

Carroll, D., 1970, Clay minerals: A guide to their x-ray identification: Geological Society of America Special Paper 126.

Deer, W. A., Howie, R. A., and Zussman, J., 1962, Rock forming minerals, volume 3, sheet silicates: Longman, London, 270 p.

Eggelton, R. A., and Chappell, B. W., 1978, The crystal structure of stilpnomelane. Part III. Chemistry and physical properties: Mineralogical Magazine, v. 42, p. 361–368.

Foster, M. D., 1962, Interpretation of the composition and a classification of the chlorites: U.S. Geological Survey Professional Paper 414–A, 33 p.

Hey, M. H., 1954, A new review of the chlorites: Minerological Magazine, v. 30, p. 277–292.

Hutton, C. O., 1956, Further data on the stilpnomelane mineral group: American Mineralogist, v. 41, p. 608–615.

Laduron, D. M., 1971, A staining method for distinguishing paragonite from muscovite in thin section: American Mineralogist, v. 56, p. 1117–1119.

Müller, G., 1966, The relationships among the chemical composition, refractive indices, and density of coexisting biotite, muscovite and chlorite from granitic rocks: Contributions to Mineralogy and Petrology, v. 12, p. 173–191.

Phillips, W. R., 1964, A numerical system of classification for chlorites and septechlorites: Mineralogical Magazine, v. 33, p. 1114–1124.

15

Tectosilicates

SILICA GROUP

Quartz

SiO$_2$
Hexagonal (trigonal)
Uniaxial (+)
n_ω = 1.544
n_ϵ = 1.553
δ = 0.009

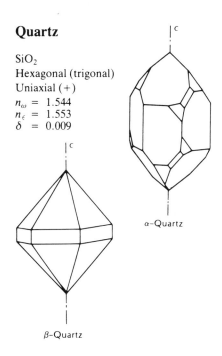

α-Quartz

β-Quartz

RELIEF IN THIN SECTION: Low positive relief.

COMPOSITION AND STRUCTURE: Quartz is a tectosilicate in which every Si is tetrahedrally bonded to four oxygens and each SiO$_4$ tetrahedra shares its four oxygens with other SiO$_4$ tetrahedra. The structure consists of interlocking spirals of SiO$_4$ tetrahedra that run parallel to the c axis. Quartz is essentially pure SiO$_2$, although trace amounts of Ti, Fe, Mn, Al, or other cations may be present.

PHYSICAL PROPERTIES: H = 7; G = 2.65; usually colorless or white in hand sample, but almost any color is possible; white streak; vitreous luster.

COLOR: Colorless in thin section and grain mount. Numerous fluid inclusions may cause it to look cloudy, particularly when seen in grain mount.

FORM: Figure 15.1. Typically anhedral granular to highly irregular in igneous and metamorphic rocks. Detrital grains are usually more or less equant. The variety chalcedony typically has a fibrous or feathery appearance, produced by numerous thin parallel grains. Chert and other microcrystalline varieties may be composed of innumerable fine more or less equant grains. Volcanic rocks may contain euhedral to subhedral phenocrysts with more or less hexagonal cross sections and which may be embayed due to resorption in the magma. Phenocrysts may crystallize as β-quartz, which is the high-temperature polymorph. On cooling below 573°C, the β-quartz inverts to α-quartz, but preserves the habit of β-quartz, which is a short hexagonal prism terminated by a dipyramid.

CLEAVAGE: Some studies have documented the presence of poor cleavage on {10$\bar{1}$1} and {10$\bar{1}$0}, but it is

Figure 15.1 Undulatory quartz with small laths of biotite. Field of view is 1.7 mm wide.

not normally seen in hand sample, thin section, or grain mount.

TWINNING: The common twins are the Dauphine twins with the *c* axis as the twin axis, Brazil twins with {11$\bar{2}$0} as the twin plane, and Japan twins with {11$\bar{2}$2} as the contact plane. While quite common, Dauphine and Brazil twins cannot be observed in thin section because the twin segments have the same *c* axis orientation. Japan twins can be observed but are rather uncommon.

OPTICAL ORIENTATION: Because quartz is uniaxial, the optic axis is the *c* axis. Elongate euhedral crystals cut from end to end are length slow. Quartz that has been deformed shows undulatory extinction.

INDICES OF REFRACTION AND BIREFRINGENCE: Quartz is one of the few minerals that shows essentially no variation in indices of refraction. Birefringence is 0.009, and interference colors range up to first-order white with a tinge of yellow in standard thin sections. Quartz interference colors can be used to provide a fairly accurate estimate of thin section thickness.

INTERFERENCE FIGURE: Quartz is uniaxial positive, but strained quartz showing undulatory extinction may show a biaxial interference figure with small separation of the isogyres. Some smoky quartz and amethyst (purple in hand sample) also may be biaxial with a small 2*V*. In unusually thick sections of quartz (> 1 mm), the isogyres may be absent near the melatope due to rotary polarization of light following the optic axis. This is not seen in conventional thin sections or grain mounts.

ALTERATION: Quartz is not readily altered and is very stable in weathering environments.

DISTINGUISHING FEATURES: Quartz is recognized by its low relief, low birefringence, and lack of cleavage or twinning. Plagioclase is biaxial, has cleavage, and is usually twinned. Orthoclase and sanidine are biaxial, have cleavage, and may show a single twin plane, or be somewhat clouded due to inclusions or incipient alterations. If not authigenic, microcline shows its distinctive tartan plaid twinning. Cordierite closely resembles quartz, but is biaxial and may show distinctive pinite alteration. Beryl is optically negative and has higher indices. Scapolite is uniaxial negative. The feldspars, cordierite, and scapolite also may be distinguished from quartz by a number

of staining techniques that can be used on thin sections.

OCCURRENCE: Quartz is one of the most abundant minerals. It is found in a wide variety of felsic to intermediate intrusive and extrusive igneous rocks such as granitic pegmatite, granite, granodiorite, quartz diorite, rhyolite, and dacite and may be found in small amounts in diorite, gabbro, syenite, and volcanic equivalents. In metamorphic rocks, it is abundant in slate, phyllite, schist, gneiss, and quartzite of various types. In sediments, quartz is a major constituent of most clastic rocks and may serve as a cementing agent. Hydrothermal vein deposits also usually contain substantial amounts of quartz, sometimes in the form of beautiful well-formed crystals, or as microcrystalline or massive varieties.

Quartz often forms intergrowths with feldspars. The fine vermicular intergrowth of quartz and plagioclase usually found at contacts between K-feldspar and plagioclase is called myrmekite, and similar intergrowths of quartz and feldspar in interstices in volcanic and intrusive rocks is called granophyre. Quartz and K-feldspar also may form a graphic intergrowth in pegmatites and granitic rocks.

Chalcedony

SiO$_2$
Hexagonal (trigonal)
Uniaxial (+)
n_ω = 1.53–1.544
n_ϵ = 1.53–1.553
δ = 0.005–0.009

RELIEF IN THIN SECTION: Low positive or negative relief.

COMPOSITION AND STRUCTURE: Chalcedony is fibrous quartz. Most analyses show water, most of which is present in the pore space between the fibers, but some is in the form of hydroxyl, which replaces oxygen. Material identified as chalcedony may contain cristobalite. The specific gravity is lower than quartz in proportion to the amount of pore space between the fibers.

PHYSICAL PROPERTIES: H = 6$\frac{1}{2}$–7; G = 2.57–2.64; hand-sample color is highly variable; white streak; dull vitreous to greasy luster.

COLOR: Chalcedony is usually colorless or pale brown, but strongly colored varieties may have pale colors in thin section or grain mount corresponding to the hand-sample color. Chalcedony may be cloudy to nearly opaque due to numerous inclusions and the innumerable grain boundaries that scatter the light. Color banding may be visible in larger masses.

FORM: Figure 15.2. The quartz fibers that make up chalcedony are typically aligned in a parallel or spherulitic fashion and have a feathery appearance. Chalcedony may form the cement in clastic sediments or may form colloform or encrusting masses. The fibers are usually elongate perpendicular to the c axis along one of the a crystal axes, and are twisted. Fibers elongate parallel to the c axis also are occasionally observed.

CLEAVAGE: There is no cleavage, but chalcedony usually fractures parallel to the fibers forming elongate grains.

TWINNING: Most quartz twins are not visible with the petrographic microscope.

OPTICAL ORIENTATION: Fibers elongate along an a axis are length fast. The fibers are typically twisted. In sections cut parallel to the length of the fibers, the optic axis is brought to a vertical position once in every 180° of twist, so there are repeated patches of extinction along the length of the fibers (Figure 15.2). In random sections, the twisting is expressed as variations in the birefringence along the length of the fibers and produces an irregular wavy extinction for the whole mass of fibers.

INDICES OF REFRACTION AND BIREFRINGENCE: The indices of refraction and birefringence are somewhat lower than for pure quartz, due to the presence of water in pores between the fibers (Figure 15.3).

INTERFERENCE FIGURE: The fine grain size usually precludes obtaining usable interference figures. Biaxial figures may sometimes be seen, due to the complications associated with the fibrous nature.

ALTERATION: Chalcedony may recrystallize to form granular quartz but is not usually altered.

DISTINGUISHING FEATURES: Low relief, feathery appearance, and low birefringence are characteristic. Chrysotile also is fibrous, but it is often pale green and has distinctly different hand-sample properties. X-ray techniques may be needed to distinguish chalcedony from fibrous cristobalite.

OCCURRENCE: Chalcedony is usually the result of low-temperature crystallization of quartz, whether in sedimentary rocks as a cement, in voids or cavities in volcanic rocks in the form of geodes or agate, or in hydrothermal deposits. It is probably formed by the crystallization of silica gels.

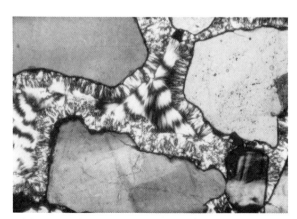

Figure 15.2 Feathery-appearing chalcedony cementing quartz and microcline grains in a sandstone (crossed polars). Field of view is 2.3 mm wide.

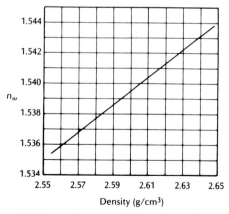

Figure 15.3 Approximate variation of n_ω of chalcedony with density, which is a function of pore space between the fibers. From Frondel (1982). Used by permission of the Mineralogical Society of America.

Tridymite

SiO$_2$
Orthorhombic (pseudohexagonal)
Biaxial (+)
n_α = 1.468–1.482
n_β = 1.470–1.484
n_γ = 1.474–1.486
δ = 0.002–0.004
$2V_z$ = 40–90°

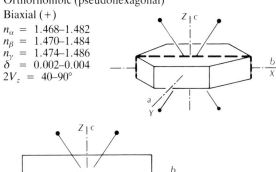

RELIEF IN THIN SECTION: Moderate negative relief.

COMPOSITION AND STRUCTURE: The structure of tridymite consists of sheets of SiO$_4$ tetrahedra with alternate tetrahedra pointing up and down. The apical oxygen of each tetrahedron is shared with a tetrahedron from an adjacent sheet, forming a framework structure with each oxygen shared by two tetrahedra. All tridymite crystallizes at high temperature as hexagonal β-tridymite but distorts slightly with cooling to become orthorhombic α-tridymite. The structure is fairly open and can interstitially accommodate varying amounts of Na, K, and Ca, with electrostatic charge balance maintained by replacing Si^{4+} with Al^{3+}.

PHYSICAL PROPERTIES: H = 7; G = 2.27; colorless or white in hand sample; white streak; vitreous luster.

COLOR: Colorless in thin section and grain mount.

FORM: Crystals are typically hexagonal plates that may be rounded or strongly embayed due to resorption. Basal sections show a hexagonal outline and other sections usually appear to be lath shaped. Tridymite may also form radiating or spherical aggregates or be anhedral granular.

CLEAVAGE: Tridymite may have poor prismatic cleavage, but it is not generally visible in thin section, nor does it control the shape of fragments.

TWINNING: The process of inversion from β-tridymite to α-tridymite may cause the crystals to become twinned on {10$\bar{1}$6}. The twins are often composed of three wedge-shaped segments. Complex lamellar twinning also may be seen in larger crystals.

OPTICAL ORIENTATION: X = b, Y = a, Z = c, optic plane = (100). Elongate lath-shaped sections show parallel extinction and are length fast.

INDICES OF REFRACTION AND BIREFRINGENCE: The indices of pure synthetic tridymite are n_α = 1.469, n_β = 1.469, and n_γ = 1.473. Substituting Na, Ca, K, and Al causes a small increase in the indices. Birefringence is low (0.002–0.004), so interference colors in thin section are no higher than first-order gray.

INTERFERENCE FIGURE: Tridymite is biaxial positive. $2V_z$ is usually around 70°, but may range from about 40 to 90°. $2V_z$ of synthetic tridymite is 36°. $2V$ may be different in different segments of twinned crystals.

ALTERATION: α-tridymite is almost always a pseudomorph (paramorph) after β-tridymite. It is also commonly pseudomorphically replaced by single crystals or fine-grained aggregates of quartz. The c axis of the replacing quartz generally does not coincide with the position of the c axis in the original tridymite.

DISTINGUISHING FEATURES: The crystal habit with wedge-shaped twins, low birefringence, and low indices of refraction with moderate relief are characteristic. Quartz has higher birefringence and is uniaxial, and cristobalite is uniaxial or essentially isotropic.

OCCURRENCE: Tridymite is found in silicic to intermediate extrusive and shallow intrusive igneous rocks. It may form phenocrysts, be part of the groundmass, or form small crystals lining vesicles or other voids. Tridymite is occasionally found in vesicles in basaltic volcanics. The associated vesicle-filling minerals are quartz, cristobalite, fayalite, sanidine, topaz, hematite, or garnet.

Tridymite is sometimes found in silicic xenoliths in basaltic volcanics and in the contact zones adjacent to basaltic intrusives. It is also found in stony meteorites.

Cristobalite

SiO$_2$
Tetragonal (pseudoisometric)
Uniaxial (−)
n_ω = 1.486–1.488
n_ϵ = 1.482–1.484
δ = 0.002–0.004

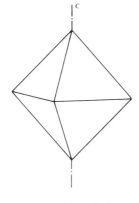

RELIEF IN THIN SECTION: Moderate negative relief.

COMPOSITION AND STRUCTURE: The structure of cristobalite is similar to tridymite and consists of sheets of SiO$_4$ tetrahedra with tetrahedra pointing alternately up and down. The apical oxygen on each tetrahedra is shared with a tetrahedron from an adjacent sheet. The difference between tridymite and cristobalite is in the position of adjacent sheets with respect to each other, and is somewhat analogous to the difference between cubic and hexagonal closest packing. As in tridymite, there is interstitial solid solution of Na, K, or Ca, and Al^{3+} may replace Si^{4+} to maintain the electrostatic charge balance. Cristobalite crystallizes as isometric β-cristobalite, but inverts to tetragonal α-cristobalite upon cooling.

PHYSICAL PROPERTIES: H = 6$\frac{1}{2}$; G = 2.32–2.34; colorless, white, or pale yellowish in hand sample; white streak; vitreous luster.

COLOR: Colorless in thin section and grain mount.

FORM: Crystals are octahedrons or cubes, which are actually pseudomorphs of α-cristobalite after β-cristobalite. In volcanic rocks, fibrous cristobalite may form spheroidal pellets or may be intergrown with tridymite or sanidine needles to form small spherulites, both with radial patterns. A fibrous variety called *lussatite* resembles chalcedony and may be intergrown with it.

CLEAVAGE: None. Fractures are commonly curved.

TWINNING: Lamellar twinning with composition planes parallel to octahedron faces {111} is common. There either may be a single set of lamellae or two sets intersecting at an angle.

OPTICAL ORIENTATION: One of the *a* axes in isometric β-cristobalite becomes the *c* axis after inversion to α-cristobalite and is therefore the optic axis. Fibers are usually length slow, less commonly length fast, and usually show parallel extinction.

INDICES OF REFRACTION AND BIREFRINGENCE: Pure synthetic cristobalite has indices n_ω = 1.487 and n_ϵ = 1.484, and natural material has indices quite close to these values. The birefringence is low (0.002–0.004) and may be essentially zero. Interference colors in thin section are no higher than first-order gray.

INTERFERENCE FIGURE: Due to fine grain size and very low birefringence, interference figures are difficult to obtain. Figures from particularly thick basal sections have broad diffuse isogyres on a gray field. Most cristobalite is uniaxial negative, but biaxial varieties with small 2*V* have been reported.

ALTERATION: Cristobalite may be pseudomorphically replaced by quartz and may itself pseudomorphically replace tridymite.

DISTINGUISHING FEATURES: Crystal habit, low birefringence, and moderate negative relief are characteristic, as is its presence in spherulites. Tridymite is biaxial and has somewhat higher negative relief. X-ray techniques are often required to identify fine-grained material.

OCCURRENCE: Cristobalite is usually found in volcanic or shallow intrusive rocks. It may line cavities in association with tridymite, topaz, garnet, fayalite, or other minerals, and it may form spherulites or be a constituent of the fine-grained groundmass. Cristobalite also has been found in meteorites and in sandstone that has been partially fused by contact with high-temperature magma or by underground coal fires. The fibrous variety lussatite is found in serpentinites and in other environments associated with chalcedony.

Opal

$SiO_2 \cdot nH_2O$
Amorphous
Isotropic
$n = 1.43$–1.46

RELIEF IN THIN SECTION: Moderate negative relief.

COMPOSITION AND STRUCTURE: Opal consists of a mixture of amorphous and crystalline silica. The amorphous silica usually consists of extremely small (0.1–0.5 nm) spherical masses that are closely packed, and the crystalline material usually consists of extremely small crystals of tridymite or cristobalite. The amount of crystalline material may be quite variable. If the silica spheres have a fairly uniform size, they can pack together in a regular manner to produce a diffraction grating that yields the beautiful play of colors seen in precious opal. Variable amounts of water may be present, either in the void space between the spheres or incorporated in some manner as part of the amorphous silica. Heating or long exposure to the air may cause dehydration and cracking.

PHYSICAL PROPERTIES: $H = 5\frac{1}{2}$–$6\frac{1}{2}$; $G = 2.00$–2.25; wide range of color in hand sample; white streak; vitreous to waxy luster. An internal play of colors may be seen in precious opal.

COLOR: Usually colorless in thin section and grain mount; also gray, brown, or other colors. Fragments in grain mount may appear turbid due to inclusions.

FORM: Opal forms colloform or encrusting masses, vein fillings, or irregular masses. It may be the cementing agent in sandstone and may replace organic material as in petrified wood.

CLEAVAGE: None. Fractures are conchoidal.

INDICES OF REFRACTION AND BIREFRINGENCE: Indices of refraction decrease with increasing amounts of water (Figure 15.4), and are usually in the range 1.44 to 1.45. Opal is usually isotropic, but some varieties may show weak birefringence due to strain. The index may change because of absorption of water or index of refraction liquids.

ALTERATION: With dehydration and crystallization, opal converts to quartz.

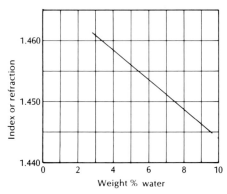

Figure 15.4 Variation of maximum index of refraction of opal with water content (Kokta, 1930).

DISTINGUISHING FEATURES: The isotropic character, moderate negative relief, and occurrence are characteristic. Volcanic glass and analcime have higher indices. Fluorite has distinct cleavage.

OCCURRENCE: Opal is common as a precipitate in voids in volcanic rocks and shallow intrusives and may replace part of the rock. It is precipitated both by groundwater and by hydrothermal solutions. Opal is found in sedimentary rocks as a cementing agent; as a replacement of shells, wood fiber, or other organic material; or as continuous beds formed by precipitation of silica gel or deposition of opaline microfossil shells and sponge spicules. It also is deposited around geysers and hot springs as geyserite or siliceous sinter.

Volcanic Glass

Noncrystalline
Isotropic
$n \cong 1.48$–1.62

RELIEF IN THIN SECTION: Low to moderate positive or negative relief.

COMPOSITION AND STRUCTURE: Volcanic glass is described in the chapter on tectosilicates for lack of a better place to put it. It is not crystalline but rather is an amorphous glass, though it frequently contains small crystals of feldspars or other silicate minerals.

PHYSICAL PROPERTIES: H = 5–6; G = 2.3–2.9; colorless, white, red, black, brown, etc. in hand sample; vitreous luster.

COLOR: Colorless, tan, reddish, or brown in thin section and grain mount. Color is darker with increasing Fe–Mg content and decreasing Si content.

FORM: Volcanic glass may be massive glassy, pumicious, vesicular, and fragmental. It commonly contains small crystallites or phenocrysts and may show flow banding.

FRACTURE: Conchoidal fractures are common in fresh glass.

INDICES OF REFRACTION AND BIREFRINGENCE: The indices of refraction increase in a systematic way with decreasing silica content from about 1.48 for rhyolitic glass to about 1.6 for basaltic glass. It is isotropic unless it has been strained.

ALTERATION: Volcanic glass devitrifies fairly rapidly, so glass older than the Cenzoic era is extremely rare. The devitrification products are clay, zeolites, feldspars, quartz, tridymite, cristobalite, and other minerals. The grain size may be too fine to distinguish individual minerals.

Palagonite is a yellowish brown, brown, or greenish material produced by hydrating basaltic glass. *Perlite* is the light-colored equivalent produced by hydrating rhyolitic glass. Both are isotropic and have indices higher than the glass from which they were derived.

DISTINGUISHING FEATURES: The isotropism and occurrence in volcanic rocks is distinctive. Opal has lower indexes of refraction.

FELDSPARS

The feldspars are by far the most abundant group of minerals. They are found in nearly all igneous rocks, most metamorphic rocks, and are an important constituent in many sedimentary rocks.

The structure consists of a corner-sharing framework of SiO_4 and AlO_4 tetrahedra with K^+, Na^+, or Ca^{2+} cations in the interstices of the framework. There is one interstice for each four tetrahedra, so the general formula for all the feldspars is MT_4O_8, where M represents the large cations and T represents the tetrahedral Al and Si cations. If M is monovalent K^+ or Na^+, then one out of every four tetrahedra must be occupied by Al^{3+} so that the valence charges balance. If M is divalent Ca^{2+}, then half of the tetrahedra must be occupied by Al^{3+}.

The composition of the common feldspars can be expressed in terms of three end members: K-feldspar (Or, $KAlSi_3O_8$), albite (Ab, $NaAlSi_3O_8$), and anorthite (An, $CaAl_2Si_2O_8$). Natural feldspars fall into either the plagioclase series with compositions between albite and anorthite or the alkali feldspar series with compositions between albite and K-feldspar (Figure 15.5). Ba also may be present and *celsian* ($BaAl_2Si_2O_8$) is a barium feldspar.

At high temperatures, the Al and Si are randomly distributed among the tetrahedral sites in the alkali feldspars. However, as temperature decreases, they tend to become ordered with Al selectively locating

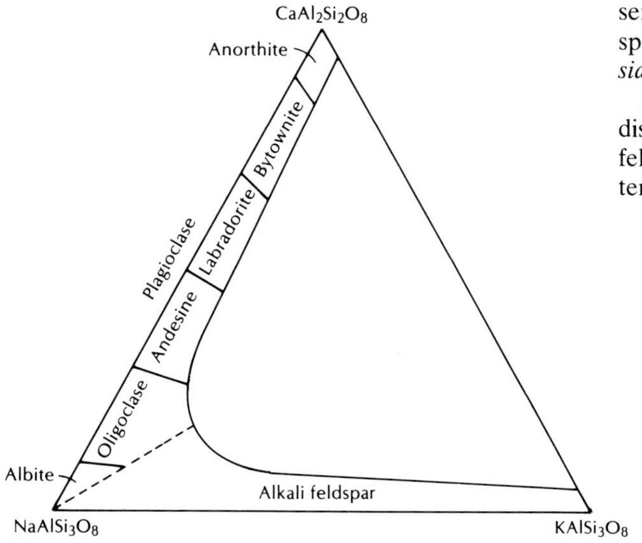

Figure 15.5 Composition range of most natural feldspars.

itself in just one of the four tetrahedral sites. The disordered and ordered varieties are given the prefix *high-* and *low-*, respectively. In the anorthite end member, both the high- and low-temperature varieties are strongly ordered with Si and Al alternating, due to the requirement that Al not occupy adjacent tetrahedra. Consequently, the optical properties of low and high plagioclase converge for anorthite-rich plagioclase, as will be seen later.

The crystal symmetry is controlled both by the size of the large cation and by the degree of ordering of Al and Si. The highest symmetry of the tetrahedral framework is monoclinic. In high K-feldspar (sanidine), this symmetry is maintained because K is large enough to keep the structure open and Al and Si are randomly distributed among the tetrahedral sites. In low K-feldspar (microcline), ordering of Si and Al in the tetrahedral sites causes a small distortion of the framework, which reduces the symmetry to triclinic. In plagioclase, smaller Na and Ca cations allow the structure to collapse or distort with a reduction in symmetry to triclinic for both high and low varieties regardless of Al, Si distribution.

Most feldspar probably crystallizes as the disordered high-temperature variety and becomes ordered in the process of cooling, provided there is sufficient time. Plutonic rocks typically contain low plagioclase and/or K-feldspar, while volcanic rocks usually contain high plagioclase and/or alkali-feldspar. Hypabyssal intrusives may have feldspars with intermediate degrees of order.

Plagioclase

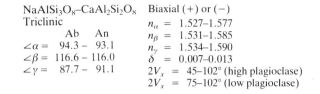

$NaAlSi_3O_8$–$CaAl_2Si_2O_8$
Triclinic

	Ab	An
$\angle\alpha =$	94.3 –	93.1
$\angle\beta =$	116.6 –	116.0
$\angle\gamma =$	87.7 –	91.1

Biaxial $(+)$ or $(-)$
$n_\alpha = 1.527–1.577$
$n_\beta = 1.531–1.585$
$n_\gamma = 1.534–1.590$
$\delta = 0.007–0.013$
$2V_x = 45–102°$ (high plagioclase)
$2V_x = 75–102°$ (low plagioclase)

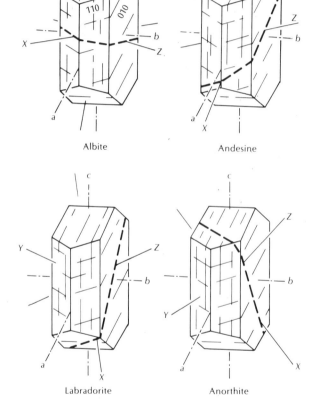

Albite   Andesine

Labradorite   Anorthite

RELIEF IN THIN SECTION: Low positive or negative relief.

COMPOSITION: At high temperatures, plagioclase shows continuous solid solution from albite (Ab) $NaAlSi_3O_8$ to anorthite (An) $CaAl_2Si_2O_8$. At low temperatures, there may be unmixing in three composition ranges. Peristerite intergrowths are found for bulk compositions between about An_2 and An_{16}, and usually consist of alternating lamellae of An_0 and An_{25} compositions. Bøggild intergrowths are found in the bulk composition range of about An_{47} to An_{58}, and apparently consists of lamellae of $\sim An_{45}$ and $\sim An_{60}$. Huttenlocher intergrowths occur for bulk compositions of about An_{67} to An_{90}, and apparently consist of lamellae of $\sim An_{67}$ and $\sim An_{95}$. These fine intergrowths are responsible for the iridescent play of colors sometimes found in hand samples of plagioclase. In thin section, the intergrowths may rarely be visible as thin diffuse lamellae intersecting the albite twinning at an acute angle, but in most cases they are too fine to distinguish, if they are present at all.

K may enter the plagioclase lattice in significant amounts only for albite-rich compositions. At high temperatures, there may be continuous solid solution between albite and K-feldspar, but at lower temperatures a solvus is present. On slow cooling, K-feldspar may exsolve from albite or, occasionally, oligoclase. Antiperthite is the term given to sodic plagioclase with exsolution blebs or lamellae of K-feldspar. Only in volcanic and shallow intrusive rocks, where the exsolution process has not been allowed to proceed, are intermediate compositions preserved, and, even then, there is often exsolution on a submicroscopic scale.

PHYSICAL PROPERTIES: $H = 6-6\frac{1}{2}$; $G = 2.63$ (Ab)–2.76 (An); white or gray in hand sample, also light green, black, pink, yellow, etc.; white streak; vitreous luster.

COLOR: Colorless in thin section and grain mount. Some plagioclase may be clouded due to partial alteration to sericite or clay and some may appear pale reddish, brownish, or gray, due to the presence of fine Fe–Ti oxide inclusions.

FORM: Figure 15.6. Plagioclase commonly occurs both as euhedral and anhedral grains. Crystals are usually tabular parallel to (010) and elongate parallel to the c or a axis. Cross sections are more or less rectangular. Cleavelandite is a variety of albite that forms very platy crystals parallel to (010).

Chemical zoning is common, particularly in volcanic and hypabassal intrusive rocks, and is expressed as a variation in the extinction angle from one zone to another. The zoning is considered normal if it changes smoothly from a more calcic core to

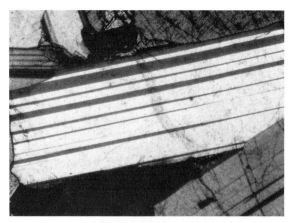

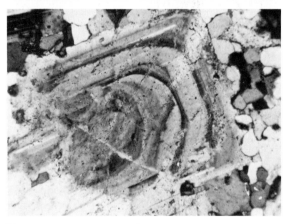

Figure 15.6 Plagioclase. (*Top*) With albite twinning. (*Middle*) A Carlsbad twin is ~N–S in the middle of the grain, parallel to the albite lamellae. Pericline lamellae are ~E–W. The dark high relief mineral to either side is orthopyroxene. (*Bottom*) Phenocryst with oscillatory zoning and sericite alteration. Width of fields of view: *top* and *middle*, 1.7 mm; bottom, 1.2 mm.

a more sodic rim. The core usually has a larger extinction angle. Reverse zoning is just the opposite. Oscillatory zoning shows alternation between calcic and sodic zones, but the overall trend is usually from a more calcic core to a more sodic rim.

"Chessboard" albite contains alternating twin lamellae somewhat like the pattern of squares on a chessboard, except the units tend to be elongate and lath shaped rather than square. It is interpreted to be produced by replacing K-feldspar or more calcic plagioclase with albite.

Intergrowth textures involving plagioclase, K-feldspar, and quartz are described in the section on alkali feldspar.

CLEAVAGE: The {001} cleavage is perfect and the {010} cleavage is good. The cleavages intersect at 93 to 94°. There is also poor cleavage on {110}, which is not seen in thin section. Cleavages tend to control fragment orientation. The cleavage in plagioclase may not be obvious unless the aperture diaphragm is stopped down to emphasize what little relief there is. In carefully made thin sections, cleavage cracks may not develop in large enough numbers to be noticeable. Cleavage is usually most noticeable in grains along the slide's edge.

TWINNING: Polysynthetic twinning is characteristic of plagioclase, although it may be absent in small grains, particularly in metamorphic rocks. The common twin laws (Figure 15.6) are as follows.

albite: (010) composition plane, polysynthetic.
pericline: twofold rotation about b axis, polysynthetic, (h01) composition plane.
Carlsbad: twofold rotation about c axis, penetration twin with (010) composition plane.

To distinguish these types of twinning, see the following discussion. Numerous other twin laws also have been found. Albite twinning is found in all compositions, but Carlsbad twins are favored in intermediate and calcic plagioclase, as is pericline twinning.

OPTICAL ORIENTATION: The optic orientation varies in a very regular manner with composition and provides the basis for several techniques of determining composition. Except by chance, none of the indicatrix axes coincides with any of the crystal axes, so extinction is inclined in almost all orientations.

INDICES OF REFRACTION AND BIREFRINGENCE: Indices of refraction show a very systematic increase with increasing anorthite content (Figure 15.7). Accurate measurement of indices—for example, with a spindle stage—can provide an estimate of the composition to within 1 or 2 percent anorthite. Birefringence varies between about 0.007 and 0.013. The change is too small to be of use in determining composition. Maximum interference colors in thin section are usually first-order gray or white. Only for very calcic plagioclase will first-order yellow colors be seen.

INTERFERENCE FIGURE: $2V$ varies systematically with composition (Figure 15.8), but it shows a distinctly different pattern for high and low varieties. $2V$ of plagioclase with an intermediate degree of ordering may fall anywhere between the curves for low (plutonic) and high (volcanic) plagioclase. The $2V$ angle by itself cannot be used to provide a reliable estimate of composition, but it can be used to resolve ambiguities involved in other techniques. Fine lamellar twinning or exsolution may make usable figures difficult to obtain in some cases. Optic axis dispersion is weak, usually with $v > r$, but also $r > v$.

ALTERATION: Plagioclase is commonly partially altered to sericite (fine-grained white mica), clay, or zeolites. The alteration may be uniformly distributed, or may be concentrated along individual twin lamellae or in the core of grains. Plagioclase also may alter to saussurite, which is a fine-grained aggregate of epidote group minerals, albite, sericite, and other minerals. Scapolite, prehnite, pyrophyllite, and calcite are other alteration products.

DISTINGUISHING FEATURES: The low relief, lack of color, biaxial character, and polysynthetic twinning distinguish plagioclase from most other minerals. Cordierite shows similar twinning, but it alters to pinite and the twins show a radial pattern in certain orientations. Pleochroic halos form around radioactive mineral inclusions (e.g., zircon) in cordierite, which also has less well-developed cleavage. Untwinned plagioclase greatly resembles quartz and may be easily overlooked. However, quartz is uniaxial, has no cleavage, and usually does not show the incipient alteration or clouding common in plagioclase. Monoclinic K-feldspars lack polysynthetic twinning. There are a number of staining techniques

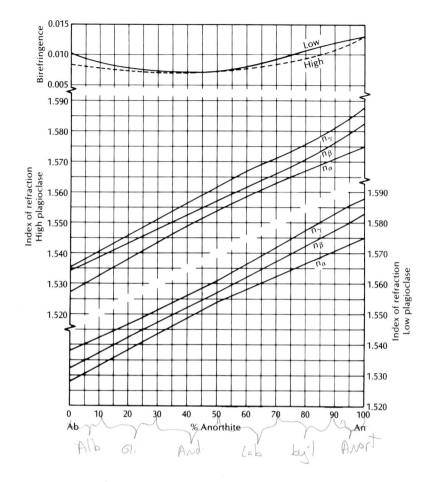

Figure 15.7 Indices of refraction and birefringence (δ) for low (plutonic) and high (volcanic) plagioclase. After Smith (1958). Used by permission of the Mineralogical Society of America.

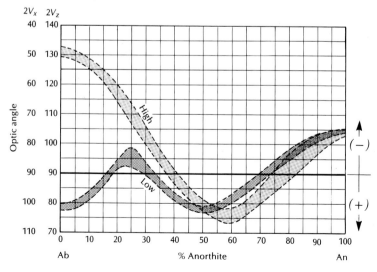

Figure 15.8 Variation of $2V$ for low (plutonic) and high (volcanic) plagioclase. Plagioclase with an intermediate degree of ordering may fall between the two trends. Based on Smith (1958), and Burri and others (1967).

that can be used to aid in the rapid identification of plagioclase, K-feldspar, and other minerals with similar appearance (e.g., Bailey and Stevens, 1960; Ruperto and others, 1964; Boone and Wheeler, 1968; Haughton, 1980).

OCCURRENCE: Plagioclase is a widespread mineral found in nearly all igneous rocks, in many metamorphic rocks, and in some sediments. Granite, granodiorite, rhyolite, pegmatite, syenite, and related rocks usually contain albite and oligoclase. Intermediate rocks such as diorite and andesite typically contain andesine. Gabbroic and basaltic rocks usually contain labradorite or, less commonly, bytownite. Anorthite is rather rare, but may form the core of plagioclase grains in mafic rocks, or be found in unusual mafic or ultramafic rocks.

The plagioclase in low-grade metamorphic rocks is usually albite (An_{0-10}), and in medium- and high-grade rocks is usually oligoclase or andesine. Anorthite contents greater than An_{50} are not common in metamorphic rocks, but fairly pure anorthite may be found in metamorphosed carbonates.

In sedimentary rocks, plagioclase occurs as detrital grains. Untwinned authigenic albite may be formed by replacement of detrital plagioclase or by crystallization due to diagenetic processes.

DETERMINING COMPOSITION: With careful measurement of optical properties it is possible to determine the composition to within 1 or 2 percent anorthite in some cases, but if knowledge of the composition is important, the optical information should be supplemented with chemical and X-ray techniques.

I. *THIN SECTION.* Determining composition in thin section depends on measurement of extinction angles and is less accurate than techniques that depend on measurement of indices of refraction.

A. *Michel–Lévy Method.* The Michel–Lévy method depends on the observation that the extinction angle measured to the (010) composition plane of albite twins varies in a systematic way with composition. Because both grain orientation and indicatrix orientation control the observed extinction angle, a number of grains are measured and the largest extinction angle is used to determine the composition. The procedure is as follows (Figure 15.9):

1. Select a grain in which the (010) composition planes between the albite twin lamellae are vertical. Usable grains have the following characteristics.

a. All the twin lamellae have essentially the same interference color between crossed polars when placed parallel to the N–S and in the 45° position. If the two sets of lamellae have different colors, either (010) is not vertical or the twins are pericline twins. If the grain is divided in half so that the lamellae on one side are all one color and on the other a different color, there is probably a Carlsbad twin separating the two segments and the Carlsbad–albite method should be used.

b. The composition planes between the twin lamellae are crisp and sharp. If they are not, (010) is not vertical and the grain cannot be used. Confirm that (010) is vertical by raising and lowering the focus with the high power objective. The position of the composition plane should not shift from side to side.

c. The extinction angles measured below should differ by no more than 4°.

2. Place the composition plane on the N–S. Rotate clockwise to get one set of lamellae extinct with the fast ray vibration direction on the N–S and note the extinction angle. With the composition plane returned to the N–S, the extinction angle for the other

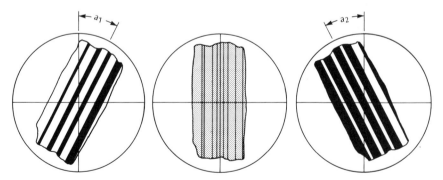

Figure 15.9 Michel–Levy method. Extinction angles a_1 and a_2 are measured to the fast ray vibration direction in their respective sets of albite twin lamellae. The two angles are averaged. A half-dozen or more grains are measured and the largest average extinction angle is used to estimate the plagioclase composition from Figure 15.10.

set of lamellae is measured by rotating counterclockwise. Take the average of the two extinction angles, provided that the difference is no more than 4°. If the difference is larger, discard the readings and look for another grain. In most cases, the extinction angle between the twin lamellae and the fast ray vibration direction is less than 45°, but for calcic plagioclase, it may be more than 45°. To check that the fast ray, and not the slow ray, direction is on the N–S in the extinction position, rotate the stage 45° clockwise from the extinction position and insert the gypsum plate (slow NE–SW). If the color in the lamellae in question decreases (usually to first-order yellow or red), then the fast ray had correctly been placed on the N–S. If it increased (usually to second-order blue), then the slow ray had been on the N–S in the extinction position, and additional rotation will be needed to bring the other set of lamellae to extinction with their fast ray on the N–S.

3. Repeat the measurements for a half-dozen or more grains. (More is better because it improves the accuracy.) Because only the maximum average extinction angle is useful, time can be saved here by quickly moving on if the first angle measured on a grain is less than the maximum previously measured on other grains.

4. The composition is determined with Figure 15.10. The maximum average extinction angle is plotted on the vertical axis and the composition along the horizontal axis. For angles less than about 18°, there are two possible compositions: one less than An_{20} (plutonic) or An_{12} (volcanic) and the other higher. The distinction between the two possibilities for plutonic plagioclase can be made based on indices of refraction and optic sign (Figure 15.7 and 15.8).

An_0–An_{20}	An_{20}–An_{35}
optically $(+)$	optically $(-)$
n_α less than 1.538	n_α greater than 1.538
($\sim n$ cement)	($\sim n$ cement)

To check the indices, find a place where a grain with maximum birefringence comes in contact with cement, either at a hole in the slide or along the edge. Rotate the part of the grain at the contact to extinction with the fast ray (n_α) parallel to the vibration direction of the lower polar (check with the accessory plate). Uncross the polars and check the index with the Becke line.

The accuracy of the method for low plagioclase is probably within ±5 percent anorthite content up to about An_{60}, and worse than that at higher An content. The accuracy is probably signficantly worse with volcanic plagioclase.

B. *Carlsbad–Albite Method.* The Carlsbad–albite method is closely related to the Michel–Lévy method, but only one properly oriented grain needs to be measured to obtain a composition. The procedure is as follows (Figure 15.11).

1. Select a grain with both albite and Carlsbad twinning, oriented so that the (010) composition plane of the albite twins is vertical. Usable grains have the following characteristics.

a. When the twin lamellae are placed in the N–S and 45° positions, the Carlsbad twin separates the grain into two segments with different interference colors, but all the albite lamellae within one segment have essentially the same interference color. The composition plane of the Carlsbad twin may be slightly irregular.

b. The twin lamellae are crisp and sharp, indicating that (010) is vertical (see Michel–Lévy method).

c. The extinction angles measured to the albite twins on any one Carlsbad segment differ by no more than 4°.

2. On the left-hand side of the Carlsbad twin, measure the extinction angle from albite twin lamellae to the fast ray direction, as described in the Michel–Lévy method. Average the two readings.

3. On the right-hand side of the Carlsbad twin, measure the extinction angles from the albite lamellae

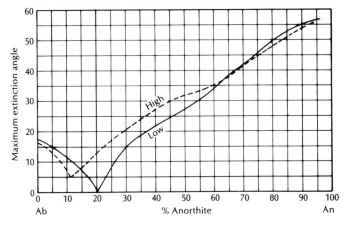

Figure 15.10 Maximum extinction angle to albite twin lamellae in sections cut perpendicular to (010) for low (plutonic) and high (volcanic) plagioclase. See text for additional discussion. Based on the curves of Tobi and Kroll (1975) (Figure 15.12).

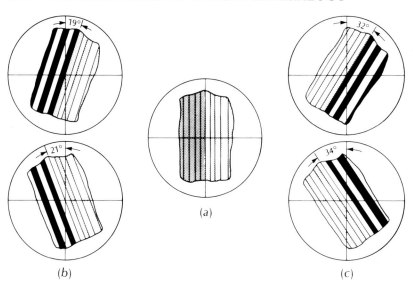

(a)

(b)   (c)

Figure 15.11 Carlsbad–albite method. (*a*) Grain with twin lamellae on the north–south. The two sides of the Carlsbad twin show different interference colors but the albite lamellae are uniform on each side. (*b*) Extinction angles measured to the albite lamellae on the left side of the Carlsbad twin. The average extinction angle in this case is 20°. (*c*) Extinction angles measured to the albite lamellae on the right side of the Carlsbad twin. The average extinction angle in this case is 33°. The composition of the plagioclase from Figure 15.12 is An_{60}.

to the fast ray direction in the same manner and average the two readings.

4. The composition is obtained from Figure 15.12a for low plagioclase and from Figure 15.12b for high plagioclase. The solid lines are for the larger of the two extinction angles found in steps 2 and 3, and the dashed lines are for the smaller angle. The point where the two lines intersect indicates the composition, which is read off the bottom of the charts.

For angles less than about 20°, there are both negative and positive values. The distinction can be made in a section cut perpendicular to the *a* axis. The angle is negative if the fast vibration direction lies in the obtuse angle between the {001} and {010} cleavages, and positive if it lies in the obtuse angle between the cleavages. However, as a partical matter, the distinction between positive and negative angles cannot be made, and lines for both positive and negative angles should be used to find intersections.

Note that the same set of readings may produce two or even three different intersections. For example, with high plagioclase, if the smaller extinction angle is 10° and the larger is 15°, intersections can be found at An_1 (where the dashed -10 and solid -15 lines cross), at An_{29} (near the top of the diagram where the dashed 10 and solid 15 lines cross), and at An_{66} (near the bottom of the diagram where the solid 15 and the second dashed 10 lines cross). For relatively albite-rich compositions, the ambiguity is resolved in the same way described with the Michel–Lévy method. For more calcic compositions, additional grains can be measured to remove the ambiguity.

The accuracy of the Carlsbad–albite method is somewhat better than the Michel–Lévy method, but

depends in part on the orientation of the grain used. The method is best if the two extinction angle measurements differ by no more than about 15° and the intersection plots in an area on the diagrams where the curves cross at a high angle. While one properly oriented grain can yield a composition, caution suggests that several be measured.

Figure 15.12 can be used to determine the composition of zones within zoned grains, provided that at least one zone is cut by a Carlsbad twin. The vertical axis of the chart shows the inclination of the *c* axis to the microscope stage. The top edge of the chart shows the extinction angles that would be found for sections cut parallel to the *c* axis, that is, parallel to (100), and the bottom edge of the chart is for sections cut at right angles to the *c* axis. One good set of readings for a single zone in a plagioclase crystal is sufficient to determine the orientation of the section through the crystal. The composition of zones that do not cross the Carlsbad twin can then be estimated by measuring a single extinction angle in the zone.

II. *GRAIN MOUNT*. There are two techniques that can be used to determine composition in grain mount: one involves extinction angles and the other indices of refraction.

A. *Schuster Method*. The Schuster method involves measurement of extinction angles on cleavage fragments in grain mount. The procedure to use is as follows.

1. Select a fragment lying flat on a cleavage surface.

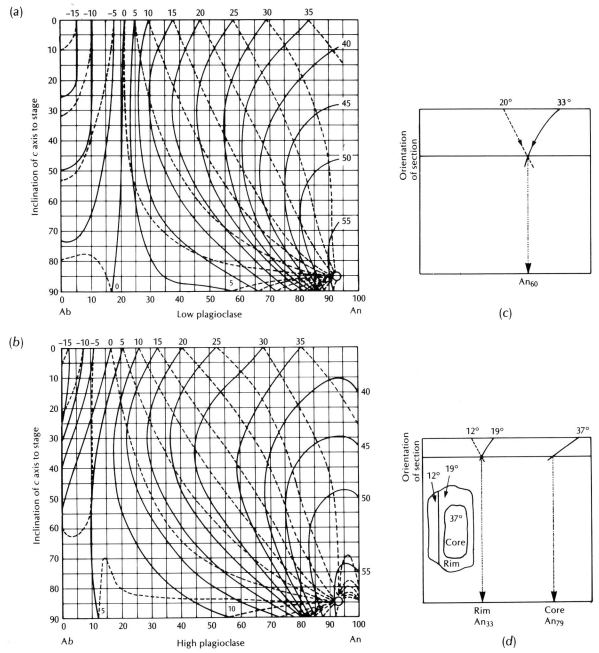

Figure 15.12 Charts for use with the Carlsbad–albite method of determining plagioclase composition. Solid lines are for the larger extinction angle and dashed lines are for the smaller extinction angle. See text for additional discussion. (*a*) Low plagioclase. (*b*) High plagioclase. (*c*) Plagioclase composition is read off the bottom of the chart below the point where the curve for the larger extinction angle (33°) intersects the curve for the smaller angle (20°). The orientation of the section is read off the left side of the diagram. (*d*) The rim is bisected by the Carlsbad twin so that the orientation of the grain is known. The core, which is not bisected by the twin, must have the same orientation; therefore, if extinction angles to albite twins can be measured (37°), the composition of the zone can be determined. From Tobi and Kroll (1975). Used by permission of the *American Journal of Science*.

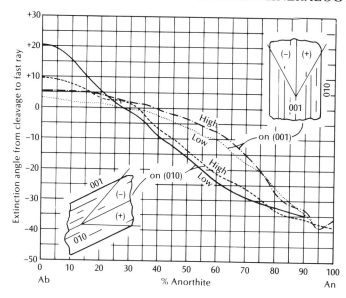

Figure 15.13 Extinction angles from cleavage to fast ray direction foɪ plagioclase fragments lying on the {001} and {010} cleavages. From Burri and others (1967). Used by permission of Birkhauser Verlag.

2. Identify which cleavage the fragment is lying on. The criteria are as follows.

a. {001} Cleavage: (1) Albite twin lamellae can be visible and show the same extinction angle to the right and to the left. (2) The albite twin lamellae are parallel to the {010} cleavage.

b. {010} Cleavage: (1) Pericline lamellae may be visible and show different extinction angles to the right and left. (2) Pericline twin lamellae may be oriented at an angle to the trace of {001} cleavage.

3. Measure the extinction angle between the trace of {010} cleavage for fragments on (001), or {001} cleavage for fragments on (010), and the fast ray vibration direction.

4. Compare the extinction angle with the appropriate curve in Figure 15.13. A single angle may indicate two possible compositions for plagioclase more sodic than An_{50} because negative and positive angles generally cannot be distinguished. The ambiguity is readily resolved by using an index oil with $n = 1.542$ and comparing the index of the fast ray (n_α' to the oil. Samples more sodic than An_{25} have n_α' less than 1.542, and more calcic samples have a higher index.

The accuracy for samples on the {001} cleavage is fairly low for plagioclase between An_0 and An_{60} because the curves are rather flat. This method is most conveniently used as a preliminary step in the Tsuboi method because it quickly gives an approxi-

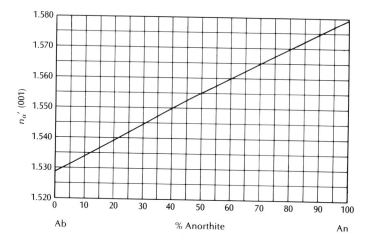

Figure 15.14 Index of the fast ray (n_α) in fragments lying on cleavages. The diagram is constructed for fragments on (001) but is applicable to fragments on (010) as well. After Morse (1968).

mate composition so that the correct oil can be rapidly selected to determine the index n_α' of the fast ray in the cleavage fragments.

B. *Tsuboi Method.* The Tsuboi method involves measuring the index of refraction of the fast ray (n_α') in cleavage flakes lying on either (001) or (010). The procedure is as follows.

1. By examination, select a fragment lying on either cleavage surface; it does not matter which.

2. Cross the polars and rotate the grain to extinction with the fast ray direction parallel to the lower polar. Use an accessory plate to determine which ray is fast (Chapter 5).

3. In plane light, compare the index n_α' with the index of the oil.

4. Repeat, using different oils, until a match is obtained.

5. Use Figure 15.14 to determine the composition. The accuracy depends on how accurately n_α' is determined. If the index is measured to ±0.001, accuracy of ±1 percent An is claimed for the range An_{20} to An_{60}, and ±3 percent An outside of that range. The accuracy for volcanic rocks is somewhat lower, probably no better than ±5 percent An.

Alkali Feldspars

$(K,Na)AlSi_3O_8$

Microcline
Triclinic
$\angle\alpha$ = 90.6°
$\angle\beta$ = 115.9°
$\angle\gamma$ = 87.7°
Biaxial (−)

n_α = 1.514–1.526
n_β = 1.518–1.530
n_γ = 1.521–1.533
δ = 0.005–0.008
$2V_x$ = ~65–88°

Orthoclase
Monoclinic
$\angle\beta$ = 116°
Biaxial (−)

n_α = 1.514–1.526
n_β = 1.518–1.530
n_γ = 1.521–1.533
δ = 0.005–0.008
$2V_x$ = 40–~70°

Sanidine
Monoclinic
$\angle\beta$ = 116°
Biaxial (−)

n_α = 1.514–1.526
n_β = 1.518–1.530
n_γ = 1.521–1.533
δ = 0.005–0.008
$2V_x$ = 0–40°[OP ⊥ (010)]
　　　0–47°[OP ∥ (010)]

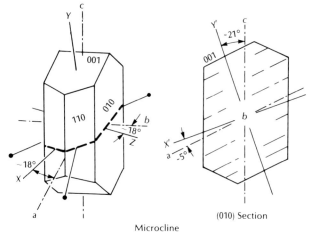

Microcline

(010) Section

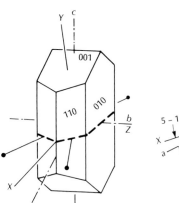

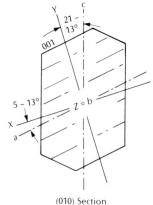

Orthoclase

(010) Section

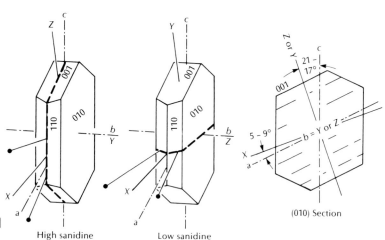

High sanidine

Low sanidine

(010) Section

Anorthoclase

Triclinic

$\angle \alpha = \sim 92°$
$\angle \beta = 116.3°$
$\angle \gamma = 90.2°$

Biaxial (−)

$n_\alpha = 1.519–1.529$
$n_\beta = 1.524–1.534$
$n_\gamma = 1.527–1.536$
$\delta = 0.005–0.008$
$2V_x = 0–55°$

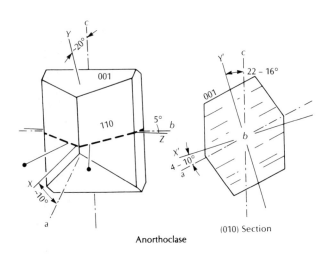

Anorthoclase

(010) Section

RELIEF IN THIN SECTION: Low negative relief.

COMPOSITION AND STRUCTURE: The alkali feldspars include compositions between K-feldspar (Or, KAl-Si$_3$O$_8$) and albite (Ab, NaAlSi$_3$O$_8$). Except near the albite end of the series, the anorthite content is usually less than 5 percent. The terminology applied to the alkali feldspars has evolved over more than two centuries, but despite, or perhaps because of, the fact that it is one of the most thoroughly studied mineral groups, the terminology remains poorly defined and subject to regular revision. The primary

problem is that the historical terminology that distinguished four varieties of K-feldspar (sanidine, orthoclase, microcline, and adularia) was based primarily on physical and optical properties, which, in some cases, were not well defined. In recent years, it has been shown that these distinctions are not always well founded on a structural or chemical basis. For example, material that has been called monoclinic orthoclase may prove to be submicroscopically twinned triclinic microcline. The distinctions between the various species used here are based on their optical properties and are as follows.

> Microcline: Optically triclinic K-feldspar characterized by its distinctive cross-hatched twin pattern. The optic plane is nearly perpendicular to (010). $2V_x$ is typically greater than 65° but lower values are possible. Note, however, that authigenic K-feldspar may be triclinic and is therefore microcline, but it can be untwinned or have only simple twins.
> Orthoclase: Optically monoclinic K-feldspar with $2V_x$ larger than $\sim 40°$ and the optic plane perpendicular to (010).
> Low sanidine: Optically monoclinic K-feldspar with $2V_x$ between ~ 40 and 0° and the optic plane perpendicular to (010).
> High sanidine: Optically monoclinic K-feldspar with the optic plane parallel to (010) and $2V_x$ between 0 and $\sim 47°$.
> Adularia: Authigenic or low-temperature hydrothermal K-feldspar with the {110} prism being the dominant form. Cross sections are diamond shaped.
> Anorthoclase: Optically triclinic sodic alkali feldspar characterized by grid twinning similar to microcline but on a finer scale, and with $2V_x$ less than 55°; found in volcanic and hypabassal intrusive rocks. The composition typically lies between Ab$_{63}$Or$_{27}$ and Ab$_{90}$Or$_{10}$.

On a structural basis, samples classified as sanidine based on the $2V$ angle typically have a high degree of disorder of Al and Si in the tetrahedral sites, and microcline samples are usually highly ordered. Samples classified as orthoclase typically have intermediate degrees of order. Anorthoclase is a triclinic sodium-rich alkali feldspar that tends to be disordered. Adularia shows a wide range of order.

Theoretically, there is complete solid solution between sanidine and high albite, which are the high-temperature, disordered phases. Anorthoclase is part of the sanidine-high albite series with compositions between Ab$_{63}$Or$_{27}$ and Ab$_{90}$Or$_{10}$, with up to ~ 10 mole percent anorthite. At lower temperatures a solvus is present, so that for the low temperature, highly ordered sequence neither microcline nor albite have more than about 15 percent Na or K, respectively. Orthoclase tends to be somewhat more sodic, with compositions as sodic as Or_{70} common. Because K-feldspar frequently crystallizes at high temperature with an intermediate composition, cooling causes exsolution of sodium-rich and potassium-rich feldspars. In microcline, low albite, and some orthoclase, the exsolution lamellae may be visible in thin section and often in hand samples as well. Exsolution may take place in other alkali feldspars, but the individual lamellae may be too fine to be recognized with the microscope. Common exsolution and intergrowth textures involving the feldspars are as follows (Figure 15.15):

Perthite: Exsolution blebs and lamellae of albite in microcline or orthoclase.
Antiperthite: Exsolution blebs or lamellae of K-feldspar in sodic plagioclase.
Cryptoperthite: Submicroscopic exsolution lamellae of one alkali feldspar in another.
Rapakivi: Rims of sodic plagioclase around a phenocryst of K-feldspar.
Antirapakivi: Rims of K-feldspar on crystals of plagioclase.
Granophyre: An intergrowth of quartz and feldspar that occurs in interstices between grains in shallow intrusive rocks, usually of granitic composition. The quartz forms small vermicular grains or blebs or may form a cuniform pattern.
Graphic and Micrographic: Intergrowth of quartz in K-feldspar crystals. The quartz forms small, more or less angular (i.e., cuniform) grains that are usually optically continuous over the entire K-feldspar crystal. Common in pegmatites and granitic rocks.
Myrmekite: An intergrowth of plagioclase and quartz often found at K-feldspar-plagioclase contacts.

PHYSICAL PROPERTIES: $H = 6-6\frac{1}{2}$; $G = 2.55-2.62$; usually white or pink in hand sample (microcline may be blue or green, sanidine and anorthoclase may be clear and glassy); white streak; vitreous luster. A play of colors may be present in some varieties (called moonstone) due to cryptoperthitic intergrowths.

COLOR: Colorless in thin section and grain mount. Clouding, particularly in orthoclase, is common.

INDICES OF REFRACTION AND BIREFRINGENCE: The indices of refraction appear to vary in a linear manner with the degree of Al–Si order for common sani-

Figure 15.15 Feldspar intergrowth textures. (*Top*) Perthite. Exsolution lamellae of albite (NE–SW) in microcline. (*Middle*) Myrmekite, consisting of irregular quartz grains in plagioclase at contact between plagioclase (right) and microcline (left). (*Bottom*) Granophyre in space between plagioclase laths in norite. Width of views of view: (*top*) and (*bottom*) 2.3 mm; (*middle*) 1.2 mm.

dine, orthoclase, and microcline (Figure 15.16). The variation appears to be independent of the amount of sodium present in the structure, whether samples are monoclinic or triclinic, or to the presence of sub-microscopic twins or exsolution lamellae. Birefringence is typically around 0.007, so interference colors in thin section are no higher than first-order white. Anorthoclase usually has indices of refraction somewhat higher than potassium-rich feldspar, and the indices approach those for albite ($n_\alpha = 1.527$, $n_\beta = 1.531$, $n_\gamma = 1.534$) as the sodium content increase.

OPTIC ANGLE: The value of $2V_x$ varies in a systematic way with degree of Al–Si order and with composition. For common sanidine, orthoclase, and microcline, the variation as a function of Al-Si order is shown in Figure 15.16. This diagram allows fairly accurate estimates of the degree of order to be made based on the $2V_x$ angle and orientation of the optic plane. The variation of $2V_x$ over the full composition range is shown in Figure 15.17. Anorthoclase typically has $2V_x$ between 0 and 55° as would be expected from Figure 15.17 because it occupies part of the field between high albite and high sanidine.

Microcline

FORM: Figure 15.18. It is commonly found as anhedral and euhedral grains in many igneous and metamoprhic rocks. Crystals are roughly tabular parallel to (010) and elongate parallel to the c or a axis. Exsolved albite may form lamellae, irregular ribbons, and blebs to form perthite.

CLEAVAGE: Perfect cleavage parallel to {001} and good cleavage on {010} intersect at 90°41′. Fragments tend to lie on the cleavage surfaces. In thin section, cleavage does not stand out because of the low relief. To spot it, look at fractured grains at the slide's edge and adjust the aperture diaphragm to accentuate the relief.

TWINNING: Twins according to the albite and pericline laws are characteristic of microcline which has inverted from monoclinic symmetry, and produce a distinctive cross-hatched or "tartan plaid" pattern. However, low-temperature hydrothermal and authigenic microcline may have only simple twins or may lack twins. The contacts between lamellae are

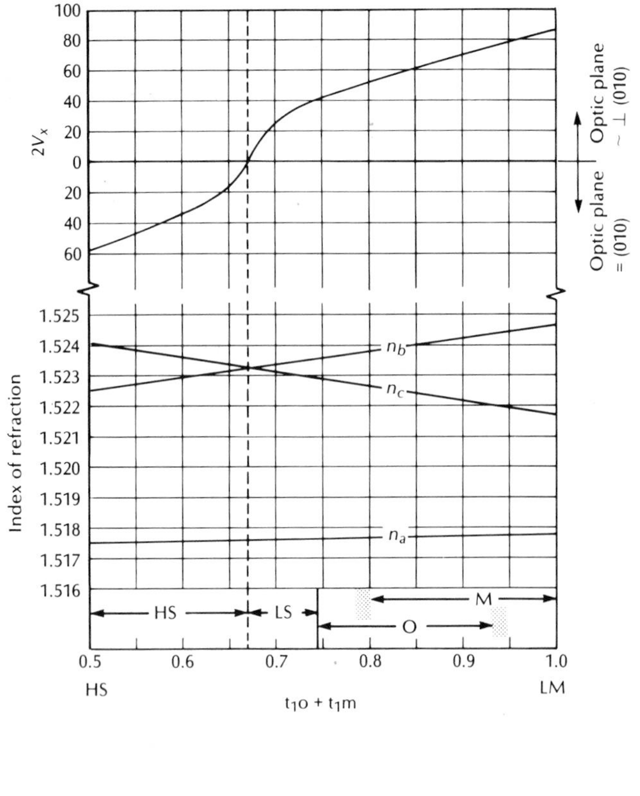

Figure 15.16 Variation of $2V_x$ and indices of refraction for common K-feldspar as a function of degree of Al–Si order. t_1o and t_1m are the occupancy of Al in the T_1o and T_1m tetrahedral sites. If Al and Si are randomly distributed among the tetrahedral sites, as in high sanidine (HS), then one-quarter of the T_1o and one-quarter of T_1m sites must be occupied by Al: therefore, $t_1o + t_1m = 0.50$. If all of the aluminum is placed in just one site (e.g., the T_1o site) then $t_1o = 1.00$, and $t_1m = 0$, so $t_1o + t_1m = 1.00$ as in low microcline (LM). The principal indices of refraction n_a, n_b, and n_c are for light vibration most nearly parallel to the a, b, and c crystal axes, respectively. The approximate range of high sanidine (HS), low sanidine (LS), orthoclase (O), and microcline (M) are indicated. After Su and others (1984).

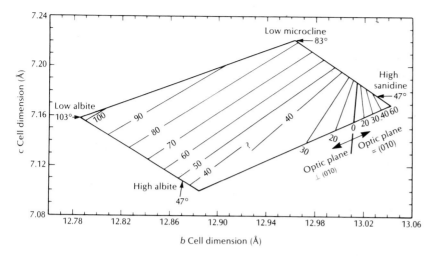

Figure 15.17 Relation of $2V_x$ to the b and c unit cell dimensions of alkali feldspars. From Stewart and Ribbe (1983). Used by permission of the Mineralogical Society of America.

diffuse, and individual lamellae appear to pinch and swell and are discontinuous. The composition plane for the albite twins is (010), and the composition plane for the pericline twins is inclined about 35° to (001) and contains the b axis. Carlsbad twins may also be present, particularly in phenocrysts from granitic rocks, and they divide crystals into two segments with a slightly irregular composition plane parallel to (010). Baveno, Manebach, and other twins also may be found.

OPTICAL ORIENTATION: The optic plane is roughly parallel with (001), $X \wedge a \cong 18°$, $Y \wedge c \cong 18°$, and $Z \wedge b \cong 18°$. Extinction is inclined to cleavages and crystal outlines in all sections, but the ubiquitous twinning makes measurement difficult. Fragments lying on the {001} and {010} cleavages show extinction angles of about 15 and 5°, respectively, when measured to the trace of cleavage. Fragments on the {010} cleavage may show extinction angles as large as 12° in cryptoperthites and angles larger than 7° probably indicate the presence of significant amounts of submicroscopically exsolved Na feldspar.

INDICES OF REFRACTION AND BIREFRINGENCE: Fully ordered low microcline has indices of approximately $n_\alpha = 1.518$, $n_\beta = 1.522$, $n_\gamma = 1.525$, and the indices appear to vary systematically with degree of Al–Si order (Figure 15.16). Sodic cryptoperthites have indices intermediate between microcline and low albite. Birefringence is low (~0.007) so interference colors in thin section are no higher than first-order white.

INTERFERENCE FIGURE: Microcline is biaxial negative with $2V_x$ usually greater than 65° and typically in the range 70 to 85° (Figures 15.16 and 15.17). However, $2V_x$ as small as 35° is possible. Optic axis dispersion is weak with $r > v$. Twinning usually makes good figures difficult to obtain. Optically positive microcline with $2V_x \cong 103°$ has been reported but the large $2V_x$ is due to the presence of submicroscopic lamellae of albite. Fragments on the {001} and {010} cleavages yield off-center optic normal (flash) and obtuse bisectrix figures, respectively.

ALTERATION: The common alteration products are sericite and clay.

DISTINGUISHING FEATURES: The grid twinning pattern is characteristic except in authigenic or low-temperature vein material. Lack of twinning is diagnostic of crystallization below ~300°C. The combination of albite and pericline twinning in plagioclase is similar to microcline twinning, but the twin lamellae are cleanly defined and do not pinch and swell. The twinning in anorthoclase is similar, but usually on a finer scale. Anorthoclase has a smaller $2V_x$ and is found in volcanic and shallow intrusive rocks.

OCCURRENCE: Microcline is found in granite, granodiorite, pegmatite, syenite, and related plutonic igneous rocks. It is not found in volcanic rocks. Microcline also is common in regional metamorphic rocks, usually of fairly high grade.

Because microcline is fairly stable in the weathering environment, it is a common constituent of

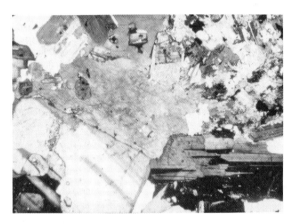

Figure 15.18 K-feldspar. (*Top*) Microcline with tartan plaid twinning. (*Middle*) Orthoclase (O) between plagioclase grains. (*Bottom*) Sanidine phenocryst in rhyolite. Width of fields of view: (*top*) and (*middle*) 5 mm; (*bottom*) 2.3 mm.

immature clastic sediments. In some cases, overgrowths of new K-feldspar may develop on clastic grains.

Orthoclase

FORM: Common as anhedral and euhedral grains in igneous rocks. Crystals are commonly elongated parallel to *c* or *a*, and roughly tabular parallel to (010). Some orthoclase is perthitic, and much of it is cryptoperthitic.

CLEAVAGE: Perfect cleavage on {001} and good cleavage on {010} intersect at 90°. Cleavages tend to control fragment orientation. Cleavage may be difficult to see in thin section because of low relief. Look at fractured grains at the slide's edge and adjust the aperture diaphragm to accentuate the relief.

TWINNING: Carlsbad twins are common and divide crystals into two segments with a slightly irregular composition plane parallel to (010). Baveno twins produce prismatic crystals elongate parallel to the *a* axis with the {021} composition plane on the diagonal through the prism. Manebach twins also are elongate parallel to the *a* axis, but the {001} composition plane is parallel to one set of prism faces. Other twins have been reported, but polysynthetic twins similar to plagioclase or microcline are not found.

OPTICAL ORIENTATION: $X \wedge a = +5$ to $+13°$, $Y \wedge c = +21$ to $+13°$, $Z = b$, optic plane is perpendicular to (010) and inclined between 5 and 13° to (001). The maximum extinction angle to the {001} cleavage is usually 5 to 6°, and is seen in sections parallel to (010) and on fragments lying on the {010} cleavage. The extinction angle may be as large as 13° in fairly sodic cryptoperthites. Fragments on the {001} cleavage and sections cut at right angles to (010) show parallel extinction to cleavage.

INDICES OF REFRACTION AND BIREFRINGENCE: Indices of refraction appear to vary in a linear manner with degree of Al–Si ordering (Figure 15.16). An increase in Na content causes a small increase in the indices, but the degree of ordering appears to be substantially more important. Birefringence is low (~0.007) so maximum interference colors in thin section are first-order white.

INTERFERENCE FIGURE: $2V_x$ varies systematically with degree of Al–Si ordering (Figure 15.16). An increase in Na content causes an increase in the $2V_x$ angle but for the compositions found in common orthoclase the effect is negligible (Figure 15.17). The convention adopted here is to use a $2V_x$ of 40° as the division between orthoclase and sanidine. Orthoclase may have $2V_x$ as large as 85° but $2V_x$ is usually less than 70°. Fragments lying on (010) or (001) yield optic normal (flash) and off-center obtuse bisectrix figures, respectively. Optic axis dispersion is $r > v$.

ALTERATION: The common alteration products are clay and sericite.

DISTINGUISHING FEATURES: Orthoclase is distinguished from sanidine based on $2V_x$ angle, and from microcline based on lack of grid twinning. Caution should be exercised, however, since some triclinic K-feldspar (particularly authigenic and hydrothermal samples) may not have obvious grid twinning. Sections that show both cleavages intersecting at right angles and fragments on the {001} cleavage should show parallel extinction. Orthoclase greatly resembles quartz, and may be easily overlooked. However, it shows negative relief ($n < n_{cement} \cong 1.538$), is often slightly clouded with incipient alteration, and is biaxial. A number of staining techniques have been devised to rapidly distinguish between K-feldspar, plagioclase, quartz, and other similar minerals (e.g., Boone and Wheeler, 1968); Bailey and Stevens, 1960; Ruperto and others, 1964).

OCCURRENCE: Orthoclase is a widespread mineral in granite, granodiorite, syenite, and related felsic rocks, and particularly in fairly shallow intrusives of Phanerozoic age. Sanidine tends to be more common in volcanic rocks, and microcline is more common in deep-seated intrusives. Orthoclase is also found in contact and regional metamorphic rocks, and as detrital grains in clastic sediments.

Sanidine

FORM: Figure 15.18. Sanidine is common as phenocrysts that are tabular parallel to (010) and somewhat elongate parallel to the a axis. Cross sections tend to be square, rectangular, or six sided. Acicular crystals are found in spherulites, and slender microlites are common in rapidly cooled rocks. Zoning is common and is expressed as variations in birefringence and extinction angle.

CLEAVAGE: Perfect cleavage on {001} and good cleavage on {010} intersect at 90°. Fragments tend to lie on cleavage surfaces. Cleavage may not be obvious in thin section due to the low relief. Look at fractured grains at the slide's edge and adjust the aperture diaphragm to accentuate relief.

TWINNING: Carlsbad twins with a composition plane parallel to (010) divide crystals into two segments. Manebach and Baveno twins may also be seen (cf. orthoclase).

OPTICAL ORIENTATION: There are two orientations possible in sanidine. In high sanidine, the optic plane is parallel to (0$\bar{1}$0) with $X \wedge a \cong +5°$, $Y = b$, and $Z \wedge c = +21°$. Low sanidine is oriented like orthoclase, with the optic plane normal to (010) and $X \wedge a = +5$ to $+9°$, $Y \wedge c = +21$ to $+17°$, and $Z = b$. The maximum extinction angle of 5 to 9° to the {001} cleavage is seen in sections parallel to (010) and in fragments lying on (010). Sections cut normal to (010) and fragments lying on (001) show parallel extinction to cleavage traces. Grains elongate parallel to the a axis are length fast.

INDICES OF REFRACTION AND BIREFRINGENCE: Indices of refraction appear to vary systematically with degree of Al–Si ordering (Figure 15.16). Indices also increase moderately with increasing Na content, but for the compositions commonly found, the effect is negligible. Birefringence is low (0.005–0.008), so maximum interference colors in thin section are no higher than first-order white.

INTERFERENCE FIGURE: The $2V_x$ angle for low sanidine is less than 40° in an optic plane normal to (010) for the convention adopted here and decreases to 0° with increasing degree of disorder in the tetrahedral sites and then opens again in (010) to as much as 47° for high sanidine. The low-sanidine orientation is more common and the $2V_x$ angle is usually small. The $2V_x$ angle in high sanidine decreases, and in low sanidine increases, with increasing Na content (Figure 15.17) but for common sanidine compositions, the effect is negligible. Cleavage fragments of high and low sanidine on (001) show optic normal (flash) and slightly off-center obtuse bisectrix figures, respectively. Provided that the $2V_x$ angle is not so small as to make the two figures indistinguish-able, the difference can be used to distinguish low from high sanidine. In thin section, the distinction can be made provided that the (010) plane can be identified, either because it is parallel to a Carlsbad twin composition plane or because crystals tend to be tabular parallel to (010). Obtain an acute bisec-trix or optic axis figure and rotate the optic plane to N–S. Return to orthoscopic illumination and note whether (010) is N–S (high sanidine) or E–W (low sanidine). Optic axis dispersion is weak with $r > v$ (low sanidine) or $v > r$ (high sanidine).

ALTERATION: Sericite and clay are the common alter-ation products.

DISTINGUISHING FEATURES: Low sanidine is dis-tinguished from orthoclase by a smaller $2V_x$ angle. High sanidine is distinguished from orthoclase and low sanidine by the orientation of the optic plane. Quartz has a similar appearance but is uniaxial and has indices higher than cement. High albite is com-monly twinned and has $2V_x$ larger than 45°.

OCCURRENCE: Sanidine is the common K-feldspar in silicic volcanic rocks such as rhyolite, rhyodacite, phonolite, and trachyte, and in dikes and other shal-low intrusives. Igneous sanidine is commonly fairly sodic (Ab_{20}–Ab_{60}) and may be cryptoperthitic. In rhyolite and obsidian, sanidine is found in spheru-lites intergrown with cristobalite. Sanidine also occurs in high-temperature contact metamophic rocks.

Clastic sediments may contain sanidine, but mic-rocline and orthoclase are more common.

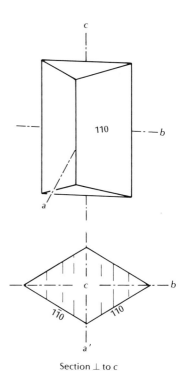

Section ⊥ to c

Figurer 15.19 Crystal habit of adularia.

or rhombic cross sections. Some may display a pat-tern of four alternating sectors with the boundaries between the sectors roughly parallel to the diagonals of the rhomb. Triclinic varieties may display microcline-type twinning but commonly do not. The optic orientation is similar to the other K-feldspars but can be quite variable.

Adularia

The term *adularia* is given to K-feldspar whose dominant form is the {110} prism (Figure 15.19). It typically forms by authigenic processes in sedimen-tary rocks or by crystallizing in hydrothermal veins. It may be either triclinic or monoclinic. If adularia is triclinic, it is properly classified as microcline, and if monoclinic it should be classified as orthoclase or sanidine, depending on its structural state. Individ-ual crystals, however, may contain domains of both monoclinic and triclinic symmetry.

Crystals characteristically show diamond-shaped

Anorthoclase

FORM: Crystals tend to be prismatic and elongate parallel to the *c* axis. Anhedral grains and microlites also are common.

CLEAVAGE: Perfect {001} and good {010} cleavages intersect at about 92°. Cleavages tend to control fragment orientation with (001) favored.

TWINNING: A complex pattern of twinning according to the albite and pericline laws are characteristic. The composition planes are {010} (albite) and {h01} (pericline). The pericline twins are usually nearly parallel to the (001) plane. The twinning in microcline is quite similar, but anorthoclase has finer lamellae, and the orientation of the pericline lamellae are distinctly different, although in anhedral grains the orientation will not be obvious. The distinction can be made in sections cut parallel to (010) or on fragments laying on (010). In anorthoclase, the twin lamellae are nearly parallel to the trace of the {001} cleavage, and in microcline the lamellae are inclined at an angle. Except in detrital grains there is rarely any confusion because microcline and anorthoclase have distinctly different occurrences.

OPTICAL ORIENTATION: The optic plane is approximately parallel to (001) with $X \wedge a \cong 10°$, $Y \wedge c \cong 20°$, and $Z \wedge b \cong 5°$. Extinction to the cleavages and crystal outlines is inclined in all sections. The maximum angle to the {001} cleavage is about 4 to 10°. Fragments lying on (001) and (010) show extinction to cleavage traces of 1 to 6°, and 4 to 10°, respectively.

INDICES OF REFRACTION AND BIREFRINGENCE: Indices of refraction increase with increasing Na content. Birefringence is low (0.005–0.008), so maximum interference colors in thin section are first-order gray or white.

INTERFERENCE FIGURE: Anorthoclase is biaxial negative with $2V_x$ less than about 55°. The complex twinning may make good figures difficult to obtain. Cleavage fragments yield off-center obtuse bisectrix and optic normal (flash) figures.

ALTERATION: The common alteration products are sericite and clay.

DISTINGUISHING FEATURES: Anorthoclase resembles microcline, but the twin lamellae are usually finer and it has a smaller $2V_x$. They also have distinctly different occurrences.

OCCURRENCE: **Anorthoclase is found in Na-rich felsic volcanics such as alkalic rhyolite, trachyte, phonolite, and equivalent hypabyssal intrusives.**

FELDSPATHOIDS

Nepheline

$Na_3K(Al_4Si_4O_{16})$
Hexagonal
Uniaxial (−)
$n_\omega = 1.529–1.546$
$n_\epsilon = 1.526–1.544$
$\delta = 0.003–0.005$

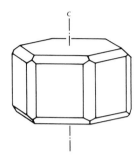

RELIEF IN THIN SECTION: Usually low positive relief, also low negative relief.

COMPOSITION AND STRUCTURE: Nepheline is the most common feldspathoid. Its structure is similar to tridymite and consists of sheets of tetrahedra. The tetrahedra point alternately up and down so that the apical oxygen of each tetrahedron is shared with a tetrahedron from an adjacent sheet. Approximately half of the tetrahedra are occupied by Al, the remainder by Si. Na and K occupy interstices between the sheets. At high temperature, there is complete solid solution between pure sodium nepheline and kalsilite ($K_4Al_4Si_4O_{16}$), but under normal magmatic conditions a solvus is present at intermediate compositions. Most nepheline has approximately 25 mole percent K and 75 mole percent Na, probably because there are three small sites and one large site that the Na and K occupy. However, smaller or larger percentages of K may be found, particularly in volcanic rocks. Small amounts of Ca also are usually found in the alkali sites. Chemical zoning may be reflected by changes in birefringence.

PHYSICAL PROPERTIES: H = 6; G = 2.55 – 2.67, white or gray in hand sample, white streak; vitreous to greasy luster.

COLOR: Colorless in thin section and grain mount. Inclusions or incipient alteration may make it cloudy, particularly in intrusive rocks.

FORM: Figure 15.20. In intrusive rocks, nepheline is usually anhedral or subhedral, and in volcanic rocks,

Figure 15.20 Nepheline (center) with plagioclase and microcline. Field of view is 5 mm wide.

it tends to form subhedral to euhedral crystals either as part of the groundmass or as phenocrysts. Crystals are usually stubby prisms with either a square or hexagonal cross section. Longitudinal sections are usually rectangular. Kalsilite exsolution lamellae may be found.

CLEAVAGE: There is fair cleavage parallel to the {10$\bar{1}$0} prism faces and the basal pinacoid {0001}, but they are not usually seen in thin section nor does cleavage strongly influence fragment shape or orientation. Fractures tend to be irregular.

TWINNING: Twins with composition planes parallel to {10$\bar{1}$0}, {11$\bar{2}$2}, and {33$\bar{6}$5} are possible but are not usually seen in thin section.

OPTICAL ORIENTATION: Nepheline is uniaxial, so the c axis is the optic axis. Longitudinal sections through euhedral crystals are length fast, with parallel extinction.

INDICES OF REFRACTION AND BIREFRINGENCE: Substituting K for Na causes a slight increase in refractive indices. Most nepheline has indices above 1.54; therefore, it shows low positive relief. Birefringence is low and maximum interference colors seen in thin section are first-order gray. Kalsilite has indices $n_\omega \cong 1.540$ and $n_\epsilon \cong 1.535$.

INTERFERENCE FIGURE: Nepheline is uniaxial negative, although slight separation of isogyres may be seen in some samples. Due to the low birefringence,

the isogyres tend to be rather broad and fuzzy on a first-order gray or white background. Calcium-rich samples may be optically positive.

ALTERATION: Nepheline may be altered to clay minerals, analcime, sodalite, calcite, and cancrinite.

DISTINGUISHING FEATURES: Nepheline resembles the feldspars but is uniaxial, lacks good cleavage and twinning, and has lower birefringence. Quartz is optically positive and lacks the clouding common in nepheline. Kalsilite is very similar but is quite rare. The distinction between kalsilite and nepheline can be made with staining techniques (e.g., Shand, 1939; Bailey and Stevens, 1960) or with X-ray diffraction techniques.

OCCURRENCE: Nepheline is a common mineral in syenite, nepheline syenite, phonolite, foidites, and related alkalic igneous rocks. It is commonly associated with K-feldspar, plagioclase, biotite, sodic and sodic-calcic amphiboles and pyroxenes, cancrinite, sodalite, melilite and leucite. Only rarely is nepheline associated with quartz, which is usually secondary. Nepheline also is found in unusual alkalic mafic rocks with olivine, calcic clinopyroxene, and monticellite. Nepheline may also be found in metamorphic rocks, presumably as the result of sodium metasomatism. Kalsilite is found in rare potassic mafic lavas.

Sodalite Group

Isometric
Isotropic

Sodalite $Na_8(Al_6Si_6O_{24})Cl_2$
$n = 1.483–1.487$

Nosean $Na_8(Al_6Si_6O_{24})SO_4$
$n = 1.470–1.495$

Haüyne $(Na,Ca)_{4–8}(Al_6Si_6O_{24})(SO_4)_{1–2}$
$n = 1.494–1.510$

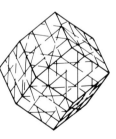

RELIEF IN THIN SECTION: Moderate negative relief.

COMPOSITION AND STRUCTURE: The structure of the sodalite group is a very open tetrahedral framework, the cavities of which contain the large anions (Cl$^-$,

SO_4^{2-}), which are surrounded by four Na^+ or Ca^{2+}. Half the tetrahedra are occupied by Al^{3+} and half by Si^{4+} to maintain charge balance. The composition of sodalite is usually close to its ideal formula, but minor substitution of Ca or K for Na, and Fe^{3+} for Al^{3+} may take place. In nosean, SO_4^{2-} occupies half of the anion sites, with the remainder vacant, and more Ca may be present than in sodalite. Haüyne differs from nosean in that it contains substantial Ca, which is balanced by filling more of the anion sites with SO_4^{2-}. *Hackmanite* is a variety of sodalite containing significant amounts of S^{2-} substituting for Cl^-, and *lazurite* can be considered a variety of haüyne containing substantial amounts of Cl^- and S^{2-} instead of SO_4^{2-}.

PHYSICAL PROPERTIES: H = $5\frac{1}{2}$–6; G = 2.27–2.33 (sodalite), 2.3–2.4 (nosean), 2.44–2.50 (haüyne); blue or gray in hand sample, also pink, yellow, or green; white streak; vitreous luster.

COLOR AND PLEOCHROISM: Sodalite and nosean are colorless, gray, or very pale blue in thin section and grain mount. Haüyne is commonly blue; also colorless or blue-green. There is no pleochroism because all are isotropic. The outer rim of grains may be dark, and the color, if present, may be patchy.

FORM: Crystals are dodecahedra, which in thin section show six-sided cross sections. Sodalite is often anhedral in plutonic rocks. Symmetrically or zonally arranged inclusions are common in nosean and haüyne. Phenocrysts may be partially resorbed and have embayed edges.

CLEAVAGE: There is poor {110} dodecahedral cleavage (i.e., six cleavages parallel to the dodecahedron faces), which is not usually obvious. It may be visible in fractured grains at the edge of slides.

TWINNING: There is twinning on {111} but, because the minerals are isotropic, twinning can be recognized only if betrayed by the external shape of crystals.

INDICES OF REFRACTION AND BIREFRINGENCE: The indices of the group generally increase with increasing amounts of Ca and SO_4. All have indices substantially below the index of cement used in thin sections. The sodalite group is isotropic, but may show weak birefringence around inclusions.

ALTERATION: Members of the sodalite group may alter to zeolites (often fibrous), clay, calcite, or cancrinite.

DISTINGUISHING FEATURES: Members of the sodalite group are distinguished by their low indices, isotropism, and occurrence. Leucite is weakly birefringent and shows a distinctive twinning pattern. Garnet has strong positive relief. Fluorite has even lower indices and good cleavage. Analcime does not show blue colors and has poor cubic cleavage, but otherwise it is quite similar. While analcime may form as a primary mineral in igneous rocks, it is usually a groundmass mineral and does not form phenocrysts. It may not be possible to distinguish the three members of the group from each other based on their optics. However, sodalite is usually colorless, lacks abundant inclusions, and is the only one found in plutonic rocks; nosean and haüyne commonly contain inclusions, and haüyne is commonly blue in thin section.

OCCURRENCE: The sodalite group is found in silica-deficient alkalic igneous rocks. Sodalite is found in both volcanic and plutonic rocks, while haüyne and nosean are generally restricted to volcanic rocks. The volcanic rocks that the sodalite group is found in include phonolites, alkali basalts, trachyte, and feldspathoidal basalt. The commonly associated minerals are nepheline, leucite, and sanidine. Sodalite is found in nepheline syenite and related rocks, and both sodalite and lazurite may be found in contact metamorphosed carbonates.

Leucite

$KAlSi_2O_6$
Tetragonal (pseudoisometric)
Uniaxial (+) (nearly isotropic)
n_ω = 1.508–1.511
n_ϵ = 1.509–1.511
δ = 0.000–0.001

RELIEF IN THIN SECTION: Low negative relief.

COMPOSITION AND STRUCTURE: Leucite is a feldspathoid with an open framework structure consisting of interlocking four- and sixfold rings of tetrahedra. The composition of most leucite is fairly close to the ideal formula, but small amounts of Na may replace K, and Fe^{3+} may replace Al^{3+}. Leucite crystallizes with an isometric structure that distorts on cooling to tetragonal symmetry.

PHYSICAL PROPERTIES: $H = 6$; $G = 2.47$; white or gray in hand sample; white streak; vitreous luster.

COLOR: Colorless in thin section and grain mount. It may be clouded due to inclusions or alteration.

FORM: Figure 15.21. Leucite commonly occurs as equant trapezohedral crystals that show eight-sided to nearly round cross sections. It also occurs as small microlites, and skeletal or anhedral grains in the groundmass of volcanic rocks. Inclusions of glass, olivine, magnetite, pyroxene, and other minerals may be present.

Figure 15.21 Leucite phenocryst. Field of view is 5 mm wide.

CLEAVAGE: Pseudododecahedral cleavage {110} is very poor and not seen in thin section, nor does it control fragment shape.

TWINNING: The transition from the high temperature iosometric structure to the tetragonal structure causes the formation of characteristic complex lamellar twins, with composition planes parallel to {110} dodecahedral faces in the isometric form. The twins are usually arranged in a more or less concentric manner, or intersect at angles of about 60°.

OPTICAL ORIENTATION: One of the a axes in the original isometric form becomes the tetragonal c axis on cooling. Crystals, however, show no elongation.

INDICES OF REFRACTION AND BIREFRINGENCE: Indices of refraction show almost no variation. Birefringence is very low, and small grains appear essentially isotropic. Maximum interference colors in thin section are lower first order gray. The gypsum plate is often needed to detect the birefringence. (Insert the plate. If the interference color is even slightly different than the color normally produced by the gypsum plate with no sample on the stage, then the material is weakly birefringent.)

INTERFERENCE FIGURE: The complex twinning and very low birefringence make interference figures difficult to obtain and interpret. A good figure will consist of very broad and fuzzy isogyres on a gray field. The isogyres may show some separation with rotation. The optic sign is usually positive.

ALTERATION: Leucite may alter to pseudoleucite, which is a mixture of nepheline and K-feldspar.

DISTINGUISHING FEATURES: The habit, low birefringence, and twinning are diagnostic. Analcime and members of the sodalite group have higher indices and lack the complex twinning. Members of the sodalite group may be bluish in thin section.

OCCURRENCE: Leucite is found exclusively in potassium-rich mafic volcanics and associated hypabyssal intrusives. It is usually associated with olivine, nepheline, sanidine, clinopyroxene, phlogopite, apatite, and sodic or sodic-calcic amphiboles. Pseudoleucite is sometimes found in alkalic plutonic igneous rocks. Because leucite is readily weathered, it is seldom found in sediments.

Cancrinite–Vishnevite

$(Na,Ca,K)_{6-8}(AlSiO_4)_6(CO_3,SO_4,Cl)_{1-2}\cdot1-5H_2O$

Hexagonal

Uniaxial (−)

n_ω = 1.490–1.528
n_ϵ = 1.488–1.503
δ = 0.002–0.025

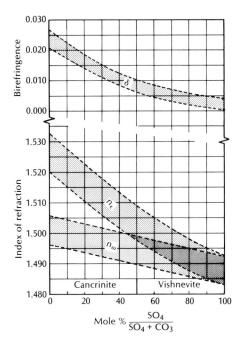

RELIEF IN THIN SECTION: Low to moderate negative relief.

COMPOSITION AND STRUCTURE: The cancrinite series has a very open framework of tetrahedra similar to the zeolites—with channels that can accommodate water molecules, and large, cagelike openings that contain the large anionic groups and the Na, Ca and K. The cancrinite–vishnevite series is defined based on the relative amounts of CO_3 and SO_4, with cancrinite being carbonate rich.

PHYSICAL PROPERTIES: H = 5–6; G = 2.32–2.51 (higher values are for cancrinite); yellow in hand sample, also white, gray, pink, or pale blue; white streak; vitreous luster.

COLOR: Colorless in thin section, sometimes pale yellow in grain mount.

FORM: Crystals are doubly terminated hexagonal prisms. Also commonly found as anhedral or subhedral grains.

CLEAVAGE: Perfect prismatic $\{10\bar{1}0\}$ cleavage (three directions at 60°) controls fragment orientation and is usually noticeable in thin sections. There also is a poor basal cleavage $\{0001\}$.

TWINNING: Lamellar twinning is rare.

OPTICAL ORIENTATION: Since cancrinite–vishnevite is hexagonal, the c axis is the optic axis. Cleavage fragments and the trace of prismatic cleavage in longitudinal sections are length fast, and show parallel extinction (cf. Figure 6.12).

Figure 15.22 Range of indices of refraction and birefringence of cancrinite and vishnevite.

INDICES OF REFRACTION AND BIREFRINGENCE: Indices of refraction and birefringence decrease with substitution of SO_2 for CO_3 (Figure 15.22). Birefringence of cancrinite is usually above 0.009, so maximum interference colors may range from first-order yellow to second-order blue. Vishnevite has lower birefringence and shows maximum interference colors of lower to middle first order. SO_4-rich vishnevite may be essentially isotropic.

INTERFERENCE FIGURE: Basal sections showing the three cleavages intersecting at 60° angles yield centered optic axis figures. Cleavage fragments and longitudinal sections showing only one cleavage direction yield flash figures.

ALTERATION: Cancrinite may be replaced by calcite.

DISTINGUISHING FEATURES: Most members of the series are distinguished from nepheline and the feldspars by stronger birefringence, and from scapolite by lower indices of refraction. Most scapolite has indices higher than the cement used in thin sections; check the Becke line. Basal sections of muscovite may resemble cancrinite, but muscovite has higher indices and is biaxial. Cancrinite and vishnevite can be tentatively distinguished based on the lower birefringence of vishnevite.

OCCURRENCE: Cancrinite-vishnevite is found in nepheline syenites and related alkalic intrusive rocks. It is usually associated with nepheline and calcite and has been interpreted to be produced by the breakdown of nepheline.

ZEOLITES

The zeolites are an important group of minerals that are found in large volumes in lacustrine and marine sediments, soils, altered volcanic rocks, low-grade metamorphic rocks, and as vesicle fillings in basalt and related rocks. Zeolites are hydrated aluminosilicates with a general formula:

$$M_x D_y (Al_{x+2y} Si_{n-x-2y} O_{2n}) \cdot m H_2 O$$

where M is usually monovalent Na or K, and D is Ca, Mg, or another divalent cation. There are over 40 different natural zeolites and over 100 synthetic zeolites with no natural counterparts. They all have a framework structure of tetrahedra that contains large channels and "cages" that contain the water and the M and D cations. The water is loosely held and may be driven off by heating, and the cations can be exchanged with other cations in solution under appropriate chemical conditions.

The occurrence in vesicles and vugs in basaltic volcanic rocks is probably best known to mineralogy students because this is one of the few occurrences in which the crystals are large enough to be examined either with a microscope or in hand sample. This, however, represents little more than a mineralogic curiosity, for the large majority of zeolite occurrences are in sedimentary rocks, altered volcanics, and low-grade metamorphic rocks, where the grain size is typically far too small to allow identification with a petrographic microscope. In the past, much of this zeolite has been dismissed as "clay", and only with the use X ray, scanning electron microscope, and other techniques has the extensive presence of zeolites in these rocks been recognized.

The zeolites described here include only those that are likely to be found in fairly large crystals and that are, in some sense, representative of some of the larger groups of zeolites.

The zeolites tend to have low indices of refraction (usually less than cement in thin section) and low birefringence. The optical properties of a given species may vary significantly depending on the degree of hydration, and there is considerable overlap of the optical properties among different species. Hence, it may be difficult to distinguish species based solely on their optical properties.

Analcime

Na(AlSi$_2$)O$_6$·H$_2$O
Isometric
Isotropic
$n = 1.479 – 1.493$

RELIEF IN THIN SECTION: Moderate negative relief.

COMPOSITION AND STRUCTURE: The structure consists of fourfold rings of tetrahedra that are linked together to form a cubic structure. Some Ca may replace Na in the ratio of one Ca^{2+} for two Na$^+$. A Ca-rich variety is called *wairakite*. The water content may vary widely, but as water content goes up, so does the proportion of Si^{4+} in the tetrahedral sites, presumably because there is less room for, and therefore fewer, Na$^+$ or Ca^{2+} cations whose charge must be balanced by substituting Al^{3+} for Si^{4+}.

PHYSICAL PROPERTIES: H = $5\frac{1}{2}$; G = 2.22–2.29; white, pink, or gray in hand sample, white streak; vitreous luster.

COLOR: Colorless in thin section and grain mount.

FORM: Crystals are typically well-formed trapezohedrons which, in thin section, yield eight-sided to nearly round sections. Also found as anhedral granules.

CLEAVAGE: There is very poor cubic {001} cleavage, which is not generally seen in thin section and which does not control fragment shape.

TWINNING: Lamellar twinning on the cube {001} and dodecahedral {110} faces has been reported in birefringent varieties. However, it is likely that what appears to be twinning is actually a complex pattern of intergrown isometric and tetragonal sectors produced during growth of the mineral.

INDICES OF REFRACTION AND BIREFRINGENCE: Most samples have indices around 1.487, but indices may be higher for K-bearing varieties, and lower for high-silica varieties. Analcime is commonly weakly birefringent ($\delta = 0.001$) and may display lower first-order interference colors. Wairakite is biaxial (+) or (−) with $n_\alpha = 1.498$, $n_\gamma = 1.502$, $\delta = 0.003$–0.004 and $2V_z = 80$–$105°$.

ALTERATION: Analcime may alter to pseudoleucite (a mixture of nepheline and K-feldspar) or other zeolite minerals.

DISTINGUISHING FEATURES: Analcime is very similar to leucite, which has higher indices, and sodalite, which may be slightly bluish. However, the feldspathoids are not found as vesicle fillings in basaltic rocks or in sedimentary and metamorphic rocks.

OCCURRENCE: Analcime is the only zeolite that is found as a primary mineral in igneous rocks. It is found in the groundmass in alkalic intrusive and extrusive units such as syenite, dolerite, alkalic basalts, and trachyandesite. As with other zeolites, analcime is found in lacustrine sediments, altered volcanics, deep marine sediments, and in vesicles, veins, and other voids in basaltic rocks. Wairakite is found in sediments and low-grade metamorphic rocks.

Natrolite

$Na_2(Al_2Si_3)O_{10}\cdot 2H_2O$
Orthorhombic (pseudotetragonal)
Biaxial (+)
$n_\alpha = 1.473$–1.490
$n_\beta = 1.476$–1.491
$n_\gamma = 1.485$–1.502
$\delta = 0.012$–0.013
$2V_z = 0$–$64°$

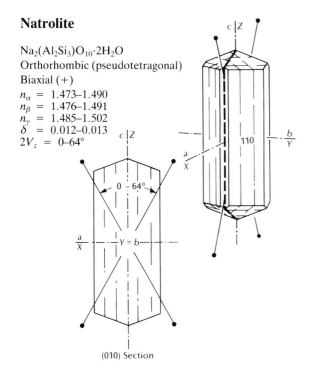

(010) Section

RELIEF IN THIN SECTION: Moderate negative relief.

COMPOSITION: There is usually some Ca and K replacing Na, and the Si:Al ratio may vary somewhat. *Scolecite* is the Ca analog of natrolite in which Ca + H_2O has replaced Na_2. *Mesolite* is an intermediate species in which the Ca: Na ratio is about 1.

PHYSICAL PROPERTIES: H = 5; G = 2.20–2.26; white or colorless in hand sample; white streak, vitreous luster.

COLOR: Colorless in thin section and grain mount.

FORM: Usually as slender prismatic crystals or fibers. Amygdules may contain radial fibrous aggregates. Cross sections of crystals are more or less square.

CLEAVAGE: Perfect {110} cleavage parallel to the two prism faces. There also is a parting parallel to {010}. Cleavage fragments are elongate parallel to the c axis.

TWINNING: Twinning on {110}, {001}, and {031} is rare.

OPTICAL ORIENTATION: $X = a$, $Y = b$, $Z = c$, optic plane = (010). In longitudinal sections, the cleavage shows parallel extinction and is length slow. Fibers and cleavage fragments also are length slow with parallel extinction. Basal sections have symmetrical extinction, but it is difficult to see since natrolite is often fibrous and has low birefringence between n_α and n_β.

INDICES OF REFRACTION AND BIREFRINGENCE: Indices of refraction increase with increasing Ca content. Maximum interference colors in thin section are first order yellow. The indices of mesolite are $n_\alpha \cong n_\beta = 1.504$–$1.508$ and $n_\gamma = 1.505$–1.509. Scolecite has indices $n_\alpha = 1.507$–1.513, $n_\beta = 1.516$–1.520, and $n_\gamma = 1.517$–1.521.

INTERFERENCE FIGURE: Most natrolite is biaxial positive with a fairly large 2V. However, uniaxial varieties have been described. Optic axis dispersion is weak with $v > r$. Cleavage fragments yield strongly off-center figures. The fibrous nature and fact that the acute bisectrix is parallel to the length of the fibers make useable figures difficult to obtain. Mesolite is biaxial (+) with $2V_z = 80°$, and Scolecite is biaxial (−) with $2V_x = 35$–$56°$.

ALTERATION: Natrolite may alter to clay or other zeolites.

DISTINGUISHING FEATURES: Stilbite usually occurs in sheaf-like rather than fibrous aggregates and is optically negative with inclined extinction. Scolecite, mesolite, and thomsonite are all fibrous, but scolecite is optically negative and has inclined extinction, and both mesolite and thomsonite have the Y indicatrix axis parallel to the length of fibers so they show both length fast and length slow character, depending on how they are oriented. In addition, mesolite shows very low birefringence (~ 0.001).

OCCURRENCE: Like the other zeolites, natrolite occurs in lacustrine and marine sediments; in vessicles, veins, and cavities in basalts, gabbros, etc., and in altered volcanic rocks.

Thomsonite

$NaCa_2(Al_5Si_5)O_{20}\cdot 6H_2O$
Orthorhombic (pseudotetragonal)
Biaxial (+)
$n_\alpha = 1.497–1.530$
$n_\beta = 1.513–1.533$
$n_\gamma = 1.518–1.544$
$\delta = 0.006–0.021$
$2V_z = 42–75°$

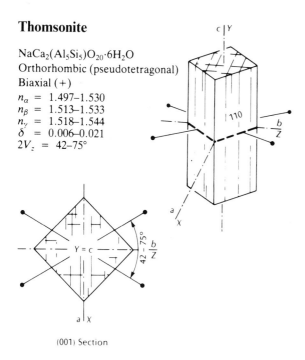

(001) Section

RELIEF IN THIN SECTION: Low negative relief.

COMPOSITION: As with the other zeolites, the ratio of Na to Ca and Al to Si may vary somewhat, and some K may replace Na or Ca.

PHYSICAL PROPERTIES: H = 5–5½; G = 2.10–2.39; white, reddish, or brownish in hand sample, also greenish; white streak; vitreous luster.

COLOR: Colorless in thin section and grain mount.

FORM: Thomsonite is usually fibrous to columnar and elongate parallel to the c axis. It is often arranged in radiating or parallel aggregates.

CLEAVAGE: Perfect cleavage on {010} and fair cleavage on {100}; intersect at 90°. Cleavage fragments are elongate parallel to the c axis and usually lie on the {010} cleavage.

TWINNING: Rare twinning on {110}.

OPTICAL ORIENTATION: $X = a$, $Y = c$, $Z = b$, optic plane = (001). The length of fragments and fibers, and the trace of cleavage in longitudinal sections show parallel extinction and are either length fast or length slow, depending on how they are oriented. Cleavage fragments lying on (010) are length slow. Basal sections are difficult to examine because of thomsonite's fibrous nature but they should show extinction that is parallel to cleavage and symmetrical to crystal outline.

INDICES OF REFRACTION AND BIREFRINGENCE: Indices of refraction increase in a general way with increasing Ca and Al. Birefringence is low to moderate (0.006–0.021). Maximum interference colors are usually middle first-order gray to yellow.

INTERFERENCE FIGURE: Thomsonite is biaxial positive with $2V_z$ between 42 and 75°. Cleavage fragments lying on (010) yield centered acute bisectrix figures. The optic plane is (001), so the melatopes lie on the N–S when the long dimension of grains and cleavage fragments is aligned on the E–W. Optic axis dispersion is moderate to strong with $r > v$. The $2V_z$ angle decreases with increasing Ca and Al content.

ALTERATION: Thomsonite may alter to clay or to other zeolites.

DISTINGUISHING FEATURES: Thomsonite is fibrous to bladed with parallel extinction and has the optic plane oriented at right angles to the length of the grains. Natrolite shows only length-slow character. Mesolite has lower birefringence. Scolecite has inclined extinction.

OCCURRENCE: Like the other zeolites, thomsonite occurs in lacustrine sediments; in vesicles, veins, and cavaties in basalt, gabbro, and serpentine; in deep marine sediments, and in hydrothermally altered volcanic rocks.

Stilbite

$NaCa_2(Al_5Si_{13})O_{36} \cdot 14H_2O$
Triclinic
$\angle\beta = 128.2°$
Biaxial $(-)$
$n_\alpha = 1.482–1.500$
$n_\beta = 1.489–1.507$
$n_\gamma = 1.493–1.513$
$\delta = 0.006–0.014$
$2V_x = 30–49°$

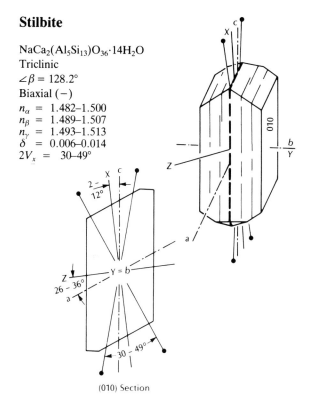

(010) Section

RELIEF IN THIN SECTION: Low to moderate negative relief.

COMPOSITION: The water content and Na:Ca and Al:Si ratios may vary somewhat, and K may substitute for Na or Ca.

PHYSICAL PROPERTIES: $H = 3\frac{1}{2}–4$; $G = 2.1–2.2$; white in hand sample, also yellowish orange or pink; white streak; vitreous to pearly luster.

COLOR: Colorless in thin section and grain amount.

FORM: Stilbite commonly forms sheaflike or radiating bundles of crystals. Individual grains are tabular parallel to (010) and elongate parallel to the c axis. A sector structure related to growth may be visible in some samples.

CLEAVAGE: A single very good cleavage parallel to {010} controls fragment orientation. A second cleavage parallel to {001} is poor.

TWINNING: Twinning on {100} is very common and may be polysynthetic or form cruciform penetration twins.

OPTICAL ORIENTATION: $X \wedge c = +2$ to $+12°$, $Y \cong b$, $Z \wedge a = +26$ to $+36°$, optic plane = (010). While the symmetry is usually considered monoclinic, the Y indicatrix axis may be inclined up to 8° from the b crystal axis indicating that the symmetry is actually triclinic for optical purposes. The optical orientation of adjacent growth sectors may be different. The trace of cleavage shows nearly parallel or slightly inclined extinction in most sections and may either be length fast or length slow. Grains are commonly elongate parallel to c, are length fast, and show a maximum extinction angle of 2 to 12° to the crystal outline.

INDICES OF REFRACTION AND BIREFRINGENCE: Indices of refraction increase somewhat with substitution of Ca for Na. Birefringence is low and maximum interference color is from first-order gray to first-order yellow. Cleavage plates show the maximum birefringence.

INTERFERENCE FIGURE: Stilbite is biaxial negative with $2V_x$ between 30 and 49°. $2V_x$ as low as 22° has been reported and $2V$ from adjacent growth sectors may not be the same. Optic axis dispersion is moderate with $v > r$. Cleavage fragments yield optic normal (flash) figures.

ALTERATION: Stilbite may alter to clay or other zeolites.

DISTINGUISHING FEATURES: The sheaflike aggregates of platy crystals and twinning are distinctive. Heulandite is optically negative and usually has lower birefringence.

OCCURRENCE: Like the other zeolites, stilbite occurs in lacustrine and marine sediments; in vesicles, veins, and cavities in basalt, gabbro, and serpentinite, and in hydrothermally altered volcanic rocks.

Chabazite

$Ca_2(Al_4Si_8)O_{24} \cdot 13H_2O$
Hexagonal (Trigonal)
Uniaxial or biaxial $(-)$ or $(+)$
$n_\alpha = 1.460–1.513$
$n_\gamma = 1.462–1.515$
$\delta = 0.002–0.010$
$2V = 0–\sim30°$

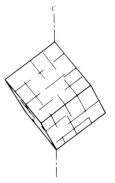

RELIEF IN THIN SECTION: Low to moderate negative relief.

COMPOSITION: There may be extensive replacement of Ca^{2+} by Na^+ or K^+, and corresponding variation in the Si:Al ratio.

PHYSICAL PROPERTIES: $H = 4\frac{1}{2}$; $G = 2.05$–2.10; white to pink in hand sample; white streak; vitreous luster.

COLOR: Colorless in thin section and grain mount.

FORM: Chabazite crystals are usually rhombohedrons that are almost like cubes since the angle between rhombohedral faces is nearly 90°.

CLEAVAGE: There are three fair to good cleavages on $\{10\bar{1}1\}$ that are parallel to the rhombohedral faces.

TWINNING: Penetration and lamellar twinning is common. Basal sections may show six segments with different optic orientation.

OPTICAL ORIENTATION: In uniaxial chabazite, the optic axis is the c crystal axis, and in biaxial varieties the acute bisectrix is probably close to the c axis. Cleavage fragments show symmetrical extinction to the nearly square outline. In thin section, extinction may range from parallel to symmetrical (cf. Figure 6.11).

INDICES OF REFRACTION AND BIREFRINGENCE: Indices of refraction decrease with increasing Na content. Birefringence is usually less than 0.005, though some samples may be as high as 0.010. Interference colors are usually first-order grays.

INTERFERENCE FIGURE: Most chabazite is uniaxial negative or biaxial negative with a small $2V$. Positive varieties are less common. Different locations on a single grain may show different interference figures. Figures are often indistinct due to low birefringence and twinning.

ALTERATION: Chabazite may alter to clay or other zeolites.

DISTINGUISHING FEATURES: The low birefringence and nearly cubic habit are distinctive but *gmelinite* and *levynite* have similar properties. Gmelinite has a basal parting and levynite may be tabular.

OCCURRENCE: Like other zeolites, chabazite is found in lacustrine and marine sediments, altered volcanic rocks; in vesicles, fractures, and voids in basalt, gabbro, and related rocks; and in hydrothermally altered rocks.

Heulandite

$Ca(Al_2Si_7)O_{18}\cdot 6H_2O$
Monoclinic
$\angle\beta = 91.5°$
Biaxial $(+)$
$n_\alpha = 1.487$–1.505
$n_\beta = 1.487$–1.507
$n_\gamma = 1.488$–1.515
$\delta = 0.001$–0.011
$2V_z = 0$–$70°$, usually $\sim 30°$

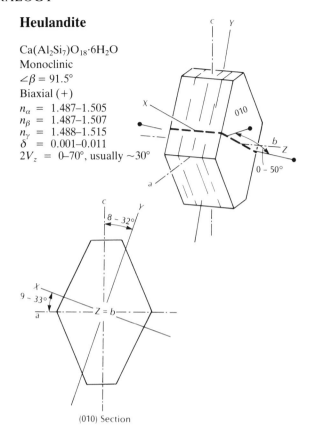

(010) Section

RELIEF IN THIN SECTION: Low to moderate negative relief.

COMPOSITION AND STRUCTURE: The structure consists of "cages" of 10 tetrahedra that are linked to form a sheetlike structure parallel to (010). K and Na may replace the Ca. The variety *clinoptilolite* has $Na^+ + Si^{4+}$ substituted for $Ca^{2+} + Al^{3+}$.

PHYSICAL PROPERTIES: $H = 3$–4; $G \cong 2.2$; white in hand sample, may be stained red or brown; white streak; vitreous to pearly luster.

COLOR: Colorless in thin section and grain mount.

FORM: Crystals are usually tabular parallel to (010) with a trapezoidal outline, or they may be nearly equant. Also found as granular aggregates.

CLEAVAGE: There is a single perfect cleavage parallel to $\{010\}$ that controls fragment orientation.

TWINNING: Not usually twinned.

OPTICAL ORIENTATION: $X \wedge a = +9$ to $+33°$, $Y \wedge c = -8$ to $-32°$, $Z = b$, optic plane is perpendicular to (010). Sections cut so that the {010} cleavage is vertical show parallel extinction, and random sections will show parallel or moderately inclined extinction to the trace of cleavage. The trace of cleavage is length fast. Sections cut parallel to (010) do not show the cleavage trace and show inclined extinction to crystal outlines.

INDICES OF REFRACTION AND BIREFRINGENCE: Indices of refraction and birefringence decrease with substitution of Na and Si for Ca and Al. Interference colors in thin section are no higher than first-order white, and are commonly lower to middle first-order gray. Clinoptilolite has $n_\alpha = 1.476–1.491$, $n_\beta = 1.479–1.493$, $n_\gamma = 1.479–1.497$, $\delta = 0.003–0.006$, and $2V_x =$ small to large. The cleavage is length slow.

INTERFERENCE FIGURE: Heulandite is biaxial positive with $2V$ usually in the vicinity of 30°, although the full range is 0 to 70°. Optic axis dispersion is $r > v$ with discernible crossed bisectrix dispersion. Cleavage fragments yield centered acute bisectrix figures. In thin section, acute bisectrix figures are found in sections parallel to (010), which show no cleavage and very low birefringence.

ALTERATION: Heulandite may alter to clay or other zeolites.

DISTINGUISHING FEATURES: Stilbite has higher birefringence and is optically negative. Laumontite has three cleavages and higher birefringence.

OCCURRENCE: Like the other zeolites, heulandite is found in lacustrine sediments, altered volcanics, hydrothermally altered rocks, and in vesicles, cavities, and veins in basalt, gabbro, serpentinite and related rocks.

Laumontite

$Ca(Al_2Si_4)O_{12} \cdot 4H_2O$
Monoclinic
$\angle\beta = 111.5°$

Biaxial ($-$)
$n_\alpha = 1.502–1.514$
$n_\beta = 1.512–1.524$
$n_\gamma = 1.514–1.525$
$\delta = 0.008–0.016$
$2V_x = 25–47°$

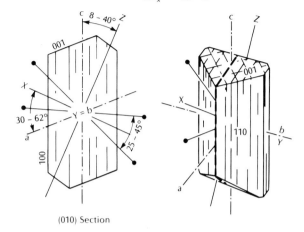

(010) Section

RELIEF IN THIN SECTION: Low negative relief.

COMPOSITION: The primary compositional variations in laumontite are in substitution of Na for Ca, and variation in the Si:Al ratio. *Leonhardite* is a partially dehydrated variety that contains $3\frac{1}{2}H_2O$, rather than $4H_2O$. Dehydration may occur on exposure to air.

PHYSICAL PROPERTIES: H $\cong$ 4; G = 2.23–2.41; colorless or white in hand sample, also yellowish or reddish; white streak; vitreous to pearly luster.

COLOR: Colorless in thin section and grain mount.

FORM: Laumontite usually forms prismatic crystals or fibrous aggregates. Grains are elongate along the c axis. It also forms granular aggregates.

CLEAVAGE: There are three perfect cleavages that produce cleavage fragments elongate parallel to the c axis. One cleavage is parallel to {010} and the other two are parallel to the {110} prism.

TWINNING: Twinning is common parallel to {100}.

OPTICAL ORIENTATION: $X \wedge a = +30$ to $+62°$, $Y = b$, $Z \wedge c = -8$ to $-40°$, optic plane = (010). The $Z \wedge c$ extinction angle of 8 to 40° is seen in sections cut parallel to (010) and fragments lying on the {010} cleavage. The slow ray is closer to the trace of cleavage (length slow). Extinction is parallel to the {010} cleavage and symmetrical to the {110} cleavages in basal sections. The extinction angle for leonhardite is usually greater than 45° to the slow ray in (010) sections (length fast).

INDICES OF REFRACTION AND BIREFRINGENCE: Indices decrease with dehydration and with increasing Na content. Birefringence is weak to moderate and maximum interference color in thin section is first-order white to first-order red. Sections parallel to (010) that show the maximum interference color also show the $Z \wedge c$ extinction angle.

INTERFERENCE FIGURE: Laumontite is biaxial negative with an intermediate $2V_x$ (25–47°). Dehydration to leonhardite causes a small decrease in $2V_x$. Optic axis dispersion is strong with $v > r$. Fragments lying on cleavages yield either flash figures or strongly eccentric figures.

ALTERATION: Laumontite may alter to other zeolites.

DISTINGUISHING FEATURES: Heulandite has only one cleavage and lower birefringence. Stilbite is length fast and has only one cleavage. Natrolite and thomsonite show parallel extinction. Chabazite forms more or less equant crystals and has lower birefringence and $2V_x$.

OCCURRENCE: Like the other zeolites, laumontite is found in lacustrine and marine sediments, altered volcanic rocks, low-grade metamorphic rocks, and in cavities and fractures in basaltic volcanics, gabbro, serpentinites and related rocks.

OTHER TECTOSILICATES

Scapolite

| Marialite | $Na_4Al_3Si_9O_{24}Cl$ |
| Meionite | $Ca_4Al_6Si_6O_{24}CO_3$ |

Tetragonal
Uniaxial ($-$)
n_ω = 1.532–1.607
n_ϵ = 1.522–1.571
δ = 0.004–0.038

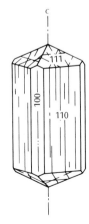

RELIEF IN THIN SECTION: Low to moderate positive relief.

COMPOSITION AND STRUCTURE: The scapolite structure contains fourfold rings of tetrahedra that are joined into a framework structure by sharing oxygens with tetrahedra from adjacent rings. It contains open cavities containing the Cl or CO_3 surrounded by four Na or Ca. There is continuous solid solution between the marialite (Ma) and meionite (Me) end members, with most samples having compositions between about Me_{20} to Me_{80}. The conventional nomenclature is marialite ($Me_{0–20}$), dipyre ($Me_{20–50}$), mizzonite ($Me_{50–80}$), and meionite ($Me_{80–100}$). Small amounts of F and SO_4 may replace the Cl or CO_3, and K may replace some Na or Ca. Compositional zoning may be expressed by variation in birefringence.

PHYSICAL PROPERTIES: H = 5–6; G = 2.50–2.78; white, gray, or greenish in hand sample, less commonly pale shades of pink, yellow, and blue; white streak; vitreous luster.

COLOR: Colorless in thin section and grain mount.

FORM: Crystals are prismatic or occasionally acicular or bladed and elongate along c. Scapolite forms anhedral grains, granular clusters, columnar aggregates, or strongly poikilitic grains enclosing associated minerals.

CLEAVAGE: There are two good cleavages parallel to the {100} prism faces, and two fair cleavages are parallel to the {110} prism faces. The cleavages can be seen intersecting at 45° angles in basal sections, but longitudinal sections show a single cleavage trace. Cleavage tends to control fragment shape, which is usually elongate parallel to the c axis.

TWINNING: None reported.

OPTICAL ORIENTATION: Scapolite is tetragonal, so the c axis is the optic axis. The trace of cleavage shows parallel extinction in longitudinal sections and is length fast. Elongate cleavage fragments have parallel extinction and are length fast.

INDICES OF REFRACTION AND BIREFRINGENCE: The indices of refraction and birefringence increase systematically with increasing meionite component (Figure 15.23). The mean index $[(n_\omega + n_\epsilon)/2]$ can be used to estimate the composition to within ±10 per-

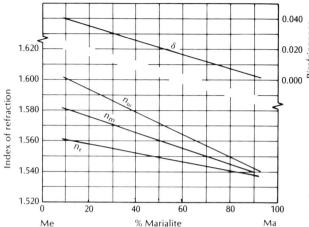

Figure 15.23 Variation of indices of refraction and birefringence (δ) with composition in scapolite. The curve of the mean index of refraction (n_m) provides the most reliable estimate of composition. After Shaw (1960), Ulbrich (1973), and Graziani and Lucchesi (1982).

cent Ma. The birefringence can be used to roughly estimate composition in thin section.

INTERFERENCE FIGURE: Scapolite is uniaxial negative, although biaxial samples with a small $2V$ ($< 10°$) have been reported.

ALTERATION: Scapolite may be replaced by aggregates containing sericite, calcite, chlorite, epidote, zeolites, or other minerals.

DISTINGUISHING FEATURES: Scapolite is distinguished from the feldspars because it is uniaxial and has different cleavage, and from quartz because it is optically negative and usually has higher birefringence. Cordierite is biaxial. Cancrinite has indices less than cement in thin section, and nepheline has lower birefringence and less well-developed cleavage.

OCCURRENCE: Scapolite is found in both contact and regional metamorphic rocks derived from calcareous sediments or from gabbroic or similar rocks. In skarn deposits and marble, scapolite is often associated with garnet, diopside, actinolite, titanite, calcite, and other calc-silicate minerals. In amphibolite, it is associated with hornblende, calcic clinopyroxene, epidote, and titanite. In altered mafic rocks it may replace plagioclase. Scapolite is also found in granulites, and less commonly in pelitic and psammitic regional metamorphic rocks. It occasionally occurs in veins cutting regional metamorphic rocks or epizonal intrusives.

Beryl

$Be_3Al_2(SiO_3)_6$
Hexagonal
Uniaxial ($-$)
$n_\omega = 1.560–1.608$
$n_\epsilon = 1.557–1.599$
$\delta = 0.003–0.009$

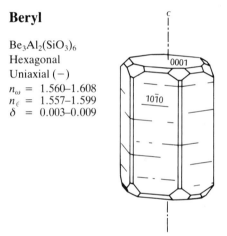

RELIEF IN THIN SECTION: Low to moderate positive relief.

COMPOSITION AND STRUCTURE: Beryl's structure consists of sixfold rings of silicon tetrahedra that are stacked in columns parallel to the c axis. The stacks of tetrahedra rings are held together laterally and vertically by bonding through Al in sixfold coordination, and Be in distorted fourfold coordination with oxygen on the outside of the rings.

The primary compositional variation is in the presence of alkali ions (Li, Na, K, and Cs are most common). Li is sufficiently small that it can substitute for both Al^{3+} and Be^{2+}, but the others are probably located in open channels in the center of the rings. The charge surplus associated with the cations in the channels is probably balanced by substituting Fe^{2+}, Li^+, Mg^{2+}, or other cations for Al^{3+}, or by leaving vacancies in the Be sites. In general, the addition of alkalis correlates with a decrease in the amount of Be, but there is substantial variability due to the variety of substitutions that are possible. Water also may be present in the channels. Trace amounts of Cr and Mn may substitute for Al and are probably responsible for the color of emerald (dark green) and morganite (pink), respectively.

Concentric zoning is common with the core of crystals usually having lower alkali content than the rim.

PHYSICAL PROPERTIES: $H = 7\frac{1}{2}–8$; $G = 2.66–2.92$ (higher with increasing alkalis); usually pale green, white, or yellow in hand sample, also blue, green, or pink; white streak; vitreous luster.

COLOR AND PLEOCHROISM: Usually colorless in thin section or grain mount, although the strongly colored varieties may be weakly colored in thicker sections or grain mounts, with pleochroism $\omega < \epsilon$; for example, ω = light blue, ϵ = blue or ω = yellowish green, ϵ = green. Longitudinal sections through elongate prismatic crystals are darker when the long dimension is parallel to the vibration direction of the lower polar.

FORM: Prismatic hexagonal crystals are relatively common. Basal sections are usually hexagonal and longitudinal sections are rectangular, although some crystals are tapered with a larger diameter at one end than at the other. Also found as tabular hexagonal prisms or as anhedral or interstitial grains. Small needles may be arranged in radiating masses. Tiny inclusions may be distributed in bands parallel to crystal faces.

CLEAVAGE: There is a fair to poor basal {0001} cleavage that is not usually seen in thin section.

TWINNING: Twinning is rare but has been reported on {$31\bar{4}1$} and {$40\bar{4}1$}.

OPTICAL ORIENTATION: Longitudinal sections parallel to the c axis are length fast and show parallel extinction.

INDICES OF REFRACTION AND BIREFRINGENCE: There appears to be a general increase in the indices and birefringence with an increase in the amount of alkalis. However, the scatter in the data is relatively large, so indices and birefringence cannot be used as a reliable guide to the amount of alkalis present. The range of indices for common beryl are $n_\omega = 1.680 \pm 0.005$ and $n_\epsilon = 1.675 \pm 0.005$, but higher or lower indices may be found and the full range is given above. The birefringence is low and yields first-order gray or white colors in thin section. Chemical zoning is usually expressed as an increase in birefringence from the center to the edge of grains.

INTERFERENCE FIGURE: In most samples, basal sections yield centered uniaxial figures. Strain in the crystal structure may produce biaxial beryl with $2V_x$ up to 17° and variable optic orientation. Grains may have biaxial cores and uniaxial rims or vice versa.

ALTERATION: The usual alteration products are sericite or clay minerals.

DISTINGUISHING FEATURES: Both quartz and nepheline resemble beryl, although beryl has slightly higher indices and relief. Quartz is optically positive, and in most cases, nepheline has lower indices than the cement used in thin section (check the Becke line).

Nepheline is also very rare in the quartz-rich granite and pegmatite in which beryl is usually found. Apatite has significantly higher relief in thin section.

OCCURRENCE: Beryl is common in granitic pegmatites and is usually associated with quartz, alkali feldspar, albite, muscovite, biotite, and tourmaline. In granite, beryl is less common but not unusual. Beryl is also found in nepheline syenite or in high-temperature hydrothermal veins associated with tin and tungsten minerals. Contact metasomatism of gneiss, schist, or carbonate rocks may rarely produce beryl.

Cordierite

$Mg_2Al_3(AlSi_5)O_{18}$
Orthorhombic (pseudohexagonal)
Biaxial $(-)$ or $(+)$
$n_\alpha = 1.521–1.561$
$n_\beta = 1.524–1.574$
$n_\gamma = 1.527–1.578$
$\delta = 0.005–0.016$
$2V_x = 40–90°$, less commonly positive $2V_z = 90–75°$

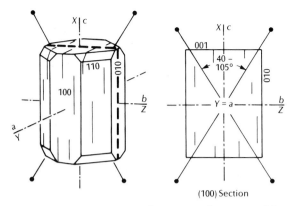

(100) Section

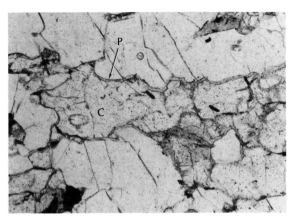

Figure 15.24 Cordierite (C) rimmed with pinite (P). Note the similarity to surrounding quartz. The slightly darker color of the cordierite is due to numerous fine opaque inclusions. Field of view is 1.7 mm wide.

RELIEF IN THIN SECTION: Low negative or positive relief.

COMPOSITION AND STRUCTURE: Cordierite's structure is essentially the same as that of beryl and consists of sixfold rings of tetrahedra that are stacked to form columns parallel to the c axis. One out of the six tetrahedra in each ring is occupied by Al, the remainder by Si. The columns of rings are tied together laterally and vertically by Mg in octahedral coordination and Al in tetrahedral coordination with the oxygen atoms on the outside of the rings. The high-temperature polymorph *indialite* has hexagonal symmetry, like beryl. A slight distortion at low temperature, caused by ordering of the Si and Al in the rings, reduces the symmetry to orthorhombic in cordierite.

The principal compositional variation in cordierite is substituting Fe^{2+} for Mg, and there is complete solid solution between the iron and magnesium end members, although most natural cordierite is magnesium-rich. Minor amounts of Mn or Ti may substitute for Mg, and Fe^{3+} may replace small amounts of Al. The channels in the center of the rings may contain variable amounts of H_2O, CO_2, or Ar and small amounts of large alkali ions, usually Na or K.

PHYSICAL PROPERTIES: H = $7-7\frac{1}{2}$; G = 2.53–2.78; usually gray, gray-blue, blue, or indigo blue in hand sample; white streak; vitreous luster.

COLOR AND PLEOCHROISM: Usually colorless in thin section or grain mount. Thicker sections or grain mounts of iron-rich samples may be light blue and pleochroic with $Z > Y > X$: X = colorless, pale yellow, pale green; Y = pale blue; Z = pale blue to violet. There may be yellow pleochroic halos around radioactive inclusions.

FORM: Figure 15.24. Commonly found as anhedral grains or irregular porphyroblastic grains, often with numerous inclusions of quartz, opaque grains, or other minerals. Less commonly as euhedral hexagonal prismatic crystals.

CLEAVAGE: There is a single fair cleavage on {010} and poor cleavages on {100} and {001}. The cleavage is not usually obvious in thin section.

TWINNING: Lamellar and cyclic twinning is common. The usual twin planes are {110} and {130}. The twinning may resemble plagioclase twins, although basal sections may display a radial or cyclic pattern with roughly hexagonal symmetry. Cordierite from high-grade metamorphic rocks is more commonly twinned than that from lower-grade rocks.

OPTICAL ORIENTATION: Most cordierite is biaxial with $X = c$, $Y = a$, $Z = b$, optic plane = (100). Indialite is hexagonal and uniaxial. Extinction is parallel to the cleavage traces and can range from parallel to either 30 or 60° to the composition planes of the twins, depending on orientation. Neither fragments nor the common habits have a dominant long dimension, and the cleavages are usually poorly developed, so the sign of elongation usually cannot be determined.

INDICES OF REFRACTION AND BIREFRINGENCE: There is a general increase in indices of refraction and birefringence with substitution of Fe^{2+} and Mn for Mg, but with considerable variability due to complications associated with variable content of the channels. The presence of water causes an increase in the indices of refraction. Dehydrating cordierite by strongly heating it may decrease the indices by up to 0.02, and dehydrated cordierite shows a very systematic increase in indices with increasing Fe and Mn content (Selkregg and Bloss, 1980).

INTERFERENCE FIGURE: Most cordierite is biaxial negative with $2V_x$ between about 40 and 90°. High water and Na content cause a decrease in the $2V_x$ angle. Some cordierite is biaxial positive with $2V_z$ between 80 an 90°, occasionally as low as 75°. Optic axis dispersion is weak, usually $v > r$, less commonly $r > v$. Indialite is uniaxial negative.

ALTERATION: Cordierite is readily altered to *pinite*, which is a fine-grained greenish or yellowish aggregate of chlorite, muscovite, and other silicates. The alteration usually develops along cracks and grain margins and may leave ragged remnants of cordierite in a matrix of pinite.

DISTINGUISHING FEATURES: Untwinned cordierite is very much like both quartz and orthoclase, and distinguishing between these minerals may give difficulty. Quartz is uniaxial, does not have alteration production like pinite, and in many metamorphic rocks displays undulatory extinction due to strain, which is unlike that displayed by cordierite. Orthoclase has more distinct cleavage and has indices that are lower than the cements usually used in thin sections, while cordierite often has higher indices (check the Becke line). Twinned cordierite may be mistaken for plagioclase, but basal sections sometimes display a radial pattern of twins, which plagioclase does not develop, and plagioclase has more distinct cleavage. Cordierite also may have fine, dustlike opaque inclusions or numerous rounded inclusions of quarts, and commonly shows yellowish halos around included zircon grains.

Indialite is uniaxial (−), with $n_\omega \cong 1.578$, $n_\epsilon \cong 1.560$, but otherwise has similar properties. *Osumilite* is a closely related mineral $[(K,Na,Ca)(Mg,Fe)_2(Al,Fe)_3(Si,Al)_{12}O_{30}]$ found in cavities in felsic volcanics, contact metamorphic aureoles, and granulite facies regional metamorphic rocks. Osumilite is hexagonal with indices $n_\omega = 1.540–1.547$, $n_\epsilon = 1.546–1.551$, and is uniaxial positive. It is usually pink or blue in thin section and may be anomalously biaxial with $2V$ up to 15°.

Thin sections may be stained to distinguish cordierite from the feldspars (Boone and Wheeler, 1968).

OCCURRENCE: Cordierite is a common mineral in medium and high-grade pelitic regional and contact metamorphic rocks. In hornfels of contact metamorphic zones, cordierite is common as porphyroblasts. In regional metamorphic rocks, it may form either porphyroblasts or anhedral grains along with quartz and feldspar. Associated minerals may include chlorite, andalusite, kyanite, sillimanite, staurolite, muscovite, biotite, and chloritoid. Cordierite is also common in certain mafic metamorphic rocks associated with anthophyllite and sometimes garnet. Cordierite is infrequently found in granite, pegmatite, members of the gabbro clan, or (rarely) in andesite and related volcanic rocks.

REFERENCES

Bailey, E. H., and Stevens, R. E., 1960, Selective staining of K-feldspar and plagioclase on rock slabs and thin sections: American Mineralogist, v. 45, p. 1020–1025.

Boone, G. M., and Wheeler, E. P. II, 1968, Staining for cordierite and feldspars in thin section: American Mineralogist, v. 53, p. 327–331.

Burri, C., Parker, R. L., and Wenk, E., 1967, Die optische orientierung der plagioklase: Birkauser Verlag, Basel, 333 p.

Frondel, C., 1982, Structural hydroxyl in chalcedony (type B quartz): American Mineralogist, v. 67, p. 1248–1257.

Graziani, G., and Lucchesi, S., 1982, Thermal behavior of scapolites: American Mineralogist, v. 67, p. 1229–1241.

Houghton, H. F., 1980, Refined techniques for staining plagioclase and alkali-feldspars in thin section: Journal of Sedimentary Petrology, v. 50, p. 629–631.

Kokta, J., 1930, On some physico-chemical properties of opal and their relation to artificially prepared amorphous silicic acid: Rozpravy České Akad., v. 40, no. 21.

Morse, S. A., 1968, Revised dispersion method for low plagioclase: American Mineralogist, v. 53, p. 105–115.

Ruperto, V. L., Stevens, R. E., and Normans, M. B., 1964, Staining of plagioclase feldspar and other minerals with F. D. and C. Red No. 2: U. S. Geological Survey Professional Paper 501B, p. B-152–153.

Selkregg, K. R., and Bloss, F. D., 1980, Cordierites: compositional control of Δ, cell parameters and optical properties: American Mineralogist, v. 65, p. 522–533.

Shand, S. J., 1939, On staining of feldspathoids and on zonal structure of nepheline: American Mineralogist, v. 24, p. 508–513.

Shaw, D. M., 1960, The geochemistry of scapolite, Part I, Previous work and general mineralogy: Journal of Petrology, v. 1, p. 218–260.

Smith, J. R., 1958, The optical properties of heated plagioclase: American Mineralogist, v. 43, p. 1179–1194.

Stewart, D. B., and Ribbe, P. H., 1983, Optical properties of feldspars. In: Ribbe, P. H., (ed.), Feldspar Mineralogy, second edition, Reviews in Mineralogy, v. 2A. p. 121–139. Mineralogical Society of America, Washington, D. C.

Su, S. C., Bloss, F. D., Ribbe, P. H., and Stewart, D. B., 1984, Optic axial angle, a precise measure of Al,Si ordering in T_1 tetrahedral sites of K-rich alkali feldspars: American Mineralogist, v. 69, p. 440–448.

Tobi, A. C., and Kroll, H., 1975, Optical determination of the An-content of plagioclases twinned by the Carlsbad-law: A revised chart: American Journal of Science, v. 275, p. 731–736.

Ulbrich, H. H., 1973, Crystallographic data and refractive indices of scapolites: American Mineralogist, v. 58, p. 81–92.

Appendix A: Sample Preparation

Grain Mount

A grain mount consists of a number of small grains of an unknown mineral placed on a glass slide, immersed in a liquid, and covered with a coverslip. The basic steps involved in preparing a grain mount are as follows:

1. Crush a sample of an unknown mineral in a mortar and pestle or other apparatus.
2. Sieve the sample and retain for examination the fraction that passes a 140-mesh sieve and does not pass a 200-mesh sieve. These grains will be between 0.105 and 0.074 mm. The finer material may be saved for X-ray examination and the coarser material recrushed if needed. Less than a gram of sample is all that is needed in most cases.
3. Sprinkle a few dozen grains of the unknown mineral on a microscope slide and cover with a piece of cover slip. One cover slip may be broken into enough pieces to prepare a half a dozen or more grain mounts. Avoid getting fingerprints on the cover slip or glass slide.
4. With a dropper, place a small amount of immersion oil next to the cover slip so that capillary action draws it under and immerses the grains. The usual error at this point is to use too much oil or to get oil on top of the cover slip.
5. Place the sample on the microscope stage and examine. Do not get immersion oils on the lenses or other parts of the microscope.

Other more sophisticated techniques that involve mounting the grains in various resins or gels also are available. Consult Hutchison (1974) for details of these procedures.

Thin Section

Thin sections are slices of rock about 0.03 mm thick that are mounted on microscope slides. A wide variety of machinery is available to speed various steps in the process and improve accuracy and consistency. The basic process is as follows:

1. A chip of rock is cut out of a hand sample with a diamond saw. The chip should be cut so that it is no larger than a microscope slide.
2. The side of the rock chip that will be glued to the slide is ground flat on a lap wheel and then polished using progressively finer and finer abrasive. The usual procedure is to start with 120-grit and then move progressively to 240-, 320-, 400-, and 600-grit abrasive. Finer abrasive may be used to produce a higher-quality polish, but this is not generally necessary for routine work. It is important at this point to ensure that the polished surface is as flat as possible. If it is not flat, the usual culprits are the uneven application of pressure while grinding, or lap wheels that are not flat.
3. The rock chip is then glued to a clean microscope slide. The details of the procedure depend on which type of cement is used. Canada balsam is the traditional cement, but a number of epoxies and other cements with superior qualities are now used for most work. The glue line needs to be thin and uniform, so be sure the slide and rock chip are free of grit and dust, and use the least possible cement. One or two drops are generally more than adequate for a conventional thin section. Bubbles trapped in the cement are a persistent problem. One technique for avoiding them is to place the cement in the middle of the rock chip and then place the slide on the cement. As the cement spreads outward it will, with luck, expel the air between slide and rock chip. If bubbles are trapped they can sometimes be worked out by judicious poking and prodding on the slide with the eraser end of a pencil.
4. Once the cement is cured, the surplus rock is cut off with a diamond trim saw.
5. The rock chip is now ground down to its final thickness of about 0.03 mm. A wide range of equipment is available to do this grinding. The simplest is just a glass plate on which water and abrasive are sprinkled. Lap wheels and other more sophisticated equipment also can be used. Usually, the final grinding to just the right thickness must be done by hand with fine (600

grit) abrasive on a glass plate. The thickness of the thin section can be determined by examining the interference color of quartz, feldspar, or other minerals whose birefringence is known (see Chapter 5).

6. If desired, the thin section can be stained for K-feldspar, plagioclase, carbonate minerals, and others. Hutchison (1974) describes a variety of staining techniques.

7. The cover slip is now cemented in place. Because there are many places to make errors, the novice thin section maker can anticipate that a frustratingly high percentage of his or her initial attempts will wind up in the trash.

Spindle Stage

The spindle stage (Figure 2.6) consists of a fine wire on which a single mineral grain is cemented. The spindle is mounted on a base plate so that it can pivot about a horizontal axis while holding the grain in immersion oil between a glass window and a cover slip. The spindle stage is attached to the microscope stage with stage clips or a mechanical stage.

A mineral grain is mounted on the spindle stage as follows:

1. Select a fragment of the mineral to be examined which is about 0.2 mm in diameter. Larger or smaller grains may be used if needed.

2. Place a very small amount of adhesive on the tip of the spindle by gently touching it to a drop of adhesive.

3. While the adhesive is still wet, touch the spindle to the grain and align the grain so that it is centered as closely as possible on the tip of the spindle. Allow the adhesive to cure.

4. Insert the spindle into the spindle stage. Place a cover glass over the grain on the cover glass supports and introduce a drop or two of immersion oil under the cover glass.

Consult Bloss (1981) for a more detailed discussion of the spindle stage.

REFERENCES

Bloss, F. D., 1981, The spindle stage: principles and practice: Cambridge University Press, Cambridge and New York.

Hutchison, C. S., 1974, Laboratory handbook of petrographic techniques: Wiley, New York, 527 p.

Appendix B: Ray Velocity Surfaces

For purposes of optical mineralogy, the indicatrix is the most convenient way of illustrating optical properties. However, ray velocity surfaces also can be used and conveniently illustrate the relationship between rays, wave fronts, and wave normals.

Isotropic Ray Velocity Surface

Consider a piece of an isotropic mineral like halite, which has a point source of light embedded in it (Figure B.1). When the light is flashed on, rays of light radiate out from the source with the same velocity in all directions. If the velocity of the light is plotted along each ray, the ends of these velocity vectors outline a sphere called the ray velocity surface whose radius is proportional to the velocity of light in the material. The wave front associated with any ray is tangent to the ray velocity surface at the point where the ray pierces the surface, so for the point source of light shown in Figure B.1 the wave

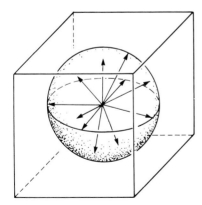

Figure B.1 Isotropic ray velocity surface. Since the velocity of light is the same in all directions, the isotropic ray velocity surface is a sphere whose radius is proportional to the velocity of the light.

fronts are spheres expanding outward from the light source. A good two-dimensional analog can be produced by dropping a pebble into a pond. Waves radiate outward from the point of impact with uniform velocity in all directions. The wave fronts, which mimic the ray velocity surface, are circular.

Uniaxial Ray Velocity Surface

In uniaxial minerals, there are two rays to contend with so two surfaces are formed: one for the ordinary ray and one for the extraordinary ray. Because the ordinary ray has uniform velocity in all directions through the crystal, its ray velocity surface is a sphere. The ray velocity surface of the extraordinary ray is a spheroid of revolution whose axis coincides with the optic axis. If the mineral is optically positive, the velocity of the extraordinary ray is less than the ordinary ray so the extraordinary ray surface is a prolate spheroid (a sphere that has been stretched out in one direction). The prolate spheroid nests inside the ordinary ray sphere and comes in contact where the optic axis (*c* axis) pierces both of them (Figure B.2*a*). If the mineral is optically negative, the velocity of the extraordinary ray is higher than the ordinary ray so the extraordinary ray surface is an oblate spheroid (a sphere that has been flattened along one axis). The ordinary ray sphere nests inside the oblate spheroid and they come in contact where the optic axis (*c* axis) pierces both of them (Figure B.2*b*). The radius of either surface is proportional to the velocity of light rays parallel to the radius.

Biaxial Ray Velocity Surface

The biaxial ray velocity surface (Figure B.3) has limited value because it is a rather complex figure. Consider light rays radiating out from the origin in

Figure B.3a to form the ray velocity surface. Every path in the X–Y plane has one ray vibrating parallel to the Z axis with velocity v_γ ($v_\alpha > v_\beta > v_\gamma$) so the slow ray forms a circular section with radius v_γ. The other ray vibrates in the X–Y plane with velocities ranging from v_α (propagating along Y) to v_β (propagating along X) so the fast ray forms an elliptical section with axes v_α and v_β.

In the Y–Z plane the fast ray always vibrates parallel to X, so it forms a circular section with radius v_α. The slow ray forms an elliptical section with axes v_β and v_γ with velocities ranging from v_β (propagating along Z) to v_γ (propagating along Y).

In the X–Z plane, one ray always vibrates parallel to Y with velocity v_β and yields a circular section.

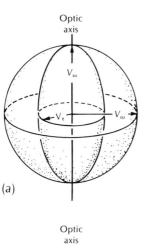

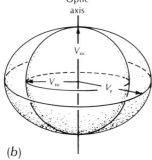

Figure B.2 Uniaxial ray velocity surfaces. (a) Uniaxial positive. The ordinary ray velocity surface is a sphere with radius v_ω, the velocity of the ordinary rays. The extraordinary ray velocity surface is a prolate spheroid whose long axis is v_ω and whose short axis is v_ϵ, the minimum velocity for the extraordinary ray. (b) Uniaxial negative. The ordinary ray sphere is nested inside the oblate spheroid defined by the extraordinary rays.

The other ray vibrates in the X–Z plane with velocities ranging from v_γ to v_α, so the second section is an ellipse with axes v_γ and v_α, so the second section is an ellipse with axes v_γ and v_α. At point I (Figure B.3b) the two sections cross and define a secondary optic axis, along which both rays have the same velocity.

Figure B.3b shows only one octant of the ray velocity surface. If the remaining parts of the surfaces are filled in, two nested surfaces like that shown in Figure B.3c are formed. The slow ray surface on the inside comes in contact with the fast ray surface at the points where the secondary optic axes penetrate the figure. A dimple on the fast ray surface marks the point where it touches the slow ray surface, which has a corresponding conical bump.

The secondary optic axes identified in the ray velocity surface do not coincide with the primary optic axes which are perpendicular to the circular sections of the indicatrix. Figure B.4 illustrates the relationship between the primary and secondary optic axes. The secondary optic axis lies in the center of the cone of light produced by internal conical refraction (Figure 7.8). The locations of the primary optic axes (OA_1) are shown in Figure B.3c. The secondary optic axes have little practical importance in optical mineralogy. Their orientation can be determined only by calculation, or graphically as shown in Figure B.4. The mineral behaves as though it is essentially isotropic only if a primary optic axis is vertical on the stage of the microscope. If the secondary optic axis is vertical, the mineral displays some birefringence, although it typically is very weak since the angle between the primary and secondary optic axes is generally quite small in most minerals. Even if the birefringence is very strong, the angle between the two sets of axes is rarely more than a few degrees.

Use of Ray Velocity Surfaces

A ray velocity surface can, in principle, be used to solve the same optical problems as an indicatrix, although the indicatrix is substantially more convenient. The relationship between rays, wave fronts, and wave normals will be illustrated for two cases: normal incidence in a principal section and inclined incidence in a principal section. Recall that wave fronts are surfaces that connect similar points on adjacent waves (Figure 1.5). The vibration direction of the light is always parallel to the wave front. The wave normal is a line at right angles to the wave front. The ray is the direction of propagation of the light energy. In isotropic materials, the ray and wave normal are parallel. In anisotropic materials however, the ray and wave normal are not generally parallel.

Normal Incidence in a Principal Section

Consider light rays traveling from the origin as shown in Figure B.5. The ordinary ray touches its ray velocity surface at point o and the extraordinary ray touches its surface at point e. Tangents to the surfaces at those points are parallel to the wave fronts for the ω and ϵ rays, respectively. Because

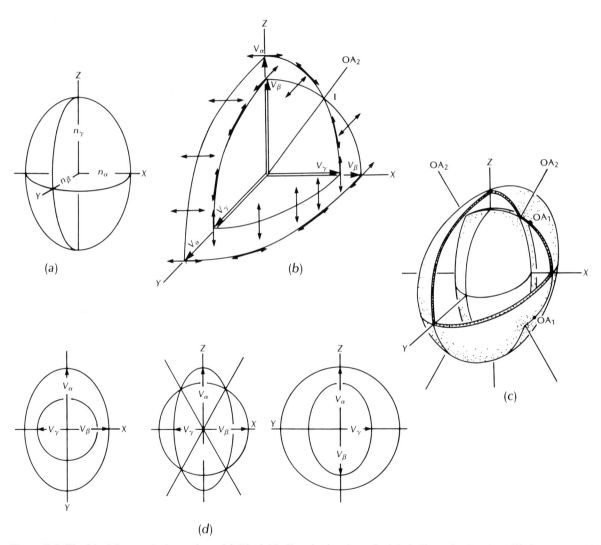

Figure B.3 The biaxial ray velocity surface. (*a*) Biaxial indicatrix showing principle indices of refraction. (*b*) One octant of the ray velocity surface showing vibration directions for light propagating within the three principle planes. OA_2 is a secondary optic axis. (*c*) Sketch showing the nested slow and fast ray surfaces. The two surfaces come in contact where they are pierced by the secondary optic axes. The points where the primary optic axes pierce the figure are labeled OA_1. (*d*) The three principal sections of the ray velocity surfaces.

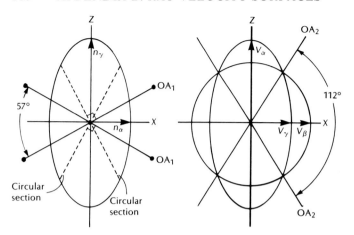

Figure B.4 Relationship between primary and secondary optic axes. (*a*) Optic plane of a biaxial indicatrix where $n_\alpha = 1.375$, $n_\beta = 2.000$, and $n_\gamma = 2.750$. $2V_x$ is 57° (cf. Figure 7.3 or Equation 7.3). (*b*) *X–Z* plane of the biaxial ray velocity surface. The angle between the secondary optic axes (OA$_2$) is 112°. The example shown here is extreme. The primary and secondary optic axes are rarely more than a few degrees apart in most minerals.

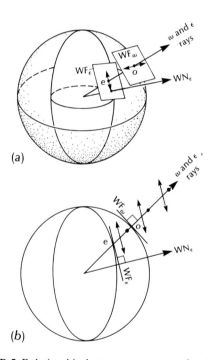

(a)

(b)

Figure B.5 Relationship between rays, wave fronts, and wave normals. Wave fronts are tangent to the points where the rays penetrate the ray velocity surfaces. The ordinary wave normal is parallel to the ordinary ray but not parallel to the extraordinary wave normal. (*a*) Perspective view. (*b*) Cross section through the ray velocity surfaces. WF$_\omega$ and WF$_\epsilon$ are the wave fronts of the ω and ϵ waves, and WN$_\epsilon$ is the ϵ wave normal.

the ordinary ray velocity surface is a sphere, the ordinary wave normal coincides with the ray. The vibration direction of light is parallel to the wave front and at right angles to the ordinary ray. This is the same behavior expected of light passing through an isotropic material. The extraordinary ray behaves differently. The wave front associated with the ϵ ray is inclined to the ray, and the ϵ wave normal and the ray are not parallel. The vibration direction of the ϵ ray is parallel to the wave front and at an angle to the ray. If the wave fronts of the ordinary and extraordinary rays are parallel, then the two rays cannot be parallel unless they are propagating at right angles to the optic axis. This observation provides the basis for explaining why two images are produced by a piece of calcite.

Consider the wave fronts passing through a calcite rhomb, shown in Figure B.6. When the light enters the calcite it is split into ω and ϵ components. Because the angle of incidence is 0°, neither wave fronts nor wave normals are bent on entering the mineral. Ray velocity surfaces are constructed at points *a* and *a'* where selected rays enter the mineral. Ray directions are found by constructing tangents to the ray velocity surfaces that are parallel to the wave fronts. For the ordinary rays, the tangents are at points *b* and *b'*, so the ordinary rays pass undeflected through the mineral with vibration direction in and out of the page as shown. Tangents for the extraordinary wave fronts are at points *c* and *c'*. The only rays that can produce these wave fronts are parallel to *a–c* and *a'–c'*, which are inclined to the ordinary ray path, and which vibrate parallel to

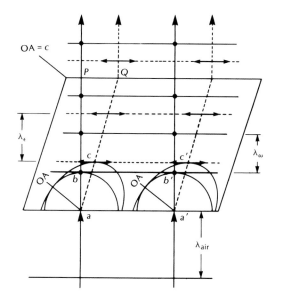

Figure B.6 Use of ray velocity surfaces to explain double refraction or calcite. See text for discussion.

the wave front, as shown. The angle between the rays has been exaggerated for clarity. It is actually 6°14′ in calcite. Recall that the light energy, and therefore the image, follows the rays. If a dot is placed at a, the ordinary ray image exits the top of the calcite at P, which is directly above a, and the extraordinary ray image exits the crystal at Q.

Inclined Incidence in a Principal Section

Inclined incidence in a principal section is illustrated for the X–Y section of a biaxial mineral in Figure B.7. Light with wavelength λ is incident at an angle of 30° to the normal. The principal section through the ray velocity surface is constructed at point O where one wave front meets the mineral. The axes of the elliptical section are λ/n_α and λ/n_γ. The radius of the circular section is λ/n_β. Recall that wavelength divided by index of refraction is proportional to velocity. Tangents are drawn to the two surfaces in the section from point O' where a second wave front impinges on the mineral. The tangents are the refracted wave fronts and the lines going through the points of tangency (a and b) are the ray paths. The wave normals (WN_a and WN_b) are perpendicular to their respective wave fronts. In this case, the slow ray and wave normal are parallel and the fast ray and wave normal are not. Note that this method of determining the refraction of light incident in one of the principal sections gives the same answers as the more mathematical approach of solving the same problem using the indicatrix (Figure 7.10).

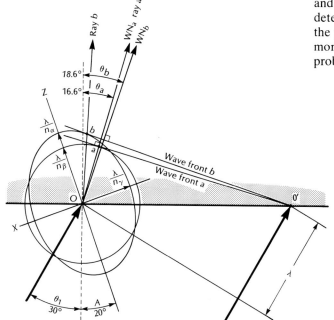

Figure B.7 Use of the biaxial ray velocity surface to determine refraction of light with inclined incidence in the X–Z principal section.

Appendix C: Identification Tables

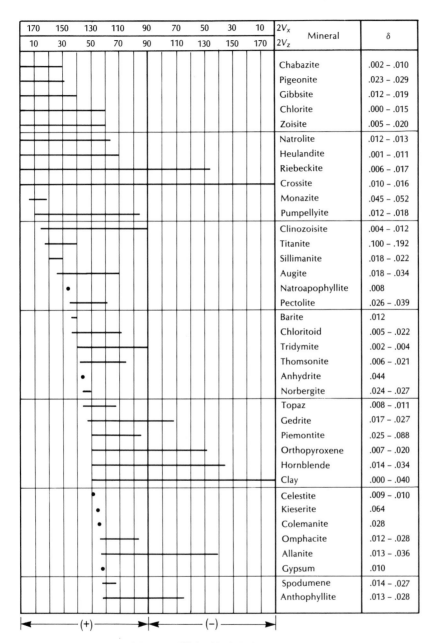

Figure C.1 $2V$ and birefringence (δ) for biaxial minerals.

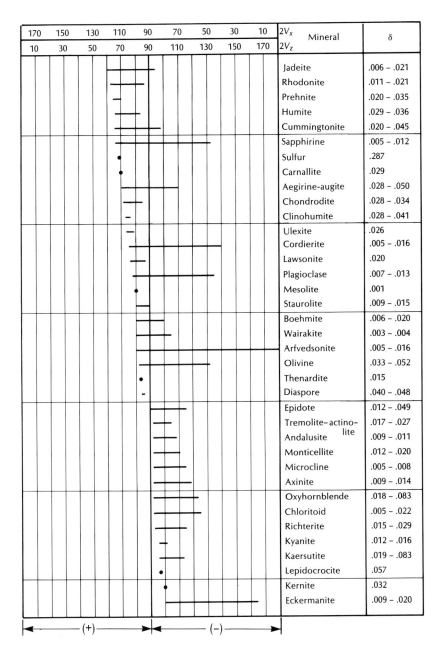

2V_x / 2V_z	Mineral	δ
	Jadeite	.006 – .021
	Rhodonite	.011 – .021
	Prehnite	.020 – .035
	Humite	.029 – .036
	Cummingtonite	.020 – .045
	Sapphirine	.005 – .012
	Sulfur	.287
	Carnallite	.029
	Aegirine-augite	.028 – .050
	Chondrodite	.028 – .034
	Clinohumite	.028 – .041
	Ulexite	.026
	Cordierite	.005 – .016
	Lawsonite	.020
	Plagioclase	.007 – .013
	Mesolite	.001
	Staurolite	.009 – .015
	Boehmite	.006 – .020
	Wairakite	.003 – .004
	Arfvedsonite	.005 – .016
	Olivine	.033 – .052
	Thenardite	.015
	Diaspore	.040 – .048
	Epidote	.012 – .049
	Tremolite–actino-lite	.017 – .027
	Andalusite	.009 – .011
	Monticellite	.012 – .020
	Microcline	.005 – .008
	Axinite	.009 – .014
	Oxyhornblende	.018 – .083
	Chloritoid	.005 – .022
	Richterite	.015 – .029
	Kyanite	.012 – .016
	Kaersutite	.019 – .083
	Lepidocrocite	.057
	Kernite	.032
	Eckermanite	.009 – .020

Figure C.1 *Continued.*

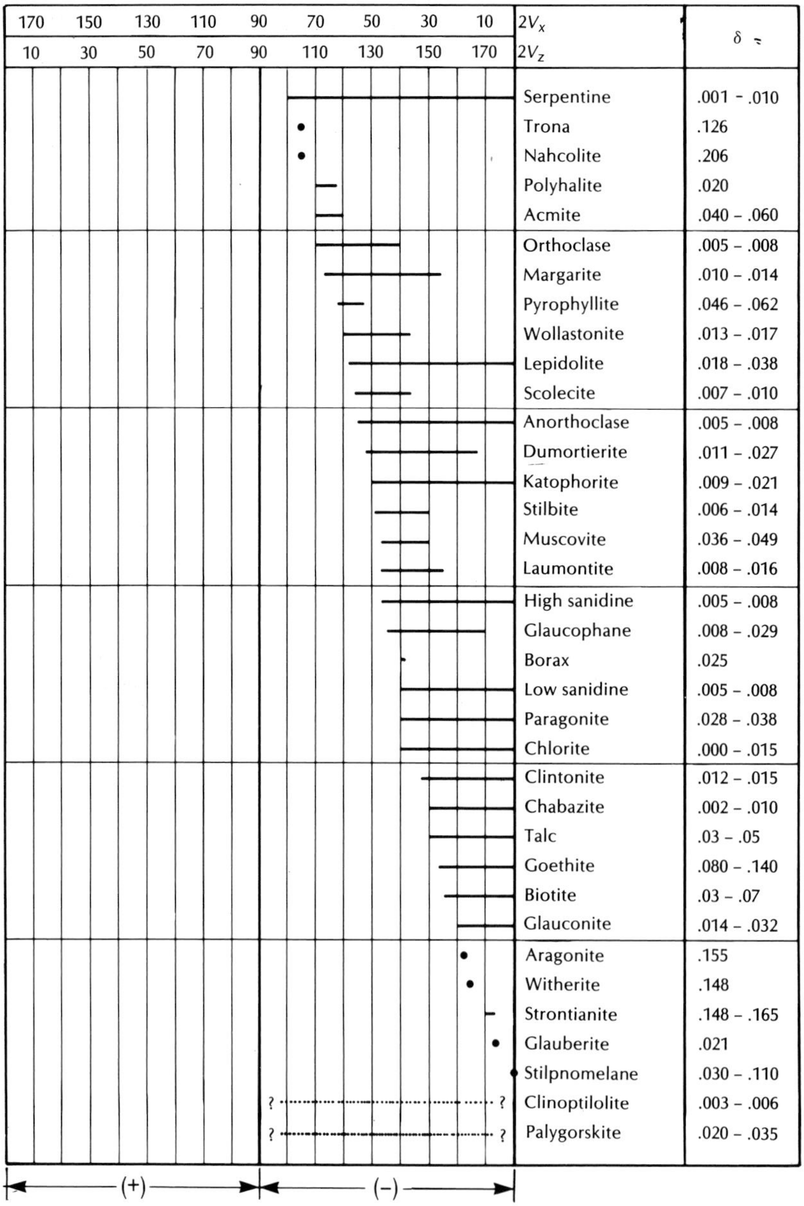

170	150	130	110	90	70	50	30	10	$2V_x$	δ
10	30	50	70	90	110	130	150	170	$2V_z$	
									Serpentine	.001 – .010
									Trona	.126
									Nahcolite	.206
									Polyhalite	.020
									Acmite	.040 – .060
									Orthoclase	.005 – .008
									Margarite	.010 – .014
									Pyrophyllite	.046 – .062
									Wollastonite	.013 – .017
									Lepidolite	.018 – .038
									Scolecite	.007 – .010
									Anorthoclase	.005 – .008
									Dumortierite	.011 – .027
									Katophorite	.009 – .021
									Stilbite	.006 – .014
									Muscovite	.036 – .049
									Laumontite	.008 – .016
									High sanidine	.005 – .008
									Glaucophane	.008 – .029
									Borax	.025
									Low sanidine	.005 – .008
									Paragonite	.028 – .038
									Chlorite	.000 – .015
									Clintonite	.012 – .015
									Chabazite	.002 – .010
									Talc	.03 – .05
									Goethite	.080 – .140
									Biotite	.03 – .07
									Glauconite	.014 – .032
									Aragonite	.155
									Witherite	.148
									Strontianite	.148 – .165
									Glauberite	.021
									Stilpnomelane	.030 – .110
									Clinoptilolite	.003 – .006
									Palygorskite	.020 – .035

(+) ———→ ←——— (–) ———→

Figure C.1 *Continued.*

Table C.1. Color in thin section

Colorless and Gray		Pink and Red	Yellow	Green	Blue and Violet	Tan and Brown
ISOTROPIC						
Amesite	Leucite	Fluorite	Allanite	Amesite	Fluorite	Allanite
Analcime	Nosean	Garnet	Collophane	Berthierine	Haüyne	Chromite
Berthierine	Opal	Hematite	Fluorite	Chlorite	Nosean	Clay
Clay	Periclase	Spinel	Garnet	Fluorite	Spinel	Collophane
Collophane	Perovskite	Volcanic glass	Limonite	Garnet	Sodalite	Garnet
Fluorite	Sodalite		Sphalerite	Greenalite		Limonite
Garnet	Sphalerite		Spinel	Hydrogrossular		Opal
Greenalite	Spinel			Perovskite		Perovskite
Halite	Sylvite			Serpentine		Sphalerite
Haüyne	Volcanic glass			Spinel		Volcanic glass
Hydrogrossular						
UNIAXIAL (−)						
Ankerite	Indialite	Anatase	Biotite	Anatase	Anatase	Anatase
Apatite	Magnesite	Corundum	Colophane	Biotite	Corundum	Ankerite
Apophyllite	Melilite	Hematite	Melilite	Chlorite	Tourmaline	Biotite
Beryl	Nepheline	Rhodochrosite	Siderite	Glauconite		Colophane
Calcite	Rhodochrosite	Tourmaline	Stilpnomelane	Stilpnomelane		Dolomite
Cancrinite	Scapolite		Tourmaline	Vesuvianite		Hematite
Chabazite	Siderite		Vesuvianite			Siderite
Corundum	Tourmaline					Stilpnomelane
Cristobalite	Vesuvianite					Tourmaline
Dolomite	Vishnevite					Vesuvianite
UNIAXIAL (+)						
Alunite	Leucite	Cassiterite	Cassiterite	Cassiterite		Cassiterite
Apophyllite	Melilite	Rutile	Melilite	Chlorite		Chalcedony
Brucite	Osumilite		Rutile	Chloritoid		Rutile
Cassiterite	Quartz		Vesuvianite	Vesuvianite		Vesuvianite
Chabazite	Vesuvianite		Xenotime			Xenotime
Chalcedony	Xenotime					Zircon
Eucryptite	Zircon					
BIAXIAL (−)						
Andalusite	Margarite	Andalusite	Andalusite	Acmite	Arfvedsonite	Acmite
Anorthoclase	Monticellite	Lepidocrocite	Anthophyllite	Actinolite	Axinite	Allanite
Anthophyllite	Muscovite	Orthopyroxene	Axinite	Aegirine-augite	Crossite	Anthophyllite
Aragonite	Nahcolite	Oxyhornblende	Geothite	Arfvedsonite	Dumortierite	Arfvedsonite
Axinite	Olivine	Sapphirine	Lepidocrocite	Biotite	Glaucophane	Axinite
Boehmite	Orthopyroxene		Olivine	Chlorite	Kyanite	Biotite
Borax	Palygorskite		Richterite	Chloritoid	Richterite	Clay
Chabazite	Phlogopite		Riebeckite	Cummingtonite-	Riebeckite	Clintonite
Chlorite	Plagioclase		Sapphirine	grunerite	Sapphirine	Gedrite
Clay	Polyhalite		Staurolite	Dumortierite		Goethite
Clinoptilolite	Pyrophyllite			Eckermanite		Hornblende
Clintonite	Richterite			Epidote		Iddingsite
Cordierite	Sapphirine			Glauconite		Katophorite
Cummingtonite-	Scolecite			Hornblende		Kaersutite
grunerite	Serpentine			Orthopyroxene		Lepidocrocite
Epidote	Stilbite			Oxyhornblende		Orthopyroxene
Glauberite	Strontianite			Sapphirine		Oxyhornblende
Glaucophane	Talc			Serpentine		Stilpnomelane
K-feldspar	Tremolite					
Kernite	Trona					
Kyanite	Wairakite					
Laumontite	Witherite					
Lepidolite	Wollastonite					

Table C.1. *Continued.*

Colorless and Gray		Pink and Red	Yellow	Green	Blue and Violet	Tan and Brown
BIAXIAL (+)						
Andalusite	Mesolite	Andalusite	Andalusite	Aegirine-augite	Arfvedsonite	Aegirine-augite
Anhydrite	Monazite	Orthopyroxene	Anthophyllite	Arfvedsonite	Crossite	Allanite
Anthophyllite	Natrolite	Piemontite	Humite group	Augite	Riebeckite	Anthophyllite
Augite	Olivine	Rhodonite	Monazite	Chlorite	Sapphirine	Arfvedsonite
Barite	Omphacite	Sapphirine	Olivine	Chloritoid		Augite
Boehmite	Orthopyroxene	Thulite	Piemontite	Cummingtonite–		Clay
Carnallite	Pectolite		Pumpellyite	grunerite		Gedrite
Celestite	Pigeonite		Riebeckite	Hornblende		Iddingsite
Chabazite	Plagioclase		Sapphirine	Monazite		Orthopyroxene
Clay	Prehnite		Sillimanite	Omphacite		Pumpellyite
Clinozoisite	Pumpellyite		Staurolite	Orthopyroxene		Sillimanite
Colemanite	Rhodonite		Sulfur	Pigeonite		Staurolite
Cordierite	Sapphirine		Titanite	Pumpellyite		Titanite
Cummingtonite–	Sillimanite			Sapphirine		
grunerite	Spodumene					
Diaspore	Thenardite					
Gibbsite	Thomsonite					
Gypsum	Topaz					
Heulandite	Tridymite					
Humite group	Ulexite					
Jadeite	Wairakite					
Kieserite	Zoisite					
Lawsonite						

Table C.2. Opaque minerals

Mineral	Color in Reflected Light[a]
Graphite	Black
Pyrite	Pale brassy yellow
Pyrrhotite	Bronze
Chalcopyrite	Brass-yellow to yellow-gold
Leucoxene	White-gray
Ilmenite	Metallic black
Hematite	Metallic black with red highlights
Magnetite	Gray
Chromite	Gray or brownish gray
Clay	Gray-earthy

[a] Produced by shining a light on the top surface of a thin section and observing through a conventional petrographic microscope.

Table C.3. Index of refraction of isotropic or nearly isotropic minerals

n	Mineral
1.43 –1.46	Opal
1.433–1.435	Fluorite
1.470–1.495	Nosean
1.479–1.493	Analcime
1.48 –1.62	Volcanic glass
1.48 –1.65	Clay
1.483–1.487	Sodalite
1.490	Sylvite
1.494–1.510	Haüyne
1.5 –1.6	Chlorophaetite
1.508–1.511	Leucite
1.544	Halite
1.58 –1.63	Collophane
1.59 –1.62	Amesite
1.62 –1.67	Berthierine
1.65 –1.68	Greenalite
1.675–1.734	Hydrogrossular
1.714–1.920	Spinel series
1.720–1.890	Garnet
1.735–1.756	Periclase
1.90 –2.12	Chromite
2.0 –2.4	Limonite
2.27 –2.40	Perovskite
2.37 –2.50	Sphalerite

Table C.4. Indices of refraction of uniaxial minerals

n_ω	n_ϵ	Mineral
UNIAXIAL (−)		
1.462–1.515	1.460–1.513	Chabazite
1.486–1.488	1.482–1.484	Christobalite
1.490–1.528	1.488–1.503	Cancrinite
1.529–1.546	1.526–1.544	Nepheline
1.531–1.542	1.533–1.543	Apophyllite
1.532–1.607	1.522–1.571	Scapolite
1.540	1.535	Kalsilite
1.560–1.608	1.557–1.599	Beryl
1.576–1.745	1.543–1.634	Stilpnomelane
1.578	1.560	Indialite
1.587	1.336	Soda niter
1.603–1.628	1.598–1.619	Carbonate–apatite
1.629–1.672	1.624–1.661	Melilite
1.631–1.698	1.610–1.675	Tourmaline
1.633–1.667	1.629–1.665	Apatite
1.658	1.486	Calcite
1.679–1.690	1.500–1.510	Dolomite
1.690–1.750	1.510–1.548	Ankerite
1.700	1.509	Magnesite
1.702–1.795	1.700–1.775	Vesuvianite
1.766–1.794	1.758–1.785	Corundum
1.816	1.597	Rhodochrosite
1.875	1.633	Siderite
2.561	2.488	Anatase
3.15 –3.22	2.87 –2.94	Hematite
UNIAXIAL (+)		
1.460–1.513	1.462–1.515	Chabazite
1.461	1.473	Tincalconite
1.508–1.511	1.509–1.511	Leucite
1.53 –1.544	1.53 –1.553	Chalcedony
1.531–1.542	1.533–1.543	Apophyllite
1.540–1.547	1.546–1.551	Osumilite
1.544	1.553	Quartz
1.559–1.590	1.580–1.600	Brucite
1.572	1.586	Eucryptite
1.572–1.620	1.592–1.641	Alunite
1.629–1.672	1.624–1.661	Melilite
1.690–1.724	1.760–1.827	Xenotime
1.702–1.795	1.700–1.775	Vesuvianite
1.920–1.960	1.967–2.015	Zircon
1.990–2.010	2.091–2.100	Cassiterite
2.61 –2.65	2.80 –2.90	Rutile

NOTE: Many phyllosilicates may be sensibly uniaxial, and the carbonates may display indices intermediate between the end-member values.

Table C.5. Indices of refraction of biaxial negative minerals arranged in order of increasing n_β

n_β	n_α	n_γ	Mineral	$2V_x(°)$
1.462–1.515	1.460–1.513	1.462–1.515	Chabazite	0–30
1.469	1.447	1.472	Borax	39–40
1.472	1.455	1.487	Kernite	80
1.479–1.493	1.476–1.491	1.479–1.497	Clinoptilolite	Small to large
1.489–1.507	1.482–1.500	1.493–1.513	Stilbite	30–49
1.49 –1.63	1.48 –1.61	1.50 –1.65	Clay	0–90
1.494	1.416	1.542	Trona	75
1.501	1.377	1.583	Nahcolite	75
1.501	1.498	1.502	Wairakite	75–90
1.512–1.524	1.502–1.514	1.514–1.525	Laumontite	25–47
1.516–1.520	1.507–1.513	1.517–1.521	Scolecite	35–56
1.518–1.530	1.514–1.526	1.521–1.533	Microcline	65–88
1.518–1.530	1.514–1.526	1.521–1.533	Orthoclase	40–70
1.518–1.530	1.514–1.526	1.521–1.533	Sanidine	0–47
1.524–1.534	1.519–1.529	1.527–1.536	Anorthoclase	0–55
1.524–1.574	1.521–1.561	1.527–1.578	Cordierite	40–90
1.530–1.603	1.529–1.595	1.537–1.604	Serpentine	Variable
1.531–1.585	1.527–1.577	1.534–1.590	Plagioclase	78–90
1.535	1.515	1.536	Glauberite	7
1.54 –1.56	1.50 –1.52	1.54 –1.56	Palygorskite	Variable
1.548–1.585	1.525–1.548	1.554–1.587	Lepidolite	0–58
1.548–1.672	1.522–1.625	1.549–1.696	Biotite	0–25
1.55 –1.69	1.55 –1.67	1.55 –1.69	Chlorite	0–40
1.562	1.547	1.567	Polyhalite	62–70
1.575–1.599	1.538–1.554	1.575–1.602	Talc	0–30
1.576–1.745	1.543–1.634	1.576–1.745	Stilpnomelane	~0
1.58 –1.65	1.56 –1.61	1.58 –1.65	Glauconite	0–20
1.582–1.620	1.552–1.580	1.587–1.623	Muscovite	30–47
1.586–1.589	1.534–1.556	1.596–1.601	Pyrophyllite	53–62
1.594–1.609	1.564–1.580	1.600–1.609	Paragonite	0–40
1.602–1.670	1.588–1.660	1.613–1.680	Anthophyllite	65–90
1.61 –1.71	1.60 –1.70	1.62 –1.73	Hornblende	35–90
1.612–1.697	1.599–1.688	1.622–1.705	Tremolite– actinolite	75–88
1.615–1.700	1.605–1.685	1.622–1.712	Richterite	64–87
1.622	1.592	1.623	Minnesotaite	Small
1.622–1.66	1.606–1.65	1.627–1.66	Glaucophane	10–45
1.625–1.652	1.612–1.636	1.630–1.654	Eckermanite	15–80
1.628–1.652	1.616–1.645	1.631–1.656	Wollastonite	36–60
1.633–1.644	1.629–1.640	1.638–1.650	Andalusite	71–88
1.635–1.710	1.627–1.694	1.644–1.722	Gedrite	72–90
1.638–1.709	1.630–1.696	1.655–1.729	Grunerite	82–90
1.642–1.648	1.630–1.638	1.644–1.650	Margarite	26–67
1.643–1.709	1.638–1.700	1.650–1.700	Arfvedsonite	0–90
1.645–1.684	1.640–1.681	1.652–1.692	Jadeite	84–90
1.646–1.664	1.638–1.654	1.650–1.674	Monticellite	69–88
1.649–1.657	1.640–1.648	1.655–1.668	Boehmite	~80
1.653–1.770	1.649–1.768	1.657–1.788	Orthopyroxene	48–90
1.655–1.662	1.643–1.648	1.655–1.663	Clintonite	0–33
1.658–1.688	1.639–1.681	1.660–1.690	Katophorite	0–50
1.660–1.701	1.654–1.694	1.668–1.705	Axinite	61–88
1.66 –1.70	1.65 –1.69	1.66 –1.70	Crossite	0–90
1.664–1.686	1.516–1.525	1.666–1.690	Strontianite	7–10
1.67 –1.78	1.65 –1.70	1.68 –1.80	Oxyhornblende	56–88
1.675–1.722	1.655–1.686	1.685–1.723	Dumortierite	13–52

Table C.5. *Continued.*

n_β	n_α	n_γ	Mineral	$2V_x(°)$
1.676	1.529	1.677	Witherite	16
1.680	1.530	1.685	Aragonite	18
1.684–1.869	1.664–1.827	1.701–1.879	Olivine	46–90
1.690–1.741	1.670–1.707	1.700–1.772	Kaersutite	66–84
1.70		1.75	Iddingsite	Variable
1.70–1.711	1.69–1.701	1.70–1.717	Riebeckite	45–90
1.700–1.815	1.690–1.791	1.706–1.828	Allanite	40–90
1.703–1.741	1.701–1.731	1.705–1.743	Sapphirine	47–90
1.708–1.734	1.705–1.730	1.712–1.740	Chloritoid	55–88
1.710–1.800	1.700–1.760	1.730–1.813	Aegirine-augite	70–90
1.719–1.725	1.710–1.718	1.724–1.734	Kyanite	78–84
1.725–1.784	1.715–1.751	1.734–1.797	Epidote	64–90
1.740–1.754	1.736–1.747	1.745–1.762	Staurolite	~90
1.780–1.820	1.750–1.776	1.795–1.836	Acmite	60–70
1.807	1.770	1.817	Tephroite	70
2.20	1.94	2.51	Lepidocrocite	83
2.22 –2.409	2.15 –2.275	2.23 –2.415	Goethite	0–27

Table C.6. Indices of refraction of biaxial positive minerals arranged in order of increasing n_β

n_β	n_α	n_γ	Mineral	$2V_z(°)$
1.460–1.513	1.460–1.513	1.462–1.515	Chabazite	0–30
1.470–1.484	1.468–1.482	1.474–1.486	Tridymite	40–90
1.474	1.466	1.495	Carnallite	70
1.475	1.469	1.484	Thenardite	83
1.476–1.491	1.473–1.490	1.485–1.502	Natrolite	0–64
1.487–1.507	1.487–1.505	1.488–1.515	Heulandite	0–70
1.500	1.499	1.502	Wairakite	80–90
1.504–1.508	1.504–1.508	1.505–1.509	Mesolite	~80
1.506	1.493	1.519	Ulexite	73–78
1.513–1.533	1.497–1.530	1.518–1.544	Thomsonite	42–75
1.522–1.526	1.519–1.521	1.529–1.531	Gypsum	58
1.524–1.574	1.521–1.561	1.527–1.578	Cordierite	75–90
1.531–1.585	1.527–1.577	1.534–1.590	Plagioclase	45–90
1.533	1.520	1.584	Kieserite	55
1.538	1.536	1.544	Natroapophyllite	32
1.55 –1.69	1.55 –1.67	1.55 –1.69	Chlorite	0–60
1.56 –1.57	1.56	1.56 –1.57	Clay	50–90
1.563–1.579	1.558–1.567	1.582–1.593	Norbergite	44–50
1.568–1.579	1.568–1.578	1.587–1.590	Gibbsite	0–40
1.576	1.570	1.614	Anhydrite	44
1.592	1.586	1.614	Colemanite	56
1.602–1.635	1.592–1.617	1.621–1.646	Chondrodite	71–85
1.602–1.670	1.588–1.660	1.613–1.680	Anthophyllite	58–90
1.603–1.615	1.592–1.610	1.630–1.645	Pectolite	35–63
1.609–1.637	1.606–1.635	1.616–1.644	Topaz	44–68
1.61 –1.71	1.60 –1.70	1.62 –1.73	Hornblende	50–90
1.615–1.647	1.610–1.637	1.632–1.670	Prehnite	64–70

Table C.6. *Continued.*

n_β	n_α	n_γ	Mineral	$2V_z$ (°)
1.619–1.653	1.607–1.643	1.639–1.675	Humite	65–84
1.623–1.624	1.621–1.622	1.630–1.632	Celestite	51
1.633–1.644	1.629–1.640	1.638–1.650	Andalusite	71–88
1.635–1.710	1.627–1.694	1.644–1.722	Gedrite	47–90
1.636–1.639	1.634–1.637	1.646–1.649	Barite	36–40
1.638–1.709	1.630–1.696	1.655–1.729	Cummingtonite	65–90
1.641–1.679	1.628–1.668	1.662–1.700	Clinohumite	73–76
1.643–1.709	1.638–1.700	1.650–1.700	Arfvedsonite	80–90
1.645–1.684	1.640–1.681	1.652–1.692	Jadeite	60–90
1.649–1.657	1.640–1.648	1.655–1.668	Boehmite	~80
1.651–1.684	1.636–1.664	1.669–1.701	Forsterite	82–90
1.653–1.770	1.649–1.768	1.657–1.788	Orthopyroxene	50–90
1.655–1.671	1.648–1.668	1.662–1.682	Spodumene	58–68
1.657–1.662	1.653–1.661	1.672–1.683	Sillimanite	20–30
1.66 –1.70	1.65 –1.69	1.66 –1.70	Crossite	0–90
1.670–1.712	1.662–1.701	1.685–1.723	Omphacite	56–84
1.672–1.753	1.664–1.745	1.694–1.771	Augite	25–70
1.674	1.665	1.686	Lawsonite	76–87
1.670–1.717	1.665–1.711	1.683–1.727	Pumpellyite	10–85
1.684–1.732	1.682–1.732	1.705–1.757	Pigeonite	0–32
1.688–1.711	1.685–1.707	1.698–1.725	Zoisite	0–60
1.70		1.75	Iddingsite	Variable
1.70 –1.711	1.69 –1.701	1.70 –1.717	Riebeckite	45–90
1.700–1.815	1.690–1.791	1.706–1.828	Allanite	57–90
1.703–1.741	1.701–1.731	1.705–1.743	Sapphirine	66–90
1.707–1.725	1.703–1.715	1.709–1.734	Clinozoisite	14–90
1.708–1.734	1.705–1.730	1.712–1.740	Chloritoid	36–72
1.710–1.800	1.700–1.760	1.730–1.813	Aegirine–augite	70–90
1.715–1.722	1.700–1.702	1.740–1.750	Diaspore	84–86
1.715–1.748	1.711–1.739	1.724–1.760	Rhodonite	63–87
1.730–1.813	1.725–1.794	1.750–1.860	Piemontite	50–86
1.740–1.754	1.736–1.747	1.745–1.762	Staurolite	80–90
1.778–1.801	1.777–1.800	1.823–1.849	Monazite	6–19
1.870–2.034	1.843–1.950	1.943–2.110	Titanite	17–40
2.037	1.958	2.245	Sulfur	69

Table C.7. Birefringence

δ	Mineral	Relief in Thin Section[a]
UNIAXIAL (−)		
0.000–0.003	Apophyllite	L+ to L−
0.000–0.011	Melilite	M+
0.001–0.007	Apatite	M+
0.001–0.020	Vesuvianite	H+
0.002–0.004	Cristobalite	M−
0.002–0.010	Chabazite	L− to M−
0.002–0.025	Cancrinite–vishnevite	L− to M−
0.003–0.005	Nepheline	L+ or L−
0.003–0.009	Beryl	L+ to M+
0.004–0.038	Scapolite	L+ to M+
0.005	Kalsilite	L+ to L−
0.008–0.009	Corundum	H+
0.015–0.035	Tourmaline	M+ to H+
0.018	Indialite	L+
0.030–0.110	Stilpnomelane	L+ to H+
0.073	Anatase	VH+
0.172	Calcite	M− to H+
0.179–0.182	Dolomite	L− to M+
0.182–0.202	Ankerite	L− to M+
0.191	Magnesite	L− to H+
0.219	Rhodochrosite	L+ to H+
0.242	Siderite	M+ to H+
0.251	Soda niter	H− to L+
0.28	Hematite	VH+
UNIAXIAL (+)		
0.000–0.001	Leucite	L+ to L−
0.000–0.003	Apophyllite	L−
0.000–0.011	Melilite	M+
0.001–0.009	Vesuvianite	H+
0.001–0.010	Chlorite	L+ to M+
0.002–0.010	Chabazite	L− to M−
0.004–0.006	Osumilite	L+
0.005–0.009	Chalcedony	L− to L+
0.009	Quartz	L+
0.010–0.021	Alunite	M+
0.010–0.021	Brucite	M+
0.013	Tincalconite	M−
0.014	Eucryptite	M+
0.036–0.065	Zircon	VH+
0.070–0.107	Xenotime	H+
0.09 –0.10	Cassiterite	VH+
0.15 –0.29	Rutile	VH+
BIAXIAL (−)		
0.000–0.015	Chlorite	L+ to M+
0.000–0.040	Clay	L− to L+
0.001–0.010	Serpentine	L− to L+
0.002–0.010	Chabazite	L− to M−
0.003–0.004	Wairakite	M−
0.003–0.006	Clinoptilolite	L− to M−
0.005–0.008	Anorthoclase	L−
0.005–0.008	K-feldspar	L−
0.005–0.012	Sapphirine	H+
0.005–0.016	Arfvedsonite	H+

Table C.7. Birefringence *Continued.*

δ	Mineral	Relief in Thin Section[a]
0.005–0.016	Cordierite	L− to L+
0.005–0.022	Chloritoid	H+
0.006–0.014	Stilbite	L− to M−
0.006–0.017	Riebeckite	H+
0.006–0.020	Boehmite	M+
0.006–0.021	Jadeite	M+
0.007–0.010	Scolecite	L−
0.007–0.013	Plagioclase	L− to L+
0.007–0.020	Orthopyroxene	M+ to H+
0.008–0.016	Laumontite	L−
0.008–0.029	Glaucophane	M+
0.009–0.013	Andalusite	M+
0.009–0.014	Axinite	M+ to H+
0.009–0.020	Eckermanite	M+
0.009–0.021	Katophorite	H+
0.010–0.014	Margarite	M+
0.010–0.016	Crossite	M+ to H+
0.011–0.037	Dumortierite	H+
0.012–0.015	Clintonite	M+
0.012–0.016	Kyanite	H+
0.012–0.020	Monticellite	M+
0.012–0.049	Epidote	H+
0.013–0.017	Wollastonite	M+
0.013–0.028	Anthophyllite	M + to H +
0.013–0.036	Allanite	H+
0.014–0.032	Glauconite	M+
0.014–0.034	Hornblende	M+ to H+
0.015–0.029	Richterite	M+ to H+
0.017–0.027	Gedrite	M + to H +
0.017–0.027	Tremolite–actinolite	M+ to H+
0.018–0.038	Lepidolite	L+ to M+
0.018–0.083	Oxyhornblende	H+
0.019–0.083	Kaersutite	H+
0.020	Polyhalite	L+
0.020–0.035	Palygorskite	L− to L+
0.020–0.045	Grunerite	M+ to H+
0.021	Glauberite	L−
0.025	Borax	M−
0.028–0.038	Paragonite	M+
0.028–0.050	Aegirine–augite	H+
0.03 –0.05	Talc	L+ to M+
0.03 –0.07	Biotite	L+ to M+
0.030–0.110	Stilpnomelane	H+
0.032	Kernite	M−
0.033–0.052	Olivine	M+ to H+
0.035–0.040	Minnesotaite	M+
0.036–0.049	Muscovite	L+ to M+
0.040–0.050	Iddingsite	H+
0.040–0.060	Acmite	H+
0.046–0.062	Pyrophyllite	L+ to M+
0.047	Tephroite	H+
0.057	Lepidocrocite	VH+
0.080–0.140	Goethite	VH+
0.126	Trona	L− to M−
0.148	Witherite	M+
0.148–0.165	Strontianite	L− to M+

Table C.7. Birefringence *Continued.*

δ	Mineral	Relief in Thin Section[a]
0.155	Aragonite	L− to M+
0.206	Nahcolite	H− to L+
BIAXIAL (+)		
0.000–0.015	Chlorite	L+ to M+
0.001	Mesolite	L−
0.001–0.011	Heulandite	L− to M−
0.002–0.004	Tridymite	M−
0.002–0.010	Chabazite	L− to M−
0.003–0.004	Wairakite	M−
0.004–0.008	Clay	L+
0.004–0.012	Clinozoisite	H+
0.005–0.012	Sapphirine	H+
0.005–0.016	Cordierite	L− to L+
0.005–0.016	Arfvedsonite	H+
0.005–0.020	Zoisite	H+
0.005–0.022	Chloritoid	H+
0.006–0.017	Riebeckite	H+
0.006–0.020	Boehmite	M+
0.006–0.021	Jadeite	M+
0.006–0.021	Thomsonite	L−
0.007–0.013	Plagioclase	L− to L+
0.007–0.020	Orthopyroxene	M+ to H+
0.008	Natroapophyllite	L+ to L−
0.008–0.011	Topaz	M+
0.009–0.010	Celestite	M+
0.009–0.013	Andalusite	M+
0.009–0.015	Staurolite	H+
0.010	Gypsum	L−
0.010–0.016	Crossite	M+ to H+
0.011–0.021	Rhodonite	H+
0.012	Barite	M+
0.012–0.013	Natrolite	M−
0.012–0.019	Gibbsite	L+ to M+
0.012–0.018	Pumpellyite	H+
0.012–0.028	Omphacite	H+
0.013–0.028	Anthophyllite	M+ to H+
0.013–0.036	Allanite	H+
0.014–0.027	Spodumene	M+

Table C.7. Birefringence *Continued.*

δ	Mineral	Relief in Thin Section[a]
0.014–0.034	Hornblende	M+ to H+
0.015	Thenardite	M−
0.017–0.027	Gedrite	M+ to H+
0.018–0.022	Sillimanite	H+
0.018–0.034	Augite	H+
0.020	Lawsonite	M+ to H+
0.020–0.035	Prehnite	M+
0.020–0.045	Cummingtonite	M+ to H+
0.023–0.029	Piemontite	H+
0.024–0.027	Norbergite	L+ to M+
0.025–0.088	Piemontite	H+
0.026	Ulexite	L−
0.026–0.039	Pectolite	M+
0.028	Colemanite	M+
0.028–0.034	Chondrodite	M+
0.028–0.041	Clinohumite	M+
0.028–0.050	Aegirine-augite	H+
0.029	Carnallite	M−
0.029–0.036	Humite	M+
0.033–0.052	Olivine	M+ to H+
0.040–0.048	Diaspore	H+
0.040–0.050	Iddingsite	H+
0.044	Anhydrite	M+
0.045–0.052	Monazite	H+
0.064	Kieserite	L− to L+
0.100–0.192	Titanite	H+
0.287	Sulfur	VH+

[a] Relief is based on comparison with cement whose index of refraction is 1.537. Cements with other indices of refraction are in common use and may yield relief that is somewhat different than shown here. The index of refraction ranges for the different relief categories are as follows:

	(+)	(−)
L = Low	1.54–1.58	1.50–1.54
M = Moderate	1.58–1.66	1.42–1.50
H = High	1.66–2.00	<1.42
VH = Very high	>2.0	

NOTE: Some phyllosilicates may be sensibly uniaxial and the birefringence of some carbonates may be intermediate between the end-member values.

Table C.8. Minerals that may display anomalous interference colors

Anatase	Clinozoisite	Pumpellyite
Apatite	Crossite	Richterite
Apophyllite	Eckermanite	Sapphirine
Arfvedsonite	Epidote	Serpentine
Borax	Glaucophane	Titanaugite
Brucite	Jadeite	Vesuvianite
Chlorite	Melilite	Zoisite
Clinohumite	Prehnite	

Table C.9. Isometric minerals that may display anomalous birefringence

Analcime
Gahnite
Garnet (grossular and spessartine)
Leucite
Sodalite
Sphalerite

Table C.10. Tetragonal and hexagonal minerals that may be anomalously biaxial

Analcime	Calcite	Quartz
Anatase	Cassiterite	Rutile
Apatite	Chabazite	Scapolite
Apophyllite	Cristobalite	Vesuvianite
Beryl	Corundum	Zircon
Brucite	Osumilite	

Table C.11. Normally birefringent minerals that may be sensibly isotropic

Allanite (metamict)	Leucoxene
Apophyllite	Limonite
Chlorite	Melilite
Clay	Serpentine
Collophane	Zircon (metamict)

Table C.12. Biaxial minerals that may be sensibly uniaxial

Anorthoclase	Gibbsite	Riebeckite
Arfvedsonite	Glauconite	Sanidine
Biotite	Goethite	Serpentine
Chlorite	Heulandite	Stilpnomelane
Crossite	Lepidolite	Talc
Clay minerals	Phengite	Zoisite
Clintonite	Pigeonite	

Table C.13. Minerals that may produce pleochroic halos in surrounding minerals

Allanite	Xenotime
Monazite	Zircon
Titanite	

Mineral Index

Minerals described in detail and the pages on which the principal descriptions are found are indicated with bold type.

Subject Index